Perspectives on Consciousness

Perspectives on Consciousness

Inquiries into Subjective Experience across Disciplines

edited by Aviva Berkovich-Ohana, Or Raphael, Eviatar Shulman,
David Rudrauf, and Idan Segev

The MIT Press
Cambridge, Massachusetts
London, England

The MIT Press
Massachusetts Institute of Technology
77 Massachusetts Avenue, Cambridge, MA 02139
mitpress.mit.edu

All artwork included in this publication is the original work of Michal Rovner.

The MIT Press would like to thank the anonymous peer reviewers who provided comments on drafts of this book. The generous work of academic experts is essential for establishing the authority and quality of our publications. We acknowledge with gratitude the contributions of these otherwise uncredited readers.

This book was set in Times New Roman by Westchester Publishing Services. Printed and bound in the United States of America.

Library of Congress Cataloging-in-Publication Data

Names: Berkovich-Ohana, Aviva, 1969- editor.
Title: Perspectives on consciousness : inquiries into subjective experience across disciplines / edited by Aviva Berkovich-Ohana, Or Raphael, Eviatar Shulman, David Rudrauf, and Idan Segev.
Description: Cambridge, Massachusetts : The MIT Press, [2025] | Includes bibliographical references and index.
Identifiers: LCCN 2025002418 (print) | LCCN 2025002419 (ebook) | ISBN 9780262547062 (paperback) | ISBN 9780262377171 (epub) | ISBN 9780262377188 (pdf)
Subjects: LCSH: Consciousness. | Neurosciences. | Cognition.
Classification: LCC BF311 .P3735 2025 (print) | LCC BF311 (ebook)
LC record available at https://lccn.loc.gov/2025002418
LC ebook record available at https://lccn.loc.gov/2025002419

EU Authorised Representative: Easy Access System Europe, Mustamäe tee 50, 10621 Tallinn, Estonia | Email: gpsr.requests@easproject.com

Contents

Figure 0.1
Michal Rovner. Silent Waters, 2022 (detail). LCD screen and video. Copyright: Michal Rovner.

Preface

Introduction

The problem of consciousness is a fundamental enigma in current scientific and philosophical debates. As it necessarily involves subjective experience, it poses difficult challenges to studies of consciousness, and persists as a theoretical and practical dilemma. Nevertheless, there is a growing consensus that an adequate explanation of consciousness requires incorporating understanding of subjective experience, as well as its biological/neural substrate.

One of the unsolved mysteries in consciousness studies is how qualia emerges from brain activity?[1] As expressed in Thomas Huxley's famous words (1866), "How it is that anything so remarkable as a state of consciousness comes about as a result of irritating nervous tissue, is just as unaccountable as the appearance of the djinn, when Aladdin rubbed his lamp." This is the well-known "explanatory gap" (Levine, 1983), or "hard problem of consciousness" (Chalmers, 1996).

In philosophy, which has debated the "mind-body problem" for millennia, many assumptions have been proposed, ranging from materialism (i.e., all is reducible to the physical) to idealism (i.e., ideas make up fundamental reality), with many options in between, including different versions of dualism. For researchers of esoteric traditions, the problem is actually reversed—how does the physical world emerges from a non-physical "substance"? That is, consciousness may not necessarily reduce to the physical.

This volume supports the premise that to produce an adequate understanding of consciousness as an embodied phenomenon, it is necessary to bring back our own subjectivity into the picture. In our view, research on consciousness cannot be approached as an entirely objective endeavor, what phenomenologists might call "the natural attitude." By considering subjectivity as an essential part of the scientific effort, we might expose ourselves to emotional and conceptual challenges, even to philosophical conundrums, yet we will nevertheless remain more faithful in this way to the phenomena we are studying; consciousness is integral to subjective experience. We thus believe that exploring our own experience, is a necessary condition to scientifically describe consciousness. As knowledge is produced for the experiencers in accord with their experience of the world, the relation between the organism and its environment remains a kind of dialogical coupling. Any perspective that purports a representation of reality that is independent of subjective experience would thereby be incomplete. As Francisco Varela said, "Any third-person perspective is always haunted by the ghost of the first."[2]

This openness to our own subjective experiences facilitated our connection, as a group of editors, fueled by friendship and a mutual curiosity and exploration, and resulted in our launching two multidisciplinary workshops on "Perspectives on Consciousness,"[3] which provide the basis for this volume. These two unique academic gatherings integrated a variety of disciplines, ranging from neuroscience, computational modeling, mathematics, and physics, through neurophenomenology, psychology, and cognitive sciences, to philosophy, history, and the study of Indian, mainly Buddhist, contemplative traditions. Scientific talks were integrated within significant spaces for meditative contemplation and for open discussion of fundamental issues. We tried to make ample room for phenomenology,[4] which contributes to our practical answer to the "explanatory gap." We gave particular importance to the study of contemplative approaches and altered states of consciousness, suggesting that they offer important insights into the nature of consciousness.

The aim of this volume is to highlight the role of subjective experience in the scientific exploration of consciousness, and to provide tentative answers to the following questions: What can the study of experience contribute to the study of consciousness? How can the study of experience be conducted in ways that may enrich science? Can personal attitudes and reflections have a positive effect on the science of consciousness? How should first-person perspectives be integrated into scientific practices?

In order to facilitate the understanding of the role of subjective experience in the study of consciousness, we first offer a historical perspective on the development of the field, which includes an introduction to the major models science has offered. This is meant to create a vast "solution space," which will help the reader appreciate the contributions in this volume.

The Problem of Consciousness: A Brief History and State of the Art

Subjective experience was a central part of consciousness research at its inception, during its introspective and deep psychology phases. It was then buried for several decades before emerging again, enriched by cognitive, neuroscientific and computational perspectives well developed in the twentieth century, and exponentially gaining in interest with the enactive view in the twenty-first century.

At the outset of modern scientific psychology in the mid-nineteenth century subjective experience was still largely equated with consciousness. At this stage, introspective scholars dominated the field, including Wilhelm Wundt (1897), Hermann von-Helmholtz (1897), and William James (1890). Around the turn of the twentieth century, depth psychology was founded on the theories of Sigmund Freud, Carl Gustav Jung, and Alfred Adler. Depth psychology explored the unconscious human experience, including the study of dreams and archetypes and an analysis of the relationship between the conscious and the unconscious, arguing that the psyche spontaneously generates symbolism, and is therefore not only instinctive but spiritual in nature (Ellenberger, 1970).

The early twentieth century, with the rise of behaviorism, saw the eclipse of consciousness research, and the elimination of subjective experience altogether, from scientific psychology (Skinner, 1953; Watson, 1924). Behaviorism emerged as a reaction to depth psychology, which often unsuccessfully struggled to make predictions that could be tested experimentally. During its peak (1920s–1950s), the basic premise was that mental processes that could not be observed objectively were outside the realm of empirical science. The study of behavior, rather than mind and its relationship with the body, was the only valid approach for natural science.

In the second half of the twentieth century, radical behaviorism was highly criticized for not examining mental processes, and was deemed unable to address central issues in human psychology. The grip of behaviorism weakened with the rise of cognitive psychology and its emphasis on information processing and the modeling of internal mental processes (Gardiner, 1985; Neisser, 1967). The latter was spearheaded by researchers in linguistics and cybernetics as well as applied psychology, who used conceptual models of mental processing to explain human behavior (Mandler, 2002). However, despite the renewed emphasis on explaining cognitive skills such as memory, perception, and language comprehension, consciousness remained a largely neglected topic for several more decades. Significant developments in cognitive science have been accomplished within the computationalist perspective (Colombo & Piccinini, 2023), which holds that the mind is an information processing or computation system. Connectionism in particular offered a framework to conceive of bridges between neural processing and cognitive functions. It fostered a philosophy of emergence: how local rules can give rise to global properties via reciprocal causality. Although some researchers suggest a fundamental opposition between classical computation and connectionism, these approaches are not mutually exclusive (Rescola, 2018). Some advantages of the connectionist approach include its applicability to a broad array of functions (e.g., sensory modalities, motion, learning and memory), rendering them mechanistically explainable. However, this view made it difficult to decipher how distributed artificial networks process information so that they may account for the compositionality of mental representations and cognitive phenomena at a higher level,[5] the so-called hard problem of consciousness, as it made consciousness appear devoid of any causal relevance. One missing ingredient in the traditional computational approach was that human cognition operates continuously in real time. Alternative approaches emphasized the need for dynamic models, which define a system according to its state at any given time, with continuous transition laws between states, and a state-space,[6] representing the totality of states the system could be in (Van Gelder, 1998).

At the same time, new techniques for large-scale investigation of human brain activity and neuropsychology emerged. The early brain imaging techniques of electroencephalography (EEG) as early as 1930s, computerized tomography (CT) in the early 1970s, and positron emission tomography (PET) in the early 1980s were followed by functional magnetic resonance imaging (fMRI) in the 1990s. More recently, there were many developments in single electrode recordings, as well as brain stimulation techniques such as trans-magnetic stimulation (TMS), and single electrodes stimulation, that provided key insights into consciousness processing.[7] When the experimental strategies of cognitive psychology were combined with these brain imaging techniques, scientists could examine how brain functions correlate with mental activities. This gave rise to the field of cognitive neuroscience, which rapidly expanded to include other fields, such as social, affective, and even contemplative neuroscience. This for the first time allowed researchers to ask direct experimental questions concerning not only sensory-motor processing, but also subtler inner experiences, such as mental imagery, attention, emotion, and more recently, the sense of self.

After this brief review of the field's history, we succinctly present current theories of consciousness. We first present theories that search for neural correlates, hinging largely on the computationalist views of the second half of the twentieth century. We then present views that emphasize subjective experience, including phenomenology, which stem from early introspectionist views, and much later regained traction in the 1980s, in the transition from traditional computationalism to connectionism, and finally the so-called enactive view. We conclude with non-physicalist theories, which are experiencing a renaissance since the turn of the twenty-first century. These twists and turns are described in some detail in the next section.

Approaches to Consciousness

The Search for Neural Correlates of Consciousness

Since the 1980s, there was a major resurgence of scientific and philosophical research into the nature and basis of consciousness, and many scholars put forth conceptual and formal proposals for theories of consciousness, drawing from Eastern and Western philosophy, experimental psychology, neuroscience, physics, and mathematics. There are different ways to group the various theories of consciousness. It is beyond the scope of this preface to mention all the theories of consciousness. However, we briefly describe several prominent neurobiological theories that are relevant to several chapters in this volume. This includes higher-order theory, global workspace, predictive processing and integrated information theory (for a thorough comparison of the major neurobiological theories of consciousness see Block [2009], and Seth and Bayne [2022]; for a brief list of these and other neural theories, see table 0.1).

- *Higher-order theory* (HOT) proposes that what makes a mental state *M* conscious is the fact that it is accompanied by a simultaneous and non-inferential higher-order (i.e., metamental) state, whose content is *M*—that is, all conscious states are reportable. Unconscious mental states lack the relevant higher-order states about them (Rosenthal, 2005).
- *Global (neural) workspace theory* (GNWT) highlights the role of a competition among processors and outputs for a limited capacity resource that "broadcasts" information for widespread access and use (Baars, 1988). The model has been further developed with proposed connections to particular neural and functional brain systems (Dehaene et al., 1998). Consciousness, in both the access and phenomenal sense,[8] occurs only when the relevant content enters the larger global network involving primary sensory areas as well as other areas associated with attention. Dehaene claims that conscious perception begins only with the "ignition" of that larger global network, whereas activity in primary sensory areas will not suffice.
- *Integrated information theory* (IIT) proposes that consciousness is a purely information-theoretical property of systems. Information integration of the relevant sort is both necessary and sufficient for consciousness regardless of the substrate in which it is realized (Tononi, 2008). An information theoretic quantity ϕ (*phi*) measures the system's degree of integrated information. IIT might imply a form of panpsychism that some find problematic (see Merkeret al., 2022). Prominent researchers such as Christof Koch have endorsed the IIT framework (Koch, 2012). In contrast to HOT and GNWT, IIT links consciousness primarily

Table 0.1
A partial list of neural theories of consciousness

Family	Specific theories	References
Higher-order theory (HOT)	The four prominent neurobiological theories of consciousness; see main text for elaboration.	Rosenthal (2005)
Global workspace (GW), and neural GW		Baars (1988); Dehaene et al. (1988)
Information integration theory (IIT)		Tononi (2008)
Predictive processing (PP)		Friston (2010); Seth and Bayne (2022)
Re-entry theories	Emphasize the importance of top down signaling in shaping and enabling consciousness	Lamme (2020)
	Reentrant cortical loops	Edelman (1989)
Neural correlates of consciousness theories	Global integrated fields	Kinsbourne (2000)
	Binding through synchronous oscillation	Singer (1999); Crick and Koch (1990); Varela et al. (2001)
	N-methyl-D-aspartate (NMDA)-mediated transient neural assemblies	Flohr (1995)
	Thalamically modulated patterns of cortical activation	Llinas and Ribary (2001)
	Comparator mechanisms that engage in continuous action-prediction-assessment loops between frontal and midbrain areas	Gray (1995)
	Left-hemisphere based interpretative processes	Gazzaniga (2004)
	Emotive somatosensory homeostatic processes based in the frontal-limbic nexus or in the periaqueductal gray	Damasio (1999); Panksepp (1998)

with posterior cortical areas, as these areas exhibit neuroanatomical properties that are supposedly well-suited for generating high levels of integrated information.

• *Predictive processing* (PP) stems from a more general perspective on the interactions between the brain, the body, and the environment, which can be used to formulate explanations and predictions about properties of consciousness. PP is based on the hypothesis that the brain implements a process of active 'prediction error minimization' that approximates Bayesian inference (Hohwy & Seth, 2020; Seth & Bayne, 2022). PP considers the brain as a predictive machine that compares bottom-up inputs with top-down expectations through cortical processing, and aims at minimizing errors of prediction (Clark, 2013). That process of minimization has been formalized within the free energy principle in the context of active inference—that is, in which organisms control their behaviors based on how anticipated action consequences match sensory evidence (Friston, 2010).

In spite of the diversity of these neuroscientific theories of consciousness[9], they share at least two elements in common (Hohwy & Seth, 2020). The first is the role of consciousness in uncertainty reduction through the generation of one unified and highly informative representational state. This point often entails a threshold at which contents become conscious (e.g., broadcasting for GWNT, optimization of signal-to-noise ratio for HOT, ϕ for IIT, learning for PP). The second common element is the high importance of top-down or recursive signaling. Yet, such materialistic neuroscientific views, although highly promising, and although they have provided remarkable breakthroughs in understanding the brain, might not in fact be able to solve the "hard problem of consciousness" (Gidon et al., 2022).

The Role of Subjective Experience

Approaches to consciousness that emphasize the role of subjective experience often try to target so-called phenomenal consciousness, or the "what-it-is-likeness" of subjective experience (Nagel, 1974). It is important to distinguish here between different "flavors" of subjective experience—for example, "phenomenal" and "access" consciousness (Block, 1995, 2007). According to Block, a mental state can be conscious if it is associated with any subjective feeling (belongs to *phenomenal consciousness*), or if the information it carries is available for reasoning and guiding behavior, even if it is not associated with any subjective feeling characterizing the concurrent phenomenal experience (in the sense of having the required sort of internal access relations, like reportability, thus termed *access consciousness*). It is still debated whether a model of consciousness that explains only access consciousness (i.e., HOT) may be the only scientifically rigorous approach (Naccache, 2018). Similarly, some claim that subjective experience cannot be studied at all within the objective framework of science and should best be ignored.[10] Resistance to integrating subjective experience into a (natural) science of consciousness can also be found on the phenomenological side of this debate. There are many philosophers in the Husserlian tradition who are unwilling to consider the validity of a naturalistic science of mind (Gallagher & Varela, 2003). Even Husserl himself objected to the project of "naturalizing" phenomenology—that is, using it as part of a psychological stance (Roy et al., 1999), as explained by Zahavi:

> For Husserl, natural science is (philosophically) naive. Its subject matter, nature, is simply taken for granted. Reality is assumed to be out there, waiting to be discovered and investigated. And the aim of natural science is to acquire a strict and objectively valid knowledge about this given realm. But this attitude must be contrasted with the properly philosophical attitude, which critically questions the very foundation of experience and scientific thought . . . Positivism has denied the existence of a particular philosophical method, and has claimed that philosophy should employ the same method that all strict sciences are using, the natural scientific method. But for Husserl this line of reasoning merely displays that one has failed to understand what philosophy is all about. (Zahavi, 2004, p. 337)

Albeit the resistance to integrating subjective experience in natural sciences on behalf of some phenomenologists, the idea has been powerfully disputed by others, such as Searle (1998, p. 1938): "The problem of consciousness and the problem of so-called qualia is the same problem; and you cannot evade the identity by treating consciousness as some sort of third person, ontologically objective phenomenon and by setting qualia on one side, because to do so is simply to change the subject." Likewise, there seems to be a growing agreement that what the study of consciousness is missing in order to be able to reduce the explanatory gap, is a good understanding of subjective experience (e.g., Flanagan, 1992). Another example comes from Antti Revonsuo (2003, pp. 2–3):"All science must start by the identification and systematic description of the *explanandum*. Only after we have an understanding of the nature of the *explanandum* can we start to construct hypotheses regarding the possible underlying *mechanisms* that might be invoked, in order to explain the phenomenon."

Francisco Varela instigated a direction of research that considered early on that in order to envision the possibility of joining phenomenology to the "consciousness endeavor" of cognitive science, a historic shift had to occur, from traditional computationalism to something more akin to connectionism. According to Gallagher and Varela (2003), this is indeed what occurred in the 1980s, with a shift away from an emphasis on reductionism to an emphasis on the notions of emergence and self-organization, asking how higher-level personal structures emerged from lower-level subpersonal, self-organizing processes. There has been a renewed interest in recent cognitive science in the ecological-embodied-enactive approach, pioneered by Varela (Varela et al., 1991); a prominent example is the work of Karl Friston and his collaborators on active inference (Friston et al., 2017). Varela's approach took up the connectionist emphasis on dynamical mechanisms and self-organizing emergence, but it further claimed that cognition is best characterized as belonging to embodied agents, situated in the world. This contrasted with classical cognitivism, which seemed to advocate a disembodied view of mind, which focuses on representational symbol processing (for criticisms of that view, see Clark [1997], Cosmelli and Thompson [2010], Gallagher [2005], and Thompson and Cosmelli [2011]).

This change of perspective on the problem of consciousness set the stage for the neurophenomenolgical (NP) research program proposed by Varela (1996), which he presented as a methodological remedy for the "hard problem" of consciousness. NP aims to "naturalize" phenomenology, by developing and applying disciplined phenomenology, and then by using first-person experience to constrain naturalistic explanations. NP thus seeks to transcend mere correlational accounts—for example, through phenomenal variants grouping[11] or mathematical modeling[12]—to progressively bridge phenomenology, cognition and the brain (Gallagher & Varela, 2003; Lutz, 2002; Roy et al., 1999; Varela & Thompson, 2003). It entails "front-loading phenomenological insights" onto the research design (Gallagher, 2003). While the NP approach cannot alone provide relevant mechanistic explanations and requires a broader interdisciplinary approach, it is still a promising way to effectively join and study in tandem irreducible ingredients of consciousness such as subjective experience and the brain. Yet, as much as NP is appealing theoretically, it is empirically challenging, and little progress has been made since Varela's original proposal (Berkovich-Ohana, 2017). As far as modeling consciousness from a neurophenomenological perspective is concerned, few models have been proposed, several of them revolving around self-consciousness and its structure (Berkovich-Ohana et al., 2020).

What seems to be needed to progress with NP is a change within the researchers themselves—that is within the scientists' experience, not just in the research program. This is captured by Petitmengin's notion of "radical neurophenomenology" (Petitmengin, 2017). Petitmengin distinguishes between two types of NP. The first one is "mild" neurophenomenology, which aims at studying correlations between rigorous first-person descriptions and recordings of brain activity, and tries to evaluate the validity of first-person descriptions through objective criteria. As such, mild-NP cannot dissolve the "hard problem" of consciousness. The second type is radical neurophenomenology, which is more ambitious and aims at investigating the process of co-constitution of the subjective and the objective poles of the problem, within lived experience, and tries to evaluate first-person descriptions through processual criteria. This would involve "on the one hand experiencing concretely the dissolution of the hard problem, and on the other hand verifying the theory of enaction

Table 0.2
A partial list of (neuro)phenomenolgical theories of consciousness

Family	Specific theories	References
Phenomenological theories	Hypothesis of subjective time—subjective time as a new dimension of physics that would allow us to have a causal effect on the world	Schooler (2015)
	Pattern theory of the self (PTS)—the self is a complex pattern of dynamically related processes, which operates as a complex system that emerges from dynamical interactions of constituent processes	Gallagher (2013); Gallagher and Daly (2018)
Neurophenomenological theories	Projective consciousness model (PCM)—a projective geometrical model of the perspectival phenomenological structure of the field of consciousness with a variational free energy minimization model of active inference	Rudrauf et al. (2017, 2020, 2022); Williford et al. (2022); Safron (2020, 2022)
	Consciousness state space (CSS)—a phenomenological dynamic model for consciousness and selfhood that relates time, awareness, and emotion within a state space	Berkovich-Ohana and Glicksohn (2014); Berkovich-Ohana and Wittmann (2017)
	Implicit–explicit gradient of consciousness as such (non-dual awareness) to be added as the *z*-axis to the existing two-dimensional (states and content) map of consciousness	Josipovic (2021)
	Minimal phenomenal experience (MPE) model—an explanation for conscious experience, taking "pure consciousness" as its entry point, based on six semantic constraints	Metzinger (2020)

in lived experience" (p. 139). This view and criterion of validation set, indeed quite radically, an ideal horizon for NP, which helps to better distinguish it from other approaches: the missing ingredient to actually solve the hard problem would be for researchers, ideally, to experience non-dual states of consciousness! Such states have the potential to "maintain and even intensify the experience of how two opposites can be true at the same time" (Zajonc, 2013, p. 86). Table 0.2 is in line with this perspective, as many of the NP theories of consciousness are inspired by states of consciousness induced by contemplative practices, including "front-loading insights" from altered states of consciousness.

Studying altered states of consciousness pushes the boundaries of science. Scientists often go to great lengths to preserve the neutrality of their research by separating it from situations in which they would be seen as being personally involved. However, scientists that have a personal practice of meditation declare that it helps them both reconfigure and stabilize the boundary between science and meditation. Wiles (2019) introduced the concept of *embodied practitioner knowledge* to describe a type of internal boundary practice that enables individuals to both separate and bridge boundaries within their own bodies. Extended to the field of consciousness studies, *embodied practitioner knowledge* refers to the many ways that scientists who are studying consciousness use their own bodies and minds to translate between science and experience. This practical epistemology seems to be set in motion within the growing number of publications in contemplative neuroscience and more recently in psychedelic research, aiming at contributing to the field of consciousness research (see more particularly Timmerman, Berkovich-Ohana, and Ott, this volume). This is indeed an exciting and challenging idea, that scientists can use their embodied experience in profitable ways to their scientific endeavor.

Non-Physicalist (Non-Local) Theories of Consciousness

While from the beginning of the early modern era in the seventeenth century, consciousness was widely regarded by philosophers and psychologists as essential or definitive of the mental, during the twentieth century a materialist perspective eclipsed that view. Materialist views of consciousness took over, first through psychophysics, then (paradoxically) through behaviorism, followed by cognitive sciences, and finally by cognitive neuroscience, with the prevalent search for the neural correlates of consciousness. Yet, since the turn of the twenty-first century, there appears to be a renewal of non-physicalist theories, which regard it as more fundamental than matter. Some of these theories include dualist metaphysical views, as well as theories based in quantum physics or mathematical modeling (see table 0.3; for a recent review see Wahbeh et al. [2022]).

Irrespective of what one might assume about the scientific value of such theories and approaches, one indisputable point about them is that they challenge the postulates of common views. They bring forth accounts of several phenomenal experiences that do not seem to fit into the frameworks of physicalist theories. Wahbeh et al. (2022) consider five consciousness phenomena: reports of individuals perceiving information from distant locations, from another person, or from the future; reports where people gain skills beyond their normal capacity, or when their brain is apparently nonfunctional. Importantly, reports of such phenomena have been extensively documented in anecdotal and experimental contexts, and seem to be ubiquitous worldwide (see Wahbeh et al. [2022] and references therein). Certainly, these reported phenomena of consciousness are in no way a valid basis to support non-local theories of consciousness. Admittedly, non-physicalist theories lack empirical evidence, and sometimes even their predictions are unclear. But what these reports illustrate is that many aspects of consciousness are both experientially pervasive and epistemically elusive, and as such invite researchers to pay attention to the more puzzling aspects of their own subjective experience, and perhaps to think outside the box of standard physicalist models of consciousness.

Importantly, the prevalence of non-physicalist theories has increased from the turn of the millennium, suggesting that we might be in the midst of a paradigm shift in the field of consciousness research, in need of adequate constraints and methodologies, and reflecting a reaction to the failure of dominant, materialist approaches that incline toward the elimination of subjective experience.

This surging interest in non-physicalist theories might stem from the huge social changes occurring from the turn of the millennium. According to one theory of social change, the stages of society's development moved from agricultural society based on basic needs (where fertilizers serves as the raw material) to industrial society with tangible needs (with nonrenewable materials as the raw material), and into the currently emerging postindustrial service society with intangible needs (where information serves as the raw material) (Wilenius, 2014). This unfolding process is about value generation, and people's needs in the emerging service society are increasingly immaterial in nature. This is in line with the

World Values Survey, a survey that has been monitoring the development of values around the world for several decades, which likewise suggests a postmaterialist trend in values (Pajarinen et al., 2013). It is also reflected in the exponential rise in contemplative practices and the surging use of psychedelics in the younger population in Western societies.

Promoting the scientific value of non-physicalist models of consciousness is challenging, but what their re-emergence suggests, in our view, is that there is a need to take subjective experience completely seriously, and that the next generation of consciousness researchers should more deeply consider their own integration as experiencers within the design of physicalist models and experimental paradigms. As Max Planck ironically said: “A new scientific truth does not triumph by convincing its opponents and making them see the light, but rather because its opponents eventually die, and a new generation grows up that is familiar with it” (Planck, 1950, p. 33).

For sure, achieving significant advancements in our understanding of consciousness, its foundations and determinants, including through NP approaches, is, by far, not a straightforward task. The development of a universally accepted and comprehensive theory of consciousness will likely necessitate extensive, collaborative, interdisciplinary efforts marked by open scientific exchange, akin to the model of large-scale international cooperation seen in modern physics (Melloni et al., 2021). The formulation of such a theory, along with the capacity to evaluate the presence, capability, or potential for consciousness in beings, such as humans with brain injuries, other animals, and even machines, is vital not just for scientific progress but also for societal considerations. The issue of consciousness bears intricate ethical and, in certain cases, medical ramifications, as it is often implicitly or explicitly regarded as a crucial criterion in determining the value and moral status of life (Mudrik et al., 2023). At this point, consciousness science is not sufficiently advanced, and no theories and methods of detection are sufficiently specific and valid, to tackle the problem in a reliable manner, thereby justifying the need to apply the precautionary principle.

Organization of the Volume and Chosen Topics

This volume brings together international scholars from neuroscience, computational biology, mathematics, physics, neurophenomenology, psychology, cognitive sciences, philosophy, and the study of Indian and Buddhist contemplative traditions. Almost all the contributions result from shared experiences, explored in at least one of two consecutive workshops, where scientific discourse was integrated with experiential sessions, including meditation (formal sitting and walking meditations), but also with Sufi whirling, guided movement, and personal sharing groups. This induced an unusual intimate and lively atmosphere of collaborative effort, stimulating togetherness toward solving a fascinating riddle, strongly departing from the competitive disciplinary approach that often prevails in scientific conferences. The result was that many of the participants approached the organizers in order to express what they felt, in quite a unique manner. A sense of a community and of insight formed, which strongly motivated this book, hoping to support further reflections and initiatives.

With this volume, we aspired to somehow reenact such a space of shared experience as material for a broader community that might find it relevant for their own reflections. Admittedly, the exercise is limited, as it is not possible to create a real interdisciplinary discourse,

Table 0.3
A partial list of non-physicalist theories of consciousness

Family	Specific theories	References
Quantum physics theories	Orchestrated objective reduction (Orch-OR) theory: Consciousness arises through quantum effects occurring within subcellular structures internal to neurons known as *microtubules*.	Hameroff (2021) Hameroff and Penrose (2014)
	Operational probabilistic theory (OPT): Reality emerges from the free-will communications of a vast number of conscious entities, that control physical bodies and exist beyond the physical world	Faggin (2021)
	Theory of double causality: A non-deterministic model of the block universe where consciousness and free will are mechanisms by which the exact path between two space-time points is decided. Future traces may be experienced as observations of coordinated systems that past observations cannot explain.	Guillemant and Medale (2019)
dualist or anti-physicalist metaphysical views	A version of panpsychism that appeals to the notion of information not only to explain psychophysical invariances between phenomenal and physically realized information spaces but also to possibly explain the ontology of the physical as itself derived from the informational.	Chalmers (1996)
	Zero-point field: A version of panpsychism where consciousness permeates the universe yet is only concentrated in certain circumstances. Unlike other panpsychist theories, it is not the "matter" that is conscious but empty space. It bares testable predictions.	Keppler (2018)
	Emergent property dualism: Treats conscious properties as arising from complex organizations of physical constituents but as doing so in a radical way such that the emergent result is something over and above its physical causes.	Hasker (1999)
	Analytic idealism: Consciousness as Nature's sole fundamental ground and that all natural phenomena are ultimately reducible to universal consciousness.	Kastrup (2019)
Mathematical theories	Interface theory of perception: Based on a mathematical structure called "conscious agents," which space and time emerging from their exchanges. Perceptions is only an interface that allows us to interact dynamically with the world. It does not represent a universal objective reality but are species specific.	Hoffman (2014, 2015, 2019)
	Triadic-dimensional vortical paradigm (TDVP): A mathematical model in which we exist in a nine-dimensional (9D) finite, quantized, volumetric, spinning reality embedded in an infinite continuity (9D+). The model requires an extra component that they dub "gimmel," which is massless and energy-less.	Close and Neppe (2015, 2020)
	Soft logic: A mathematical model that extends classic logic to represent contradictory situations. It is based on soft numbers, which are part real and part multiples of zero (i.e., numbers on the zero-axis). Soft logic suggests a new mathematical framework that bridges between inner experience (i.e., conscious) and the outer world, and models the interaction between them.	Klein and Maimon (2016, 2020)

related to phenomal experience in meditation, within a book. To mitigate the issue, we use several devices attempting to bring the reader closer to a similar lived experience, of engaging with ideas in an experiential manner.

First, we asked all the contributors to write in a new and relatively uncommon mode of writing for scientific discourse, which combines personal autobiographical elements and their own humanely inspirational stories, with rigorous, but accessible, scientific contributions. While challenging for some, most experienced this as rewarding and sometimes even transformative.

Second, we included four experiential chapters (Fulder, Mayseless, Shamgar and Soloveitzik, Raphael) that are meant to generate experiences that may allow the reader to enter the question of consciousness in fresh, profitable ways. These, we suggest, may be read more like poetry, leaving aside the analytical mind, while opening up to the sensations, images, and emotional impact that these contributions might evoke. We warmly suggest engaging with the experiential exercises that are suggested in these chapters. This could be an opportunity to become acquainted with elusive, yet highly relevant, subjective phenomena, and will also help in approaching the other chapters and the book as a whole with the mindset of the workshop.

Third, we included in the book images by world renowned artist, Michal Rovner, as her artistic work deeply moved the editors as it so beautifully touches upon the essence of the human condition. Her images transcend time and space, and create a feeling of yearning to dissolve the tension between aloneness and togetherness, between the mundane and the spiritual, between order and chaos, between beauty and disharmony; perhaps this is a yearning to transcend the regular boundaries of consciousness. We warmly suggest to the reader to take contemplative breaks with these images, to evoke an altered state and a shift in attitude while reading the volume.

Fourth, instead of grouping the chapters into sections—for example, differentiating contributions based on fields, such as scientific, philosophical, or experiential—we decided to present the chapters in alphabetical order of author's surname. This reflected both our inability to do justice to some contributions by placing them into a specific section, and our intention to create a mélange of ideas and approaches, as we had experienced in the workshops, without locking ideas into predefined boxes. We hoped that this would produce a freer flow of ideas between disciplines, enriching and echoing each other.

In the next section, we briefly describe the chapters in the order they appear in the book and situate them within the relevant theories and approaches introduced above.

A Short Description of the Contributions

In chapter 1, Yochai Ataria presents a strong enactive view, in sharp contrast to the cognitive representationalist view. Personally, Ataria provides a brave testimony to the painfulness of the "hard problem," and from his experience as a phenomenologist, to the helplessness of phenomenology to provide "real" progress like neuroscientific findings. Yet, on the other hand, phenomenology seems to be the only thing that is "real" out there. In his chapter, Ataria forcefully uses phenomenological criticism to expose faults in the current cognitive-nonscientific approach, claiming that even with the ultimate technology, it would still be impossible to know everything about subjective experience. Ataria explains that the error in assuming that our ability to understand subjective experience rests entirely on technological progress runs deeper than usually presumed. These assumptions are then presented and refuted with phenomenological arguments. Ataria concludes that scientific research of consciousness that relies solely on technology leads to reduction *ad absurdum*, so that in order to gain a better understanding of subjective experience we must inevitably return to phenomenology.

In chapter 2, Aviva Berkovich-Ohana integrates neurophenomenology and contemplative science in the study of the self, as classically advocated by Varela (1996). For Berkovich-Ohana, the connection between personal exploration and scientific study of consciousness is very explicit, through insight gained in deep meditative experiences and dreams. She generously shares how these insights guide her research that investigates how altered states of consciousness change the basic minimal and narrative senses of self, focusing on the flexibility of self-boundaries and its implications for human well-being. She concludes that the integration of subjective experience into scientific studies might allow a more comprehensive exploration of consciousness as a multidimensional and dynamic phenomenon, and calls for openly "front-loading" intuitions gained during altered states onto the scientific design.

In chapter 3, Ema Demšar and Urban Kordeš phenomenologically explore the horizons of subjective experience. For Demšar, this core interest is driven by the potential to discover something so intimate, yet perhaps previously unnoticed. Kordeš shares that although trained as a natural scientist, first a physicist and later a cognitive scientist, he came to realize that the only channel where consciousness can be empirically reached is by studying subjective experience. Demšar and Kordeš argue that although the epistemic acts involved in observing and reporting experience do not merely retrieve information from an observer-independent stream of consciousness, they necessarily contribute constructively to the eventually generated phenomenological data. Yet, these horizons of attending to experience are part of experience—and can be phenomenologically investigated. Moreover, this—they conclude—opens the way toward joining findings of different first-person studies, to contribute to an integrative "map" of the phenomena of lived experience.

In chapter 4, Georges Dreyfus attempts to disentangle consciousness from the self, in light of Buddhist philosophy, and provides an understanding of what Buddhists mean when they speak of no-self. He shares that his former training as Buddhist monk and scholar in the Tibetan tradition led him to the study of consciousness and the self. He was surprised to realize that the insights gained through his studies and practice could be of use in the scientific study of consciousness and inform scientific research. The chapter explore what our senses of self can include, offering a simple threefold typology that distinguishes between minimal, core, and extended senses of self. Dreyfus proposes that it is not the minimal self that is the target of basic Buddhist practice, but the core self, which involves a strong boundary between oneself and others. This demonstrates his conviction that to deepen our understanding of consciousness we need to take into account the views and analyses of subjectivity provided by traditions such as Buddhism, which combine observation of subjectivity with rigorous philosophical analyses.

Chapter 5 is the first *experiential contribution* to this volume, in which Stephen Fulder presents consciousness in the context of the aspiration for liberation, as viewed by Buddhist teachings. He explores what a more liberated awareness is like, what are its sources and how it emerges in daily life experience. Fulder explains that when we use consciousness to explore the nature of consciousness, whether or not we use equipment, the mindset that we use will inevitably color the questions, the observations, and the interpretations. Conceptual thinking inevitably results in the mindset that consciousness is a concept. This reification bears a cost, including the sense that consciousness is unknowable from any "external" position. This, in turn, can invite observation from within. Meditation is a way of exploring, with penetrating insight and kindness, the nature of consciousness, and how it contributes to building our known selves and the known world. Fulder provides a basic step-by-step guidance on meditation, in order to release consciousness from its contracted perception of assumed forms including time and space. The readers are warmly encouraged to practice with these instructions for a few minutes.

In chapter 6, Shaun Gallagher presents what he calls indirect phenomenology, in relation to prereflective self-awareness and intersubjective interaction. He starts by sharing autobiographic details of his ventures into an interdisciplinary mode of philosophy, which was influenced by the work of philosophers like Merleau-Ponty. His strategy as a scientist is to combine classic phenomenology, neurology, and psychology, in order to answer "easy problems" of consciousness (what he calls "indirect phenomenology"). He shares his encounter with Varela, which led him to appreciate neurophenomenology, as well as with some other key figures in the field of consciousness research, which shaped his view in relation to the issue of this chapter. He then argues and explains why in order to understand consciousness in relation to social interaction, we need to consider not just brain mechanisms, but also the extensive and enactive processes of bodily engagement in the social environment, where the starting point should be prereflective self-consciousness.

In chapter 7, Yuval Noah Harari argues that the study of consciousness is not just a scientific enterprise but very much also an ethical and political one. Noah Harari maintains that from a political and ethical perspective, consciousness equals the capacity to suffer. This is why the questions of "who is conscious?", "who can feel something?", "who can suffer?" are extremely important political questions. If cows, chickens, or fish suffer, then this means that these animals are ethical subjects, which we must take into consideration when deciding how we treat them. The chapter ends by emphasizing the distinction between intelligence and consciousness, and with examination of ethical and political attitudes to artificial intelligence (AI), whereby computers and robots already surpass human intelligence in some domains. Noah Harari provocatively asks the questions: Should robots have rights? Should AI be authorized to make ethical decisions?

In chapter 8, Dominique Lamy and Nitzan Micher provide a methodological contribution, surveying the main intricacies associated with experimental discrepancies in research evaluating unconscious vision. This relates to the larger context of the experimental approach adopted by many cognitive psychology and cognitive neuroscience researchers when investigating consciousness: to target the "easy problem" of the transition between unconscious and conscious vision. "In this sense," they write, "what we learn about subjective experience from investigating unconscious processes is akin to what we learn about an object from its shadow." Yet, even this limited endeavor has yielded discrepancies. In the chapter, Lamy and Micher focus on the methodological issues that may account for many of these discrepancies, which inform current research on the function of conscious vision. Lamy's personal reflection is a beautiful testimony of why we need an interdisciplinary approach (and scientific encounters) to the study of consciousness.

In chapter 9, Oded Maimon and Moshe Klein present a new type of mathematical logic for consciousness, which they call "soft logic." They share that this theory stems from "revelations from our direct subjective experience." This led them to "fearless curiosity to research the foundation of consciousness, and to suggest a new science of mathematics for this purpose." In soft logic, a new kind of number is introduced, which contains two parts—a zero-axis part and a real-axis part. The zero axis represents inner experience, or consciousness, and the real axis represents the outside, physical world. Soft logic is then suggested as a new mathematical framework that bridges between inner experience and the outer world, and models the interaction between them. It also develops a new concept of modeling and dealing with the uncertainty of time and space. Maimon and Klein end by describing possible applications and further directions to develop this theory, including processes in which the consciousness of many individuals alters together—for example, during collective rituals. This chapter provides an example of the recent rise in non-physicalist theories of consciousness, stemming from subjective exposure to altered states of consciousness.

In chapter 10, Rafi Malach provides a compelling neuroscientific approach to consciousness. Malach emphasizes that the tight link between the physical processes in our brains and conscious and unconscious processes is one of the most rigorously established scientific truths. He explains that, while neuroscientists cannot solve the magic of the mind–brain link, they currently undergo an "exceedingly challenging and thrilling journey to uncover the principles and rules that underlie the association between physical brain states and the myriad of conscious experiences." Malach shares a personal account of his struggle with these issues as a neuroscientist and his attempts to explore them experimentally. His work has led him to suggest certain principles that seem to be key to understanding consciousness, including cooperative and strongly recurrent group activity—the joining together of neurons into a cohesive group. In contrast, slow, irregular, spontaneous waves of activity—correspond to subconscious process that might lead to creative insights.

In chapter 11, Zohar Maliniak explores possible parallels between the world of quantum mechanics (QM) and human experience. After outlining basic principles of David Bohm's interpretation of QM, Maliniak goes on to share some of his experiences from personal and collective experiential explorations of consciousness, and presents their parallels to Bohm's views on QM. He then suggests that there might be a "quantum order" for human consciousness, raising the possibility that the human brain, as the physical counterpart of the human mind, might also exhibit quantum effects. Zohar encourages philosophers and neuroscientists to adopt a "quantum" viewpoint in the study of the brain, suggesting that this will enable the framing of new research questions regarding consciousness.

In chapter 12, which is a partly academic and partly experiential contribution, Ofra Mayseless discusses the concept of spirituality, within the larger context of altered states of consciousness, from a positive psychology perspective. She starts by intimately sharing her trauma-induced spiritual awakening, which led her to integrate spirituality and academic work. In the chapter, she discusses some of the known types of spiritual experiences. She then presents scientific support of the connections between spiritual experiences and beneficial outcomes, and suggests some questions that future research dealing with spiritual experiences should tackle. Finally, Mayseless offers an experiential part, where she invites readers to connect to spiritual experiences they may have had and to become aware of their significance. The readers are again warmly invited to take a contemplative break in reading, and take part in this experiential exercise, which created a huge shift in the atmosphere and enhanced intimacy in the workshop.

In chapter 13, Ulrich Ott offers reflections on the importance of understanding mystical experience as part of an exploration of consciousness, states that have also fascinated him personally for many years. According to his presentation, while meditation is studied widely, it often relates to the generation of different positive cognitive potentials, which may be valuable but possess a restricted scope. Mystical states far exceed this, creating new states of mind that are characterized by deep joy and powerful experiences that transcend space and time, involve a strong sense of unity, and even reach events of non-duality. After discussing some of the methods for inducing such states in different religious traditions and cultures, Ott surveys some of the research on the possible neural correlates of consciousness, ending by outlining certain benefits such deep meditative practices may afford, while remaining aware that there are also risks involved.

In chapter 14, Giuseppe Pagnoni and Fausto Taiten Guareschi explore the relevance of Karl Friston's active inference framework to better understand the processes engaged in contemplative practices. This account contributes to the understanding of the meditation-induced unlearning process that many spiritual traditions point to as necessity to escape the limitations imposed by a well-established but ossified cognitive structure. Similarly, it explains the playfulness, openness, and immediacy that some adults maintain in life, so essential for any creative enterprise. They offer a clear and accessible presentation of the framework and connect it to Zen traditions. In particular, they emphasize the role of the drive for epistemic value or curiosity over that for utility as embedded in active inference to illuminate the gain of intimacy with oneself that can be achieved through such practices. Along the way, they propose a critical analysis of popular claims about meditation and their ramifications for humanity. On a personal note, Pagnoni writes "I feel that contemplative practice, by fostering a greater intimacy with one's own subjective experience, may facilitate access to an intuitive space that is often the shadowy breeding ground of novel perspectives."

In chapter 15, David Rudrauf discusses how his own approach to understanding and modeling consciousness in a scientific perspective was influenced by aspects of Francisco Varela's ideas. He emphasizes the often overlooked importance of mathematical modeling in Varela's neurophenomenological project. Mathematical modeling is expected to operate regarding both neural and phenomenological data, in order to define mutual constraints and offer a formal bridge or generative passages between the two domains. He also discusses how his own development of the projective consciousness model constitutes an attempt at contributing to the mathematization of phenomenology as a moment of the neurophenomenological project. In his personal reflection, he shares that the clear and somewhat obvious intuition of the three-dimensional projective structure of consciousness came to him in an altered state of consciousness.

In chapter 16, Shantena Augusto Sabbadini explores the question of whether QM has something to say about consciousness. On the personal level, Augusto Sabbadini asserts that for him, phenomenological experience is primary. All the rest is "about" it. Moreover, matter seems to arise from consciousness. In the chapter, he presents the quantum measurement problem, which lies at the heart of QM, and some of the possible interpretations to QM that propose solutions to this problem. He then claims that although the solution to the quantum measurement problem does not require consciousness, QM, and specifically quantum indeterminacy, offers a parallel to consciousness (and to life at large), as all are comprised out of "acts of creation," or unpredictable individual events.

In chapter 17, Stefan Schmidt discusses how rigorous work with experienced meditators could help resolve salient issues in psychology and neuroscience, concerned with concepts such as free will. On the personal level, he frankly shares that his approach is based on challenging some of the basic assumptions of science held by many people, like the primacy of matter. He begins with considerations about the puzzling and somehow circular entanglement between scientific research in general and consciousness, emphasizing that the latter always plays an irreducible role in the very foundations of science. He focuses on his work on the famous Libet experiment, and shows how the use of rigorous first-person methodologies with experienced meditator yields insightful interpretations of the experiment and its results, both from a psychological and neuroscientific standpoints, that are a departure from more standard accounts. The latter tend to reduce voluntary acts to deterministic unconscious processes, while his work suggests a more complex picture, thanks to the fine-grained descriptions afforded by his neurophenomenological approach.

In chapter 18, another experiential contribution, Or Raphael suggests that not one, but many different consciousnesses are studied in science. He suggests that the study of consciousness will not (and should not) reduce to the study of one specific consciousness only, and calls for the use of more rigorous definitions of consciousness when studying it. Finally, he explains why his definition of consciousness leaves it outside the realm of scientific study, and suggests two practices that shed light on the limits of human ability to know anything objectively, and on the possibility of using subjective experience as a means for gaining objective knowledge. Again, readers are highly recommended to try out the practice.

In chapter 19, Idan Segev introduces recent studies pertaining to the biophysical processes underlying conscious experiences, taking pain as his main example. On the personal level, Segev shares his childhood "naive" desire to alleviate human suffering and pain, which led him to neuroscience. In this chapter, Segev discusses new toolsets that enable experiments to validate, very convincingly, the tight correlations between the spiking (electrical) activity of specific groups of neurons and specific conscious experience. Importantly, he shows that through recent developments of optogenetic and related experimental tools, it has become possible to causally induce (or block) specific conscious experiences (e.g., fear) by activating (or inactivating) specific groups of neurons. Although these impressive advances have important clinical, ethical, and technological implications, he claims that the "big leap" between biophysical mechanisms operating in the brain and consciousness remains as large as ever. He maintains that the puzzle of consciousness delineates the limits of the explanatory power of science but that, on the other hand, it highlights the richness of the physical material that goes hand in hand with consciousness.

In chapter 20, Yaron Senderowicz claims that the first-person perspective "from within" is the most essential feature of consciousness. He personally shares how subjective experience intrigues him—"I realized quite soon that although I have special access to my conscious experiences, the content of all my conscious thoughts, intentions, and activities could be grasped only by attending to the shared world that includes me, the experiencing subject, as one of things in it." In his method Senderowicz combines phenomenological reflection and conceptual analysis. Senderowicz explains why the first-person view does not constitute a single cognitive capacity but rather several interrelated ones. He concludes that the similarities, interrelatedness, and substantive differences between these cognitive capacities should guide an interdisciplinary study of consciousness.

Chapter 21 is the last experiential contribution. In this chapter, Anat Shamgar and Tom Soloveitzik present their artistic process as performing artists of movement and sound. They discuss the characteristics of composition, which is a practice through which they make their art, and its relation to the unknown and the emerging: "The quest is to free the body-mind from inhibitions or expectations. Composition is questioning the ways one thinks . . . it is rather an expedition to the unknown which will surface, and validify itself since it has emerged in such a process." Shamgar and Soloveitzik then expand on some of the elements that comprise performance, including time, space, memory, practice, and communication. This chapter is written with a metaphoric, poetic language, as if from an altered state, encouraging the reader to enter into an altered state, into the "now." Echoing Pagnoni and Taiten Guareschi's meditation-induced unlearning process (chapter 14), which their guided movement enabled in the participants, they describe such process as "an interest and focus in the transformation of sensing and 'making sense,' seeing and forgetting, drawing, and erasing." The practice they offer provides an example of the fear one needs to overcome when going beyond the regular boundaries of consciousness.

In chapter 22, Eviatar Shulman offers a rich discussion of certain relevant theories from Indian philosophy to a discussion of consciousness. Steering away from the more commonly discussed ideas of Buddhist philosophy, he addresses the general philosophical and metaphysical context in which Hindu and Buddhist ideas of consciousness and the mind were generated. Contextualizing his views in relation to prevalent ideas of consciousness in Western philosophy of mind and its more practical application in brain-science, he argues for the need to take consciousness thoroughly seriously, which thus must relate to a dualist, idealist, or panpsychist framework. While he avoids the attempt to engage extensively with these views here, Shulman delves into classical Indian philosophical texts, mainly from the *Upaniṣads* and the *Sāṃkhya* school, which speak of the manner in which physical reality derives from and is generated through consciousness. With this he provides an Indian version for the so-called hard problem of consciousness, which hopes to facilitate and encourage further research along these lines.

In chapter 23, Narayanan Srinivasan describes his personal journey from function to phenomenology, in the exploration of consciousness and cognition. Starting from computational and functionalist views of the mind, Srinivasan describes studying the effects of attention on perception, intention, and time perception, initially measured objectively using reaction time and accuracy. With time, coming closer to study subjective experience, the measures used in his studies moved toward reports about experience. He then presents studies of meditation-related effects of attention on conscious experience (visual and temporal), based on the global workspace theory proposed by Bernard Baars. He also presents his work on volition, demonstrating that volition influences our actions and perception. The empirical studies he presents contribute to the literature on how different cognitive processes influence what we experience, how practices like meditation influence cognition and consciousness, and how conscious states like intentions influence subsequent perception. He also shows that our minds operate at multiple hierarchical spatiotemporal scales influencing our sense of agency. Srinivasan concludes that the strength of the current scientific approaches used to study conscious perception is the ability to design controlled lab experiments, while the weakness is that the rich structure of conscious experience is still not adequately captured by the dependent variables that are currently in use, strongly pointing to the need to incorporate phenomenology and the structure of conscious experience.

In chapter 24, Christopher Timmermann argues in favor of the introduction and development of neurophenomenological approaches in psychedelic drug research, as a paradigmatic context for exploring the relationships between "objective" and "subjective" data. He shares that his experiences have provided a sort of recursive process, by inspiring the development of research questions, so that the research has then served as a process to reflect upon and validate intuitions resulting from these experiences. In this chapter, he emphasizes the current lack of rigorous first-person methodologies in psychedelic drug research, and discusses, through his own work, how they could contribute to strengthen the scientific quality of such research. He also considers the contribution that this combination of neurophenomenology and psychedelic research could bring to the endeavor of reducing the explanatory gap between mental and physical descriptions, as well as ethical and practical consequences.

In chapter 25, Kenneth Williford discusses the problem of understanding and modeling prereflective self-consciousness (PRSC), which he presents as the hypothesis that consciousness is always (at least) prereflectively conscious of itself, in a manner that cannot be eliminated from the very notion of consciousness but does not reduce to the way consciousness is conscious of other objects. As a philosopher, Williford asserts that one's access to one's own subjective experience contributes to the scientific study of consciousness in a completely indispensable way, as it is our primary source of data about the phenomenon of consciousness and its various aspects, structures, and possibilities. After introducing the PRSC hypothesis from a conceptual and historical perspective, he discusses how the projective consciousness model might offer some account of it. In particular, he argues that the central projective geometrical component of the model might account both for the subjective perspective of consciousness and phenomenological aspects of PRSC, notably through the concept of duality in projective geometry for the latter. Importantly, this chapter has references to almost every other paper in the volume, which makes for a nice "epilogue" by recapitulating many of the central themes of the volume.

Concluding Remarks

This volume places subjective experience in the forefront, both from a theoretical and scientific perspective, and with regard to the nature of the contributions and their writing style. Here, the personal and the scientific and philosophical are intertwined, and the personal views and experiences reported by the contributors, which express their embodied perspective on consciousness, are not divorced from their science. This approach thus accepts that personal reflections and intuitions provide insights into how subjective experience can be "front-loaded" into scientific questions and design. This is because, for this subject matter, we believe we cannot genuinely separate the science from the scientist, or the scientist from their environment, reflecting in our approach some of the assumptions of the enactive approach to consciousness.

We hope that this effort can serve as a joint statement of a foundational standpoint on the value and intrinsic relevance of subjective experience, and as a basis for further inquiry, because we have only just begun. We expect that the future of the science of consciousness will be less dominated by the search for neural correlates, and adopt more philosophical elements and even training, which may help to combine a personal, yet rigorous, exploration of consciousness, perhaps even beyond its regular state and boundaries, with scientific methods and knowledge. Such a dual integrative work of exploration should foster a sensitivity to the different manifestations and qualities of consciousness acquaintance, and cultivate an awareness to their various theoretical layers and ramifications. Along the way, investigators might come experientially into contact with some of the subconscious and subtle imaginary and intuitive aspects of subjective experience, and perhaps even employ altered states along their scientific voyage. While much of this remains to be seen, it seems clearer each day that science cannot keep ignoring the robust fact of subjective experience, and that it must search for intelligent ways to integrate it into its inquiry. Whatever the outcome will be, we believe that the science of consciousness cannot avoid grappling with the circularity of consciousness as a process of self-discovery.

Notes

1. *Qualia* is defined as individual instances of subjective, conscious experience.
2. From the lecture of Emmi Varela at the Mind & Life's Contemplative Research Conference on November 5–8, 2020.
3. This took place in May 2019 and February 2020 (https://avivabo.edu.haifa.ac.il/Organization-of-Conferences).
4. By *phenomenology* we intend the branch of philosophy that studies subjective experiences, with the specific methodology established by Edmund Husserl.
5. Mental representation is a hypothetical entity that is presumed to stand for a perception, thought, memory, or the like, during cognitive operations.
6. A state-space representation is a mathematical model of a physical system as a set of input, output and state variables related by first-order differential equations or difference equations. The state-space is the space in which the variables on the axes are the state variables, and all the possible states of the system can be represented as a state vector within that space.
7. Consciousness studies were boosted by studying other animals such as nonhuman primates and rodents.
8. "Phenomenal" consciousness refers to its "what-it-is-likeness," hence a "minimal sense" of subjective experience without requiring reportability. In contrast, "access" consciousness corresponds to reportability, and thus requires processing of information, as well as attention or meta-cognition (e.g., Block, 2007).
9. We use Van Gulick's (2021) analysis of the problem as a thread for the preface.
10. The name most commonly associated with such a reductionist view is Daniel Dennett, most classically in his classic monograph *Consciousness Explained* (1991). A more robust eliminativist framework has been articulated by Paul and Patricia Churchland (1981, 1988).
11. This is elaborated with examples in Aviva Berkovich-Ohana's contribution (this volume).
12. This is elaborated in David Rudrauf's contribution (this volume).

References

Baars, B. J. (1988). *A cognitive theory of consciousness*. Cambridge University Press.

Berkovich-Ohana, A. (2017). Radical neurophenomenology: We cannot solve the problems using the same kind of thinking we used when we created them. *Constructivist Foundation, Special Issue on Neurophenomenology: Building a Science of Experience, 12*, 156–159.

Berkovich-Ohana, A., & Glicksohn, J. (2014). The consciousness state space—a unifying model for consciousness and self. *Frontiers in Psychology, 5*, 1–19. https://doi.org/10.3389/fpsyg.2014.00341

Berkovich-Ohana, A., Dor-Ziderman, Y., Trautwein, F.-M., Schweitzer, Y., Nave, O., Fulder, S., & Ataria, Y. (2020). The hitchhiker's guide to neurophenomenology—The case of studying self-boundaries with meditators. *Frontiers in Psychology, 11*. https://doi.org/10.3389/fpsyg.2020.01680

Block, N. (1995). On a confusion about a function of consciousness. *Behavioral and Brain Sciences, 18*(2), 227–247.

Block, N. (2007). Overflow, access, and attention. *Behavioral and Brain Sciences*, 30(5–6), 530–548.

Block, N. (2009). Comparing the major theories of consciousness. In M. Gazzaniga (Ed.), *The cognitive neurosciences IV* (pp. 1111–1123). MIT.

Chalmers, D. J. (1996). Facing up to the problem of consciousness. *Journal of Consciousness Studies, 2* (3):200–219.

Churchland, P. (1988). Can neurobiology teach us anything about consciousness. In N. Block, O. Flanagan, & G. Güzeldre (Eds.), *The nature of consciousness: Philosophical debates* (pp. 127–140). MIT.

Churchland, P. (1981). Eliminative materialism and the propositional attitudes. *Journal of Philosophy, 78*(2), 67–90.

Clark, A. (1997). *Being there: Putting brain, body, and world together again.* MIT Press.

Clark, A. (2013). Whatever next? Predictive brains, situated agents, and the future of cognitive science. *Behavioral and Brain Sciences, 36*(03), 181–204. https://doi.org/10.1017/S0140525X12000477

Close, E. R., & Neppe, V. M. (2015). Translating fifteen mysteries of the universe by applying a nine-dimensional spinning model of finite reality: A perspective, the standard model and triadic dimensional-distinction vortical paradigm, part I. *Neuroquantology, 13*(2), 205–217.

Colombo, M., & Piccinini, G. (2023). *The computational theory of mind.* Cambridge University Press.

Cosmelli, D., & Thompson, E. (2010). Embodiment or envatment? Reflections on the bodily basis of consciousness. In S. John, O. Gapenne, & E. Di Paol (Eds.), *Towards a new paradigm for cognitive science.* (pp. 361–385). MIT Press.

Crick, F., & Koch, C. (1990). Towards a neurobiological theory of consciousness. In *Seminars in the neurosciences* (Vol. 2, pp. 263–275). Saunders Scientific Publications.

Damasio, A. R. (1999). *The feeling of what happens: Body and emotion in the making of consciousness.* Harcourt Brace.

Dehaene, S., Kerszberg, M., & Changeux, J. P. (1998). A neuronal model of a global workspace in effortful cognitive tasks. *Proceedings of the National Academy of Sciences, 95*(24), 14529–14534.

Dennett, D. C. (1991). *Consciousness explained.* Little Brown & Co.

Edelman, G. M. (1989). *The remembered present: A biological theory of consciousness.* Basic Books.

Ellenberger, H. (1970). *The discovery of the unconscious: The history and evolution of dynamic psychiatry.* Basic Books.

Faggin, F. (2021). Consciousness comes first. In E. F. Kelly & P. Marshall (Eds.), *Consciousness unbound: Liberating mind from the tyranny of materialism* (pp. 283–322). Rowman & Littlefield Publishers

Flanagan, O. J., (1992). *Consciousness reconsidered.* MIT Press.

Flohr, H. (1995). Sensations and brain processes. *Behavioural Brain Research, 71*(1–2), 157–161.

Friston, K. (2010). The free-energy principle: A unified brain theory? *Nature Reviews Neuroscience, 11*(2), 127–138.

Friston, K., FitzGerald, T., Rigoli, F., Schwartenbeck, P., & Pezzulo, G. (2017). Active inference: A process theory. *Neural Computation, 29*(1), 1–49.

Gallagher, S. (2003). Phenomenology and experimental design. *Journal of Consciousness Studies, 10*(9-10), 85–99.

Gallagher, S. (2005). *How the body shapes the mind.* Oxford University Press. https://doi.org/10.1093/0199271941.001.0001

Gallagher, S., & Varela, F. J. (2003). Redrawing the map and resetting the time: Phenomenology and the cognitive sciences. *Canadian Journal of Philosophy, 33*(sup1), 93–132.

Gallagher, S. (2013). A pattern theory of self. *Frontiers in Human Neuroscience, 7*, 443.

Gallagher, S., & Daly, A. (2018). Dynamical relations in the self-pattern. *Frontiers in Psychology, 9*, 664. https://doi.org/10.3389/fpsyg.2018.00664

Gazzaniga, M. S. (2004). Consciousness and the cerebral hemispheres. In M. S. Gazzaniga (Ed.), *The cognitive neurosciences* (pp. 1391–1400). MIT Press.

Gidon, A., Aru, J., & Larkum, M. E. (2022). Does brain activity cause consciousness? A thought experiment. *PLOS Biology, 20*(6), e3001651. https://doi.org/10.1371/journal.pbio.3001651

Gray, J. A. (1995). The contents of consciousness: A neuropsychological conjecture. *Behavioral and Brain Sciences, 18*(4), 659–676. https://doi.org/10.1017/S0140525X00040395

Guillemant, P., & Medale, M. (2019). A unique but flexible space–time could challenge multiverse theories. *Annals of Physics, 409*, 167907.

Hameroff, S., & Penrose, R. (2014). Consciousness in the universe: A review of the "Orch OR" theory. *Physics of Life Reviews, 11*(1), 39–78.

Hameroff, S. (2021). "Orch OR" is the most complete, and most easily falsifiable theory of consciousness. *Cognitive Neuroscience, 12*(2), 74–76.

Hasker, W. (1999). *The emergent self.* Cornell University Press.

Hoffman, D. D. (2014). The origin of time in conscious agents. *Cosmology, 18*, 494–520.

Hoffman, D. D. (2015). *Consciousness and the interface theory of perception* (pp. 1–33). Edge Foundation.

Hoffman, D. (2019). Do we see reality? *New Scientist, 243*(3241), 34–37.

Hohwy, J., & Seth, A. (2020). Predictive processing as a systematic basis for identifying the neural correlates of consciousness. *Philosophy and the Mind Sciences*, *1*(II). https://doi.org/10.33735/phimisci.2020.II.64

Huxley, T. H. (1866). On the advisableness of improving natural knowledge. *Fortnightly Review*, *3*(17), 626–637.

James, W. (1890). *The principles of psychology: Vol. 1*. Dover. http://psychclassics.yorku.ca/James/Principles/index.htm.

Josipovic, Z. (2021). Implicit–explicit gradient of nondual awareness or consciousness as such. *Neuroscience of Consciousness*, *2021*(2), niab031.

Kastrup, B. (2019). *Analytic Idealism: A consciousness-only ontology*. Dissertation, Radboud University Nijmegen.

Keppler, J. (2018). The role of the brain in conscious processes: A new way of looking at the neural correlates of consciousness. *Frontiers in Psychology*, *9*, 1346.

Kinsbourne, M. (2000). How is consciousness expressed in the cerebral activation manifold? *Brain and Mind*, *1*(2), 265–274.

Klein, M., & Maimon, O. (2016). The mathematics of soft logic. *Materials Science and Engineering*, *157*(1), 012025. IOP conference series.

Klein, M., & Maimon, O. (2020). The dynamics in the soft numbers coordinate system. *Journal of Advanced Mathematics*, *18*, 2347–1921.

Lamme, V. A. F. (2020). Can we see without knowing that we see? Challenges for two theories of consciousness 1. In M. Overgaard, J. Mogensen, A. Kirkeby-Hinrup (Eds.), *Beyond neural correlates of consciousness* (pp. 62–86). Routledge.

Levine, J. (1983). Materialism and qualia: The explanatory gap. *Pacific Philosophical Quarterly*, *64*(4), 354–361.

Llinas, R., & Ribary, U. (2001). Consciousness and the brain: The thalamocortical dialogue in health and disease. *Annals of the New York Academy of Sciences*, *929*(1), 166–175.

Lutz, A. (2002). Toward a neurophenomenology as an account of generative passages: A first empirical case study. *Phenomenology and the Cognitive Sciences*, *1*(2), 133–167.

Mandler, G. (2002). Origins of the cognitive (r)evolution. *Journal of the History of the Behavioral Sciences*, *38*(4), 339–353. https://doi.org/10.1002/jhbs.10066

Melloni, L., Mudrik, L., Pitts, M., & Koch, C. (2021). Making the hard problem of consciousness easier. *Science*, *372*(6545), 911–912.

Merker, B., Williford, K., & Rudrauf, D. (2022). The integrated information theory of consciousness: a case of mistaken identity. *Behavioral and Brain Sciences*, *45*, 1–52.

Metzinger, T. (2020). Minimal phenomenal experience: Meditation, tonic alertness, and the phenomenology of "pure" consciousness. *Philosophy and the Mind Sciences*, *1*(I), 1–44.

Mudrik, L., Mylopoulos, M., Negro, N., & Schurger, A. (2023). Theories of consciousness and a life worth living. *Current Opinion in Behavioral Sciences*, *53*, 101299.

Naccache, L. (2018). Why and how access consciousness can account for phenomenal consciousness. *Philosophical Transactions of the Royal Society B: Biological Sciences*, *373*(1755), 20170357.

Nagel, T. (1974). What is it like to be a bat? *The Philosophical Review*, *83*(4), 435–450.

Neisser, U. (1967). *Cognitive psychology*. Appleton-Century-Crofts.

Pajarinen, M., Rouvinen, P., & Ylä-Anttila, P. (2013). Services: A new source of value. *ETLA Briefs*, *11*, 1–8. http://pub.etla.fi/ETLA-Muistio-Brief-11.pdf

Panksepp, J., Knutson, B., & Pruitt, D. L. (1998). Toward a neuroscience of emotion. In M. F. Mascolo, & S. Griffin (Eds.). *What develops in emotional development*? (pp. 53–84). Springer.

Petitmengin, C. (2017). Enaction as a lived experience: Towards a radical neurophenomenology. *Constructivist Foundations*, *12*(2), 139–147.

Planck, M. (1950). *Scientific autobiography, and other papers with a memorial address on Max Planck by Max von Laue* (Trans. from German by Frank Gaynor). Philosophical library.

Revonsuo, A. (2003). The contents of phenomenal consciousness. *Psyche*, *9*, 1–12.

Rosenthal, D. M. (2005). V. Consciousness, interpretation, and higher-order-thought. In P. Giampieri-Deutsch (Ed.), *Psychoanalysis as an empirical, interdisciplinary science: Collected papers on contemporary psychoanalytic research* (pp. 119–142). Verlag der Österreichischen Akademie der Wissenschaften (Austrian Academy of Sciences Press).

Roy, M. J., Petitot, F., Pachoud, B., & Varela, F. J. (1999). Beyond the gap: An introduction to naturalizing phenomenology. In *Naturalizing phenomenology* (pp. 1–83). Stanford University Press. https://cepa.info/2034

Rudrauf, D., Bennequin, D., Granic, I., Landini, G., Friston, K., & Williford, K. (2017). A mathematical model of embodied consciousness. *Journal of Theoretical Biology*, *428*, 106–131.

Rudrauf, D., Bennequin, D., & Williford, K. (2020). The moon illusion explained by the projective consciousness model. *Journal of Theoretical Biology*, *507*, 110455.

Rudrauf, D., Sergeant-Perthuis, G., Belli, O., Tisserand, Y., & Serugendo, G. D. M. (2022). Modeling the subjective perspective of consciousness and its role in the control of behaviors. *Journal of Theoretical Biology*, *534*, 110957.

Safron, A. (2022). Integrated world modeling theory expanded: Implications for the future of consciousness. *Frontiers in Computational Neuroscience*, *16*, 642397.

Safron, A. (2020). An integrated world modeling theory (IWMT) of consciousness: Combining integrated information and global neuronal workspace theories with the free energy principle and active inference framework; toward solving the hard problem and characterizing agentic causation. *Frontiers in Artificial Intelligence*, *3*, 520574.

Schooler, J. (2015). Bridging the objective/subjective divide—Towards a meta-perspective of science and experience. In T. Metzinger, & J. M. Windt (Eds.), *Open MIND: 34(T). Frankfurt am Main: MIND Group*. https://doi.org/10.15502/9783958570405

Searle, J. R. (1998). How to study consciousness scientifically. *Philosophical Transactions of the Royal Society A: Mathematical, Physical and Engineering Sciences*, *353*, 1935–1942.

Seth, A. K., & Bayne, T. (2022). Theories of consciousness. *Nature Reviews Neuroscience*, *23*, 439–452. https://doi.org/10.1038/s41583-022-00587-4

Singer, W. (1999). Neuronal synchrony: A versatile code for the definition of relations? *Neuron*, *24*(1), 49–65.

Thompson, E., & Cosmelli, D. (2011). Brain in a vat or body in a world? Brainbound versus enactive views of experience. *Philosophical Topics*, *39*(1), 163–180. https://doi.org/10.5840/philtopics201139119

Tononi, G. (2008). Consciousness as integrated information: A provisional manifesto. *The Biological Bulletin*, *215*(3), 216–242.

Van Gelder, T. (1998). The dynamical hypothesis in cognitive science. *Behavioral and Brain Sciences*, *21*(5), 615–628.

Van Gulick, R. (2021). Consciousness. In E. N. Zalta (Ed.), *The Stanford encyclopedia of philosophy* (Winter 2021 ed.). https://plato.stanford.edu/archives/win2021/entries/consciousness/

Varela, F. J. (1996). Neurophenomenology: A methodological remedy for the hard problem. *Journal of Consciousness Studies*, *3*(4), 330–349.

Varela, F. J., Thompson, E., & Rosch, E. (1991). *The embodied mind: Cognitive science and human experience*. MIT Press.

Varela, F., Lachaux, J. P., Rodriguez, E., & Martinerie, J. (2001). The brainweb: Phase synchronization and large-scale integration. *Nature Reviews Neuroscience*, *2*(4), 229–239.

Varela, F. J., & Thompson, E. (2003). Neural synchrony and the unity of mind: A neurophenomenological perspective. In A. Cleeremans (Ed.), *The unity of consciousness: Binding, integration, and dissociation* (pp. 266–287). Oxford University Press.

Wahbeh, H., Radin, D., Cannard, C., & Delorme, A. (2022). What if consciousness is not an emergent property of the brain? Observational and empirical challenges to materialistic models. PsyArXiv Preprint. https://doi.org/10.31234/osf.io/6tzw4

Watson, J. B. (1924). *Behaviorism*. The People's Institute Publishing Co., Inc.

Wilenius, M. (2014). Society, consciousness and change—An inquiry into Pentti Malaska's futures thinking. *Futures*, 58–67. https://doi.org/10.1016/j.futures.2014.05.005

Wiles, K. (2019). Embodied practitioner knowledge: Reconfiguring the boundary between science and meditation. *Science as Culture*, *28*(4), 449–469. https://doi.org/10.1080/09505431.2018.1561659

Williford, K., Bennequin, D., & Rudrauf, D. (2022). Pre-reflective self-consciousness & projective geometry. *Review of Philosophy and Psychology*, *13*, 1–32.

Zahavi, D. (2004). Phenomenology and the project of naturalization. *Phenomenology and the Cognitive Sciences*, *3*(4), 331–347. https://doi.org/10.1023/B:PHEN.0000048935.94012.4e

Zajonc, A. (2013). Contemplative pedagogy: A quiet revolution in higher education. *New Directions for Teaching and Learning*, *2013*(134), 83–94.

Figure 0.2
Michal Rovner. Culture Plate Flower, 2005. Print on archival paper. Copyright: Michal Rovner.

1 Unapologetic Phenomenological Approach to the Study of Consciousness

Yochai Ataria

Personal Reflections

How I Lost All Interest in the Study of Consciousness and Survived

Prologue: In 2008, when I was still a MA student, I came across an essay by David Chalmers. At the time, I had not heard about either Chalmers or consciousness, and I eagerly rushed to share this mind-blowing revelation with my favorite lecturer, only to hear that this was old news. Still, it felt like a bomb had just been dropped on me; there is something called the hard problem—the subjective experience—and science does not even know how to approach it. For quite a while, my life revolved around this problem. Only after a ten-year journey, I realized that in many ways, Chalmers held me back and the eloquent and dazzling article he wrote is nothing but a cliché, one that is particularly hard to shake off.

Games: With every passing year, I feel how people around me have more and more to say about the hard problem of consciousness, while my words seem to slip through my fingers. Many aspire to make a career for themselves in this field, and the words pile up in a steaming heap. I have to confess that I too take part in this game—what a waste.

Attempt 1: contradiction. We are not Turing machines; we do not engage in information processing. We do not represent the world, consciousness is not inside, and consciousness does not operate within the known laws of physics. I am overwhelmed with aphorisms; I know too much. Worse, I am too self-aware, and it holds me back. I am almost tempted to say that those who proclaim that not a single important word has yet been said about the study of consciousness are correct. But that is also a hollow platitude, one that is perhaps even more dangerous than Chalmers's cliché.

Attempt 2: construction. Consciousness originates in nothingness, or perhaps consciousness is the total thrownness into the world. All or nothing. We've made zero progress.

Attempt 3: phenomenology. While I am writing this text, consciousness slips through *my fingers. I listen to Ludovico Einaudi's piano, the black coffee does its thing; I am inundated yet empty. Glenn Gould's magic lies in the moments of silence between the sounds. Consciousness is in the moments of silence, but also in the relentless thoughts that do not stop for a second, and the fingers—we mustn't forget about the fingers. All the while, I keep thinking: This is lame, pathetic, who still has the time for more reflection?*

Mutiny: I smile to myself, and it hurts too. This is a joke and the joke is on you. I'm making a career out of it. Standing behind podiums and proclaiming that consciousness is not in the brain. Be brave and just say "I have no idea." Enough with this bullshit!

Cold facts: A friend sends me photos of neurons—cold facts, so he insists. And that's actually enlightening, I can't argue with the fact that this is real scientific progress. Maybe one day it will help us understand something about consciousness. Yet for now I prefer to cross out all these words that have been said. And as for the question, "what is my interest in consciousness?" My answer is I no longer have any interest.

Gratitude: I know that when I go into the Mediterranean Sea for the first time every year, I am alive. Unplugged by any doubt. I know for a fact that while my kids sneak into our bed for a cuddle at night, living becomes somewhat possible. The touch of Adi's hand in the morning when she makes herself coffee and I make sandwiches for the kids—that's poetry and consciousness and that's life itself at its peak. In moments like these, I am grateful that I am a conscious being—to experience the touch of the hand, to cherish these little moments. There are other moments, ones in which consciousness stupefies me.

Overview

Let us suppose we possess the technology that can examine brain activity at any level we choose—from neurons to networks. Under these conditions, could we, simply by observing this brain activity, offer an accurate account of the subjective experience? Many leading neuroscientists answer this question affirmatively. In this chapter, I defend a negative response to this question: even with the ultimate technology, it would still be impossible to know everything about the subjective experience. I present new arguments in support of this position—embracing a phenomenological perspective. The phenomenological criticism not only exposes faults in the current cognitive-nonscientific approach but also allows us to rethink the problem of consciousness—giving the subjective experience the respect it deserves.

Introduction

Let us suppose we possess technology that can examine brain activity at any level we choose—from neurons to networks to any new conceptual level that emerges through scientific analysis. Under these conditions, could we, simply by observing this brain activity, offer an accurate account of the subjective experience? Many leading neuroscientists answer this question affirmatively—to a large extent, this is the predominant view in the scientific world. Such scholars view *the so-called* hard problem of consciousness as merely a technical problem—one that does not deviate from the methodology currently practiced in cognitive neuroscience.

At first glance, Chalmers appears to present the opposite view: the subjective experience is beyond what techno-science can deliver using the current methods. In essence, however, Chalmers's claims are rooted in the same mode of thinking as that of neuroscientists. For example, he insists that there is a close correlation between consciousness and information processing: "Even though cognitive processes do not conceptually entail facts about conscious experience, consciousness and cognition do not float free of one another but cohere in an intimate way" (Chalmers, 1995, p. 213).

Essentially, both Chalmers and Christof Koch—two leading figures in the consciousness-brain discourse—set out from the same assumptions; that is, both operate within the framework of cybernetics, a scientific discipline "concerned with the role that the concept of information plays in the sciences of communication, control, and statistical mechanics" (Mitter, 2019, p. xxi). Notably, according to the father of cybernetics, Norbert Wiener, the gap between machine and organism is not significant, and as a rule he recommends avoiding words like "soul" or "life" altogether, instead suggesting that:

> in connection with machines . . . there is no reason why they may not resemble human beings in representing pockets of decreasing entropy in a framework in which the large entropy tends to increase. . . . The machine, like the living organism, is . . . a device which locally and temporarily seems to resist the general tendency for the increase of entropy. (Wiener, 1989, pp. 32–34)

In this chapter, I defend a negative answer to the opening question: even with the ultimate technology, it would still be impossible to know everything about the subjective experience. I present here new arguments in support of this position. Indeed, the error in assuming that our ability to understand subjective experience rests entirely on technological progress runs deeper than is usually presumed: it is rooted in assumptions embraced by philosophers like Chalmers as well as neuroscientists such as Koch. The following is a partial list of these assumptions, all of which will be refuted in this chapter:

1. Technology serves as a neutral tool that uncovers preexisting information.
2. There is a clear gap between the inner world and the external world—the subjective experience located inside.
3. The subjective experience is intentional content (see **glossary**); hence we can explore it as we can explore any other object.

This chapter discusses how phenomenologists undermine the above assumptions. Accordingly, each of the following sections focuses on one of these assumptions. Phenomenological criticism not only exposes faults in the current cognitive-nonscientific approach but also allows us to rethink the problem of consciousness—giving the subjective experience the respect it deserves.

Assumption 1: Technology Serves as a Neutral Tool to Uncover Preexisting Information

By offering people the best conceivable technology, we have misguided them. Naively, we tend to regard technology as some kind of neutral tool. Technology is an inherent part of the scientific revolution, and advances such as the telescope and the microscope, so we believe, simply improve our ability to see nature as it is. While this is indeed true in a significant number of cases, technology is not merely a neutral tool that allows us to explore the world. As Ihde stresses (2009, 2010), at least to some extent technology does not merely expose existing truth, it creates reality. In other words, the tools we use in our research (such as telescopes, fMRI, etc.) do not simply reveal what exists "out there" in the world; to a large extent, they create a certain kind of reality. For example, "mental illness" is a classification of certain phenomena that were once understood completely differently, and what brain research "finds" is not mental disorders but rather the definitions of this era for phenomena that we barely understand. This is similar to how a documentary film does not merely present the complete truth about a historical event, but instead constructs it—not just because the creator cannot avoid imposing their perspective on reality, but also because social biases are deeply embedded within technologies (as seen, for example, in the development of cameras and film, where there was a bias towards representing white faces over black faces).

An examination of one of the most prominent technologies in the field of neuroscience today, functional magnetic resonance imaging (fMRI), illustrates this notion. When observing colored "pictures" of the brain, we tend to believe that the fMRI is a kind of camera that simply presents brain activity. This, however, is far from true. When a brain area is more active, it requires more oxygen. To meet this increased demand, blood flow to the active area increases—fMRI scans allow us to trace this activity (which can occur at a delay of 3–8 seconds—an eternity in terms of correlative research) by detecting changes in blood oxygenation. This is the sum total of what an fMRI accomplishes, yet popular fascination with it "is reflected in countless articles in the press speculating on potential applications, and seeming to indicate that with fMRI we can read minds better than direct tests of behavior itself" (Logothetis, 2008, p. 869). Scientists seem to be quite happy with this illusion and even endeavor to uphold it.

Interestingly, while the findings can be visualized in a myriad of ways, scientists consistently present them as an image—preferably colored. These images give the impression, at least to some people, that we are able to glimpse into our soul (to use nonscientific terminology), as though we are seeing consciousness in action, our subjective experience while it is happening. According to Hagner (2006), neuroscientists make sure to maintain this illusion. However, the fMRI is not a camera—while photography creates images through exposure to light, a "picture" of brain activity is the outcome of computerized processes assisted by remarkably statistical tools. Thus, what the viewers actually "see" is nothing but the *authoritative interpretation of cognitive neurology concealed as "objective" visual dynamic images*. The fMRI shows us neither "reality" nor our subjective experience.

When we look through a microscope, we believe that if only our eyes were stronger, we would be able to see the microscopic world. However, the fMRI does not unveil a hidden world in the same way as a simple microscope does. The fMRI is not an improved eye that exposes a glimpse of reality. We tend to relate to the imaging process as though it is reality, but imaging, as its name suggests, is not reality, objectively apprehended or defined. From the outset, this imaging is the result of a certain system of beliefs. With this in mind, we may suggest that the so-called hardcore data tell us more about the assumptions of the researcher than the brain of the person inside the fMRI machine.

Indeed, the same can be said about the telescope. As Shapin (2018) argues, it was not just the telescope itself, but also Galileo's interpretation of what he saw that brought about the change in worldviews. Galileo's interpretation, more than the findings themselves, lies at the heart of the "scientific revolution."

The fMRI is just one example among many; the same analysis applies to different kinds of technologies. Therefore, even with the most sophisticated technology, what we may discover is probably a certain scientist's interpretation, which reveals and affirms their own preconceptions and the ones of their discipline—molecular biology, quantum physics, neural networks, and more (Marom, 2015).

We can take a step further, arguing that the fMRI tell us something deeper about the neuroscientific paradigm as a whole. If our brain activity is the territory and the colored-dynamic pictures are the maps, neuroscientists prefer the map over the territory. One can even suggest that neuroscientists deny the very concept of reality and prefer to operate within the simulacra (see **glossary**)—a pure representation that has lost connection with reality. Treating the fMRI as *the real thing* is strong evidence that we prefer Plato's Cave over reality—we are chained to our own beliefs.

The fMRI (which is of course only an example) creates the illusion that it tells us something important about brain activity and conscious experience whereas, in fact, it simply exposes our fetish for colored pictures. Indeed, this is one of our obsessions in the modern age: if we cannot take a picture of something, it does not exist. In Heidegger's words:

> When we reflect on the modern age, we are questioning concerning the modern world picture . . . world picture, when understood essentially, does not mean a picture of the world but the world conceived and grasped as picture. What is, in its entirety, is now taken in such a way that it first is in being and only is in being to the extent that it is set up by man, who represents and sets forth. (Heidegger, 1977, pp. 128–130)

Assumption 2: There Is a Clear Gap between the Inner World and the External World

In the current age, with all our knowledge regarding the intimate link between brain activity and subjective experience, it would be nonsensical to argue against such a linkage. Yet, it is one thing to argue in favor of *correlation,* and a completely different matter to argue that the subjective experience can be *reduced* to brain activity. It seems to me that this debate begins on the second floor, taking for granted a much deeper assumption: subjective experience *is located inside*. Please note that this assumption relies on far more basic postulates: (1) there is a clear boundary between the internal world and the external world; and (2) the subjective experience can be located "somewhere." Once we accept these assumptions, the brain appears to be the optimal candidate for the location of the consciousness.

In his *Meditations*, Descartes (1641/2008) casts doubt on *everything* yet concludes that he cannot doubt the fact that he doubts. Based on this process, Descartes determines that he is, first and foremost, a *thinking thing*. Notably, however, Descartes never really doubts that the process of thinking is located inside our body—to be more specific, within our brain.

The Cartesian mode of thinking stands at the very core of scientific thought. Like Descartes, modern scientists normally assume that the subjective experience can be located inside, specifically within our brain. Heidegger's notion of *being in the world* presents an alternative way to address this issue: "The clarification of being-in-the-world showed that a mere subject without a world 'is' not initially and is also never given. And, thus, an isolated *I* without the others is in the end just as far from being given initially" (Heidegger, 1996, p. 109). According to Heidegger, we are not locked inside. Rather, we are thrown into the world from the very outset. Merleau-Ponty clarifies this notion: "What is meant when we say that there is no world without a being in the world? Not that the world is constituted by consciousness, but rather that consciousness always finds itself already at work in the world" (Merleau-Ponty, 2012, p. 456).

Essentially, Heidegger replaces Descartes's *thinking thing* with coping and availableness. To be in the world is always to be thrown into a certain situation that calls for action, enabling us to go in a certain direction and making it harder for us to choose other alternatives. To be in a situation is to confront certain problems. In the process of coping with specific problems in a certain situation, there is simply no need to model the world within our head but rather to act upon it and, more importantly, with it: "It is more accurate to say that I see according to it, or with it, than that I see it" (Merleau-Ponty, 1964, p. 164).

To understand in greater depth how we rely on the world to solve problems, think of how we put together jigsaw puzzles: when we choose a piece, we do not turn it round in our "mind." We place it in the relevant area and begin to "play with it," turning it around until it fits into place, just as we park between two cars when we do not have a lot of room to maneuver. By means of feedback, we succeed in solving actual problems in the real world. As Rodney Brooks, a pioneer researcher in the field of robotics, famously said, "The world is its own best model—always exactly up to date and complete in every detail" (1990, p. 5). The world always enables feedback and shapes the way we work, guides us while coping with problems and solving them "online." The world is here "for us." Hence, the best way to solve problems in the world is to act in the world itself—there is no need for a central system representing the world in all its complexity.

This is a critical notion because it changes our perspective—we are not in our heads but rather in the world. To pin down this notion, let us think for a moment about a baseball player who catches a flying ball:

> The information processing perspective requires the player to formulate an estimate of various initial conditions of the ball's flight—the force of the impact, the angle of the trajectory, that kind of thing—then to create and analyze an internal model of the path along which the ball will likely move, then to use that model to guide and adjust motor movements continuously in time in order to intercept the ball. That is all well and good if we functioned as computers do, but [we can give] . . . a simpler account: to catch the ball, the player simply needs to keep moving in a way that keeps the ball in a constant visual relationship with respect to home plate and the surrounding scenery (technically, in a "linear optical trajectory"). This might sound complicated, but it is actually incredibly simple, and completely free of computations, representations and algorithms. (Epstein, 2016)

This example offers an alternative to the representative approach in terms of cost-benefit analysis: an approach that relies on the world itself. Continuing this line of thought, Dennett (1996) claims that the most important source of our excess intelligence is our custom of projecting as many of our conscious aims as possible onto the environment. We "extract" our brains to the world around us, building a variety of external means that can store, process, and represent the meaning anew.

In a system of this sort, there is no clear distinction between movement, perception, and expectations. It is impossible to separate them, and it is impossible to describe the cognitive system as a process of information processing (the sandwich model) that takes place between movement and perception, between input and output.

As was noted above, according to Descartes we are first of all *thinking things*, in a sense this notion has shaped large strands of Western culture in general and the scientific approach to the study of consciousness in particular. Essentially, once we begin with the *thinking thing*, we end up looking for something that does not necessarily exist. Heidegger turns this notion on its head: we are not a *thinking thing* but rather thrown into the world to begin with. Hence, there is no need for calculation and modeling, only coping—taking care of our environment:

> The closest kind of association is not mere perceptual cognition, but, rather, a handling, using, and taking care of things which has its own kind of "knowledge." Our phenomenological question is initially concerned with the being of those beings encountered when taking care of something. (Heidegger, 1996, p. 67)

I do not seek here to compare the Cartesian and Heideggerian approaches. At least at this stage, my only goal is to raise some doubts concerning what most of us tend to accept as obvious—namely, that all the *important stuff* takes place within us, or more specifically, that the subjective experience occurs inside our brain.

Assumption 3: Subjective Experience Is Intentional Content

Many findings lead us to believe that there is a direct link between subjective experience and brain activity. Among the most prominent is a recording of neurons that "fire" in response to lines at different angles (Hubel & Wiesel, 1962, Fig. 1.1). This experiment is so simple and elegant and the findings so obvious that, at least allegedly, there is hardly any point in arguing with it: the neuron (or group of neurons) fires in response to a stimulus in the world, directly resulting in our subjective experience. At this point one can ask: *What more could we want?* However, the phenomenological approach undermines the very foundations of experiments such as these.

According to the phenomenological approach, we are thrown into a world that is already filled with meaning and perceive it as such. The whole is not only greater than the sum of its parts but also precedes them: "The perceptual 'something' is always in the middle of some other thing, it always belongs to a 'field'" (Merleau-Ponty, 2012, p. 4). For example, when we look at Edvard Munch's *The Scream* our existential experience precedes every detail in the painting (and of course, this is not to suggest that we can make do without the details).

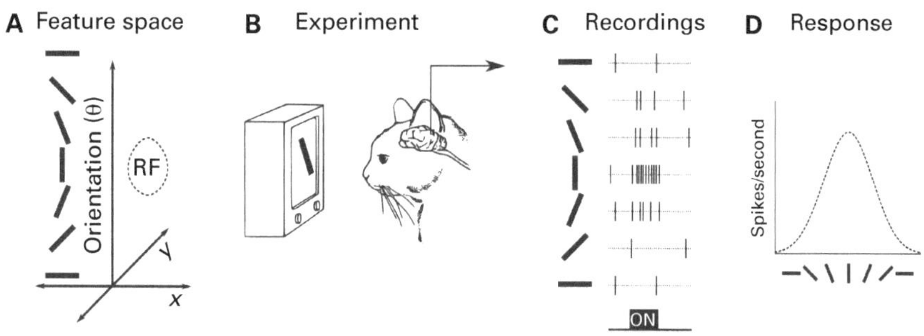

Figure 1.1
Hubel and Wiesel experiment. Responses of a complex cell in the right striate cortex (layer IV A) to various orientations of a moving black bar. I would like to thank James A. Bednar for allowing me to use this image from the paper in the *The Neuroscientist* entitled "Cortical Maps" (2016).

When we really think about the experiment in which neuronal firing is detected in response to a line that appears at a certain angle (Hubel & Wiesel, 1962), we realize that there is a fairly strong premise of a distinct and sharp separation between figure and background, and consequently that we can address the figure separately from the background. However, perception always takes place in a particular context, The question, then, is to what extent the background can be treated as "noise", that is, a mood. When concentrating on a firing group of neurons, we simply forget it. Not only do we lack the ability to perceive a straight line unless it appears against a certain background, the background itself is not passive—it changes and shapes the object (a white line on a red background is not the same as a white line on a blue background and so on).

Since subjective experience is total and holistic, it cannot be broken down into fundamental elements. Clearly, we cannot deduce from this statement that neurons firing in a particular area of the brain are irrelevant to the subjective experience; however, this is also true of the weather, our race, our gender, the state of bacteria in the gut, the political climate, our existential anxieties, and so on. We cannot ignore any of these components, but we should also avoid reducing subjective experience to one unit or the sum of several units—just as Picasso's *Guernica* cannot be broken down into the units depicted in the painting. In any case, the phenomenological critique of this experiment is more profound. To understand it, we must first examine the notion of mood.

Who among us does not have days when the world feels just too much, days when we ask "*what is the point of all this?*" or days when we feel that we no longer want to play the game? Sometimes we can point to an incident at work or at school that made us feel that way. However, more often than not, when we look closer, we realize that nothing special happened. No one said anything notable nor did we witness any unusual event. Yet still, this day was what Pink Floyd called "one of these days." On (one of) these days, to share my own subjective experience, sitting in a chair feels uncomfortable, it is hard to concentrate on what you are reading, the hours pass without being able to focus on anything, clothes feel strange on your body, you are simultaneously hungry and don't want to eat anything, people around you are annoying, you cannot do what usually comes so effortlessly, your body feels heavy, and your mind slows down to a halt, so you drink more coffee in an attempt to recover but end up even more ill at ease in your body. You just want to sleep, but at that point your body is already too caffeinated and it is not really an option. In short, nothing really "flows." Heidegger calls this a mood; essentially, we are always thrown into the world in a certain state of mind, a certain mood:

> Mood assails. It comes neither from "without" nor from "within," but rises from being-in-the-world itself as a mode of that being. But thus, by negatively contrasting attunement with the reflective apprehension of the "inner," we arrive at a positive insight into its character of disclosure. Mood has always already disclosed being-in-the-world as a whole and first makes possible directing oneself toward something. Being attuned is not initially related to something psychical, it is itself not an inner condition which then in some mysterious way reaches out and leaves its mark on things and persons. (Heidegger, 1996, p. 129)

The way we are thrown into the world shapes our existential experience. A person suffering from depression is thrown into the world differently—their field of possibilities is more limited, everything is further away and less accessible, the future is more threatening, the world is closing in on them. Likewise, a person suffering from post-trauma is thrown into a more dangerous, more frightening, menacing world. In that sense, the subjective world precedes the objective world.

With this in mind, let us pose the following question: Can we treat the subjective experience as an object (figure) that appears against a certain background, or is the subjective experience nothing other than how we are thrown into the world—a certain mood? The shift from content (which always appears against a certain background) to mood is a critical one, because from the perspective of mood we can no longer think of the subject separately from the world:

> Self and world belong together in the single entity, the Dasein [being there]. Self and world are not two beings, like subject and object, or like I and thou, but self and world are the basic determination of the Dasein itself in the unity of the structure of being-in-the world. (Heidegger 1982, p. 297)

When we are asked to describe the experience, *presumably* there is conscious content (or unconscious content that could be made conscious) that we must try to describe—in a sense, we are adopting the metaphor of a spotlight turned on our inner world. Yet subjective experience is not an object against a certain background but rather a manner of being thrown into the world; it is mood rather than content.

We tend to think about subjective experience as content that appears against a background, but in the case of subjective experience, *the figure-ground structure is fundamentally wrong*. Subjective experience is not specific content that appears against a particular background, because the subjective experience itself already includes the background.

Brentano is absoultely right when he says that, "Every mental phenomenon includes something as object within itself, although they do not do so in the same way" (Brentano, 1995, p. 68). However, that is not to say that the subjective experience can be reduced to (intentional) content, rather within the totality of our subjective experience there is also content that is part of the subjective experience but not of the subjective experience itself. Indeed, there is no doubt that almost "every mental phenomenon includes something as object within itself . . . in love loved, in hate hated, in desire desired" (Brentano, 1995, p. 68). At the same time, it is quite clear that if we think about love, hate, and other emotions in their full meanings, these are ways of *being in the world*: when we are in love, the intentional structure is dramatically changed and the whole world is perceived differently.

Subjective experience can be described in terms of the world's accessibility or situatedness, but it must include the world in its totality, because *being* is already, and always, in the world from the outset:

> I am given to myself. I am given, which is to say I find myself already situated and engaged in a physical and social world; I am given to myself, which is to say that this situation is never concealed from me, it is never around me like some foreign necessity, and I am never actually enclosed in my situation like an object in a box. (Merleau-Ponty, 2012, p. 377)

The subjective experience is not a certain line on a certain background but rather a way of being in the world, which in turn cannot be reduced to objective content. Within the totality of the subjective experience we can focus on something, yet this is only one element within the whole structure of *being in the world*.

With this in mind, clearly to even begin discussing the correlation between brain activity and subjective experience, neuroscientists must start thinking in terms of moods and abandon the attempt to reduce the subjective experience to content.

Concluding Remarks

Many neuroscientists believe that with suitable technology they will be able to explain consciousness. This chapter demonstrates that these researchers, like most philosophers, base their theories on erroneous assumptions:

1. Current neuroscientific technology does not reveal the truth but rather creates reality. There is currently too much room for interpretation and personal predispositions, and therefore many studies in the field mainly reveal the specific researcher's system of beliefs.
2. In the field of neuroscience, the Cartesian starting point—consciousness is located in the brain—is taken for granted. However, this is not necessarily the case. The boundaries between inside and outside are not so rigid, and it is possible that consciousness leaks into the world itself. Thus, consciousness is being thrown into the world to begin with.
3. Neuroscientists relate to the subjective experience as certain content, similar to a figure that appears on a background. However, based on in-depth phenomenological research, it is unclear that the subjective experience is content that can be separated from the background. Thus, to understand the subjective experience it is necessary to stop thinking about content and start thinking in terms of moods.

The current scientific approach, which considers technology above all else and believes in reduction *ad absurdum*, will not help us gain a better understanding of subjective experience. Before we ask whether the most advanced technologies will enable us to know everything about subjective experience, we should take a step back, asking ourselves a much more basic question: *What is the nature of subjective experience and how can it be explored?* The phenomenological approach is not without its errors and problems, but at least it sincerely tries to start moving in that direction.

Glossary

Intentional content: The way in which the one presents to oneself the intentional object. The easiest way to think about intentional content is by going back to the very starting point of the modern discourse regarding this topic. Franz Brentano notes that,

> Every mental phenomenon is characterized by what the Scholastics of the Middle Ages called the intentional (or mental) inexistence of an object, and what we might call, though not wholly unambiguously, reference to a content, direction toward an object (which is not to be understood here as meaning a thing), or immanent objectivity. Every mental phenomenon includes something as object within itself, although they do not do so in the same way. In presentation, something is presented, in judgment something is affirmed or denied, in love loved, in hate hated, in desire desired, and so on. This intentional inexistence is characteristic exclusively of mental phenomena. No physical phenomenon exhibits anything like it. We can, therefore, define mental phenomena by saying that they are those phenomena which contain an object intentionally within themselves. (Brentano, 1995, p. 68)

Simulacra: The classic definition of this concept is provided by Jean Baudrillard:

> By crossing into a space whose curvature is no longer that of the real, nor that of truth, the era of simulation is inaugurated by a liquidation of all referentials—worse: with their artificial resurrection in the systems of signs, a material more malleable than meaning, in that it lends itself to all systems of equivalences, to all binary oppositions,

to all combinatory algebra. It is no longer a question of imitation, nor duplication, nor even parody. It is a question of substituting the signs of the real for the real, that is to say of an operation of deterring every real process via its operational double, a programmatic, metastable, perfectly descriptive machine that offers all the signs of the real and short circuits all its vicissitudes. Never again will the real have the chance to produce itself—such is the vital function of the model in a system of death, or rather of anticipated resurrection, that no longer even gives the event of death a chance. A hyperreal henceforth sheltered from the imaginary, and from any distinction between the real and the imaginary, leaving room only for the orbital recurrence of models and for the simulated generation of differences. (Baudrillard, 1994, pp. 1–2)

References

Brooks, R. (1990). Elephants don't play chess. *Robotics and Autonomous Systems*, *6*, 3–15.

Chalmers, D. (1995). Facing up to the problem of consciousness. *Journal of Consciousness Studies*, *2*(3), 200–219.

Descartes, R. (1641/2008). *Meditations on first philosophy.* (M. Moriarty, Trans.). Oxford University Press.

Descartes, R. (1998). *Discourse on method and meditations on first philosophy.* (D. A. Cress, Trans.). Hackett Publishing Company.

Epstein, R. (2016). The empty brain. *Aeon*. https://aeon.co/essays/your-brain-does-not-process-information-and-it-is-not-a-computer.

Hagner, M. (2006). *Der geist bei derarbeit. Historische untersuchungen zur hirnforschung.* Wallstein.

Heidegger, M. (1982). *The basic problems of phenomenology* (A. Hofstadter, Trans.). Indiana University Press.

Heidegger, M. (1996). *Being and time.* (J. Stambaugh, Trans.). New York Press.

Hubel, D. H., & Wiesel, T. N. (1962). Receptive fields, binocular interaction and functional architecture in the cat's visual cortex. *The Journal of Physiology*, *160*(1), 106–154.

Husserl, E. (1989). *Ideas pertaining to a pure phenomenology and to a phenomenological philosophy—Second book: Studies in the phenomenology of constitution* (R. Rojcewicz, & A. Schuwer, Trans.). Kluwer.

Ihde, D. (2009). *Postphenomenology and technoscience.* Suny Press.

Ihde, D. (2010). *Heidegger's technologies.* Fordham University Press.

Logothetis, N. K. (2008). What we can do and what we cannot do with fMRI. *Nature*, *453*(7197), 869–878.

Marom, S. (2015). *Science, psychoanalysis, and the brain.* Cambridge University Press.

Merleau-Ponty, M. (1964) *The primacy of perception.* Northwestern University Press.

Merleau-Ponty, M. (2012). *Phenomenology of perception* (D. A. Landes, Trans.). Routledge.

Mitter, S. K. (2019). Foreword. In N. Wiener (Ed.), *Cybernetics* (pp. xxi–xxxiv). MIT Press.

Shapin, Steven. The scientific revolution. University of Chicago press, 2018.

Wiener, N. (1989). *The human use of human beings.* Free Association Books.

2 Meditation and Self-Consciousness: Neuroscience, Neurophenomenology, and Possible Implications

Aviva Berkovich-Ohana

Personal Reflections

For me, consciousness is an ever-unfolding mystery. I have spent much of my adult life exploring consciousness through various contemplative practices. Thus, the connection between personal exploration and scientific study of consciousness is very clear to me. My exploration has included different types of meditation practices, which yielded unique consciousness states, and I have also experienced many non-volitional "grace-like states," such as guided dreams.

Through discussions with many contemplative (neuro)scientists, I learned that deep meditative states often spark interest in the scientific study of consciousness. There are many documented examples of dreams inspiring scientific work, such as Descartes' dreams revealing the basis of the scientific method, Kekulé's discovery of benzene's structure, Mendeleev's vision of the periodic table, and Bohr's discovery of the atom's structure. Interestingly, Einstein said his career was an extended meditation on a dream he had as a teenager.

I want to briefly share one meditative experience and one dream that inspired my scientific inquiry, partly described in this chapter. The first inspired the scientific questions guiding my work, and the latter inspired the scientific method I employ.

The meditative experience took place in 2004 during a Vipassanā (see ***glossary****) meditation retreat. Although I started practicing in 1995, this experience was novel to me. During one sitting, I slipped into an unprecedented state of consciousness. There was no sense of time, space, thoughts, emotions, or body. It was serene and whole, like a complete void. This half-hour experience transformed my life, inspiring me to start a PhD in contemplative neuroscience and shaping my current work on the boundaries of consciousness and selfhood.*

The dream occurred in 2008 during a Mind and Life Summer Research Institute workshop. I knew little about Varela and his work until a movie about his life moved me deeply. That night, Varela visited my dream-kitchen, reassuring me that my simple setup was sufficient. This dream provided intuition and reassurance for using neurophenomenology as a research tool and adopting the enactive approach. The neurophenomenological studies we have conducted over a decade with adept meditators exemplifies Varela's call for a framework of two-way causation in studying consciousness, integrating upward causation (consciousness as an emergent property of neural activity) with downward causation (the agency of mental states).

Albert Einstein said, "The intuitive mind is a sacred gift and the rational mind is a faithful servant. We have created a society that honors the servant and has forgotten the gift." To me, the scientific work presented here, as well as the workshops and this entire volume, are a step to rectify this forgetfulness.

Overview

One traditional view held by philosophers of mind is that any conscious state entails some sense of self, either the basic minimal self or the narrative self. In this chapter, I describe how altered states of consciousness, specifically deep meditative states, alter both modes of sense of self and the implications of such alterations for human flourishing.

To these ends, I start by introducing a simple conceptualization of the dual sense of self: the minimal and the narrative modes. This framework is adopted due to its simplicity, enabling a fruitful dialogue with cognitive neuroscience by grounding philosophical conceptualizations in cognitive processes and underlying neural mechanisms. Next, I review key neuroscientific findings on meditation-related alterations in the neural mechanisms supporting the narrative self, which collectively suggest that long-term practitioners rely less on the narrative self and related processing. I then describe how neurophenomenology has contributed to the exploration of contemplation-induced alterations in the minimal self. Here, I detail our novel neurophenomenological research paradigm, which involves collaborating with long-term meditative practitioners who can volitionally produce self-dissolution experiences in a lab setting. These studies demonstrate that such experiences (1) can be phenomenologically described and studied in laboratory settings, (2) produce a replicable neural effect, and (3) are best explored using neurophenomenology. These studies introduce the notions of self-boundaries and flexibility, both of which are crucial for personal well-being, psychological growth, and relationships with others, which are the topics of the final section.

I conclude by calling for the integration of subjective experience into the scientific study of consciousness as a source of insight, inspiration, and a means to expand the scientific discourse around consciousness:

> A human being is a part of the whole called by us universe; a part limited in time and space. He experiences himself, his thoughts, and his feelings as something separate from the rest—a kind of optical delusion of his consciousness. (Albert Einstein in a letter to Robert S. Marcus, February 12, 1950, quoted in *The New York Times*, March 29, 1972)

Introduction

In psychological and cognitive literature, deep meditative states and similar conditions are often termed religious, spiritual, and sacred experiences, with the most prevalent being mystical experiences (see **glossary**; James, 1958; Stace, 1960) and altered states of consciousness (see **glossary**; Pahnke & Richards, 1972). Comparing the dimensions of mystical experiences and altered states of consciousness is beyond this chapter's scope. However, these experiences share common qualities, notably an alteration in the regular sense of self, where self-consciousness is partly or completely lost. This alteration is often referred to as a dissolution of self-boundaries, frequently experienced as unity or "selfless" experiences (Limanowski & Friston, 2020).

This chapter describes the effects of contemplative training on the sense of self. It introduces a simple conceptualization of the dual sense of self: the minimal and narrative modes, and their underlying neural mechanisms. Key neuroscientific findings on meditation-related alterations in the neural mechanisms supporting the narrative self are reviewed. The chapter also discusses how neurophenomenology has contributed to exploring contemplation-induced alterations in the minimal self. Finally, the possible implications of meditation-induced alterations of the sense of self for human flourishing and interconnectedness are considered.

The Dual Sense of Self and Underlying Neural Networks

An increasing number of publications in philosophy, psychology, and neuroscience investigate the sense of self, referring to subjects' consciousness of themselves. The concept of self is highly ambiguous, including different aspects such as somatosensory, agentive, narrative, and social components (Gallagher, 2013). Here, I adopt a prevalent framework in philosophy of mind that views the self as a two-fold phenomenon, allowing a fruitful dialogue with cognitive neuroscience by grounding philosophical conceptualizations in cognitive processes and underlying neural mechanisms.

This conceptualization is inspired by James's (1958) differentiating between the self as "I" (the subjective knower) and the self as "me" (the self-concept and autobiographical identity). In this framework, two types of self-experience can be distinguished: the "minimal self" and the "narrative self" (Gallagher, 2000). The minimal self is a foundational level of selfhood based on embodied experience, involving a sense of agency, ownership, and first-person perspective. It establishes a dynamic boundary, distinguishing our lived bodies, actions, and phenomenal interiority from the external world. The narrative self is the construction of one's personal identity extended in time, shaped by social and cultural factors, and involves high-level functions such as language and autobiographic memory.

The two philosophical notions of self-constitution, the narrative and the minimal self, support two types of cognitive processes: self-related and self-specific processes, respectively (Christoff et al., 2011). These processes are confirmed in cognitive neuroscience and related to different brain regions and networks. Self-related processing supports the narrative self by attributing or evaluating stimuli with respect to one's perceptual image or mental concept of oneself, involving regions like the default-mode network (DMN) (Raichle et al., 2001), and specifically its midline strucutres. The DMN, active at rest but less so during goal-directed tasks, is involved in introspection-related functions such as self-reflection, episodic memory, future planning, and personal moral reasoning. Self-specific processing supports the embodied self by specifying the self as an embodied subjective knower and agent. Studies of self-specific processing are less common due to the difficulty in manipulating these processes in the lab. Much work relies on virtual reality paradigms and experimental setups designed to manipulate specific facets of the minimal self, including the neural basis of the senses of agency and ownership. The literature converges on the temporal parietal junction (TPJ) for exteroceptive multisensory integration, and the insular cortex and posterior cingulate cortex (PCC) for interoceptive bodily processing (for an integrative review, see Park and Blanke [2019]).

The Narrative Self: Contemplative Neuroscience and Alterations in Self-Related Network

This section demonstrates that contemplative neuroscience supports the hypothesis that meditative practices alter the functioning of the narrative self, as suggested by Buddhist approaches: "It is generally agreed that each of us has a sense of self: the (perhaps somewhat elusive) feeling of being the particular person one is . . . But the Indian traditions suggest otherwise. Indian philosophical investigations of the self begin with the suspicion that the sense of self that everyone seems to have might be importantly mistaken—indeed that this might be the cause of our being bound to the wheel of Saṃsāra"[1] (Siderits et al., 2011, p. 3).

The accumulating knowledge concerning the networks supporting different types of self, turned out to be highly informative for contemplative neuroscience. Specifically, the DMN discovery provided the basis for a new era in the field's trajectory, inspiring a transition from studying attention and emotion-regulation process, into the core changes of the sense of self as an underlying mechanism of action of meditation (Brewer & Garrison, 2014; Tang et al., 2015). Given Buddhist philosophy's emphasis on realizing the illusory nature of the self, a key hypothesis is that long-term meditation reduces self-related processing, indicated by reduced midline DMN activity and altered connectivity.

Indeed, research using blood oxygenation level–dependent functional magnetic resonance imaging (BOLD-fMRI) shows that meditative practices involve connectivity and functional alterations in the DMN, both short-term (state) and long-term (trait) effects. Studies reveal that the DMN shows relatively little activity in meditators compared to novices, indicating diminished self-referential processing (Tang et al., 2015). I have reviewed these alterations in more detail elsewhere (Berkovich-Ohana et al. 2020b), and here provide only a brief overview.

State effects include direct measurement of activity during meditation-related practices. Such studies mostly revealed that regions of the DMN show relatively little activity in meditators compared to novice control participants (reviewed by Tang et al. [2015]). This can be largely interpreted as indicating diminished self-referential processing during various meditation types. As the posterior node of the DMN, the PCC, was related, using real-time fMRI neurofeedback, to "identifying with attributes of ourselves," or "being caught up in experience"(Brewer & Garrison, 2014), the findings of reduced DMN activity in meditation practitioners as a state and trait effect (Berkovich-Ohana et al., 2016) lend support to the theory that Buddhist-related meditation achieves its positive outcomes through a process of dis-identification from the content of ones' thoughts (Dahl et al., 2015).

Trait effects are often measured using cross-sectional design—that is, comparing meditators from different traditions and with varying levels of expertise to a control group, ideally novices. Reports of such resting-state (trait) alterations in functional connectivity in meditators include both reduced and increased functional connectivity among various DMN nodes, as well as altered functional connectivity between DMN nodes and other regions, including the cognitive control network, sensory regions, and the orbitofrontal cortex (for a recent review, see Fox et al. [2016]). These findings are hard to interpret collectively, and mostly demonstrate the malleability of the DMN.

Another way to study trait effects of meditative practices is to compare structural changes between meditators and control participants. While fMRI studies largely show that meditation practice is associated with reduced DMN activity, the related structural effects in the DMN are less clear (Fox et al., 2014). Specifically, several neuroanatomical studies investigated PCC gray matter thickness and reported contradictory results. Using MRI, we showed that gray matter in the PCC, previously related phenomenologically with the experience of attachment to one's experience (Garrison et al., 2013), reduced in meditators compared to controls (Berkovich-Ohana et al., 2020b). This correlated significantly with reduced PCC activity during both visual task and resting state, indicating a tight structure-function relationship. Interestingly, the observed gray matter reduction in the PCC was significantly and negatively correlated with both meditation expertise, and with positive affect, supporting the Buddhist intuition that the more one practices, the more one detaches from identification with experience, and the higher the experienced well-being.

These findings suggest that long-term meditation reduces reliance on self-related processing, providing physiological support for the notion that Buddhist meditations help transcend the narrative self. As Goleman wrote, "The goal of all meditative paths is to transform the meditator's consciousness. In this process, the meditator dies to his past self and is reborn to a new level of experience" (1988, p. 112).

The Minimal Self: Harnessing Neurophenomenology to Study Self-Specific Processing

In this section, I provide some overview of our ongoing neurophenomenological projects with deep meditative states in the lab. This line of work was novel and unique at the time as not much previous work focused on neurophenomenology with adept meditators (but see, e.g., a case study by Lehmann et al. [2001]). This work sparked a new wave of studies on deep meditative states by Sacchet, Laukkonen, and colleagues (e.g., Laukkonen et al., 2023; Yang et al., 2023).

The inspiration to study subtle facets of the sense of self by combining contemplative practices and harnessing neurophenomenology was given by Varela et al. (1991, p. 33): "We believe that mindful awareness practices can provide a natural bridge between cognitive science and human experience. Particularly impressive to us is the convergence that we have discovered among the main themes concerning the self and the relation between subject and object." Neurophenomenology is a research program which aspires to bridge the gap between, and integrate, first-person and third-person approaches to understanding the mind (Varela, 1996). It does so by suggesting a methodological framework allowing these two irreducible phenomenal domains to relate and reciprocally support the investigation of one another (for more examples of applications, see Schmidt's and Timmerman's chapters, this volume).

As previously mentioned, studying self-specific processes is a more complex issue than employing the network approach to study the DMN. The investigation of cases in which the embodied self appears to be radically disrupted, including psychopathologies and neurological disorders or full-body illusions, involve disruptions to a single feature of the minimal self, such as the sense of agency (e.g., thought insertion), or body ownership (e.g., full-body illusions). By contrast, deep meditative states may involve a global dissolution of sense of self (Millière et al., 2018). Unlike other altered states (e.g., dreamless sleep, drug-induced ego-dissolution, near-death experiences), deep meditative states can be non-chemically and volitionally manipulated in the lab.

In a series of pilot studies employing a small number of proficient contemplative practitioners, we harnessed an innovative magnetoencephalography (MEG) neurophenomenological setup. In the first pioneering study, we asked sixteen proficient meditators to enter volitionally into states of selflessness. Based on semi-structured interviews, we evaluated their subjective experience during those deep meditative states, and place them into groups that served to guide the neural analyses, and provided a proof-of-concept that such states could be volitionally and reliably produced in the lab (Berkovich-Ohana et al., 2020a). This study showed that meditators can volitionally enter stable altered states of consciousness and that neural architectures for the narrative and minimal self overlap in regions but differ in oscillatory signatures (Dor-Ziderman et al., 2013).

A subsequent case study was done in collaboration with Stephen Fulder (see contribution to this volume), a unique participant, owing to both his accumulated years of meditative experience (> 35 years), as well as his ability to articulate subtle aspects of experience. Fulder could volitionally dissolve self-boundaries in the lab, as confirmed by phenomenological descriptions (Ataria et al., 2015). These phenomenological insights were "front loaded" onto the neural analysis in two ways. First, the gradually descending states of self-boundary dissolution guided us toward performing a regression neural analysis, yielding a gradual reduction in high beta power (27 Hertz) in regions supporting the minimal self, specifically the PCC and TPJ (Dor-Ziderman et al., 2016). Second, the close phenomenological link between the experience of self-boundary dissolution and the selflessness states previously recorded with proficient contemplative practitioners allowed us to test the neural results (obtained in the previous data set) on this independent MEG data set. Equipped with the neural information related to self-boundaries from Fulder's data, we could regress these specific regions and frequency of interest on the larger sample. Reassuringly, the same neural measures (27 Hertz beta power in PCC and TPJ) showed a significant reduction in the entire group (Dor-Ziderman et al., 2016). This suggests that self-boundaries are flexible and that manipulating the minimal self is similar to manipulating self-boundaries. The robustness of neural regions related to both interoception and exteroception was evident, though neural data showed lower resolution compared to phenomenological data.

Building on these insights, we developed a larger study where Fulder trained forty-six meditators to manipulate their self-boundaries in the lab. This study aimed to replicate previous results and advance understanding of self-consciousness and its neural mechanisms. It targeted cognitive mechanisms related to self-boundary flexibility, such as agency, ownership, and first-person perspective, and investigates both boundary dissolution and maintenance. Phenomenologically, the results replicated and extended the previous results, showing that the group could modulate their self-boundaries in the lab, both in terms of maintenance and dissolution (Nave et al., 2021). The results support the enactive view of the minimal self, linking its generation to sensorimotor activity and attention-demanding processes. Neural results showed that participants could modulate their self-boundaries in the lab. The same neural regions and frequencies of interest were replicated for boundary dissolution (27 Hertz reduction over bilateral PCC), especially in participants experiencing deeper self-dissolution, termed "full dissolvers." A self-flexibility index, calculated from the amplitude difference between boundary dissolution and maintenance, showed significant negative correlations with phenomenological categories like attention and agency. This demonstrates the value of combining phenomenology and contemplative training for scientific characterization of processes underlying the minimal self (Trautwein et al., 2024).

These projects show that studying deep meditative states enhances our understanding of the minimal self and introduces the concepts of self-boundaries and flexibility. These notions are important for personal well-being and relationships, which is the topic of the next section.

Alterations in the Sense of Self: The Implications to Well-Being and Social Process

In this section, I discuss the relevance of contemplation-induced alterations in the sense of self and their contribution to psychological concepts such as human flourishing, happiness, and connectedness. Both personally and academically, it is crucial to bridge the basic science previously presented with psychological concepts relevant to educators, psychotherapists, and the general educated public. In this regard, my work echoes that of Professor Ofra Mayseless (this volume).

The links between an altered sense of self and human flourishing are evident in humanistic, transpersonal, and positive psychology, which focus on optimal functioning. Investigators in these fields widely agree that human psychological development, particularly concerning the sense of self, often concludes relatively early in adult life, typically between the ages of eighteen and thirty years. To transcend this conventional development, it is proposed that systematic practice for experiencing mystical or altered states may be fundamental for "unfreezing" development, enabling individuals to attain postconventional levels:

> At the preconventional and conventional stages (which together include over 90 percent of people in Western nations), people operate from their social programming, use conventional linear thinking, and create an increasingly solid independent sense of self. One's goal is to learn to operate ever more effectively in the world. Progress is defined as being more in charge of one's life by learning how to predict, measure, and explain the world. These stages are about increased differentiation; in other words, the creation of an increasingly solid and separate sense of self. When, or if, people enter the postconventional stages . . . they begin to move toward greater integration. They more and more recognize their connection with others rather than their individual separateness, gravitating ultimately toward experiencing "oneness with everything. (Hartman & Zimberoff, 2008, p. 23)

There is a general agreement among transpersonal psychologists and philosophers that characteristics of mystical or altered states—chiefly the reduction of self-boundaries and the experience of unity—can become permanent through long-term training, leading to a shift in the regular sense of self (Hartman & Zimberoff, 2008). Maslow (1971) famously termed this shift "self-transcendence," conceptualizing it as the pursuit of benefits beyond the purely personal, seeking communion with the transcendent through mystical experiences or altered states, and identifying with something greater than the individual self, often through service to others.

The theory of self-based psychological functioning by Dambrun and Ricard (2011) links self-structure and happiness. Integrating Western psychology and philosophy with Eastern traditions, this model suggests that happiness lies on a continuum. At one end is a self-centered mode, where the self is experienced as solid and independent. At the other end is selflessness, where the self is flexible, dynamic, and interconnected. A self-centered mode is associated with decreased ability to care for others, loneliness (Cacioppo et al., 2017), and drives intolerance of unpleasant emotions, experiential avoidance, reactivity to internal experiences, anxiety, and depression (Dambrun & Ricard, 2011). In contrast, selflessness is linked to altruism, kindness, respect, empathy, compassion, and harmony, and is empirically associated with meditation practice (Dambrun, 2016; Hanley et al., 2017). More recently, a set of three papers bridge the more nuanced view of the self, as described in the self-pattern theory (Gallagher, 2013), with Buddhist philosophy (Gallagher et al., 2023), enhanced well-being due to higher self-flexibility (Giommi et al., 2023), and even to self-liberation, describing a set of transformations by which the self-pattern may transform with meditative practice to a selfless-pattern of functioning (Berkovich-Ohana et al., 2024).

Finally and importantly, contemplative practices have been shown to enhance global identity, a concept that includes self-definition as part of humanity and self-investment in solidarity and concern. This enhanced global identity, in turn, promotes pro-environmental attitudes and behaviors, including environmental protection and support for climate policy (Loy et al., 2022).

Concluding Reflections

In this chapter, I aimed to achieve two goals. First, I wanted to share the fruits of my ongoing scientific research on the sense of self, following an arc from phenomenological philosophy to cognitive neuroscience, through contemplative training effects on neural activity and phenomenology, and culminating in transpersonal psychology. The second goal was to demonstrate how my subjective experience shaped my scientific search by "front loading" intuitions gained during altered states onto the scientific design. Similar themes are present in other chapters of this volume (e.g., Rudrauf, Schmidt). In doing so, I aimed to achieve one of the bridging principles of "mild neurophenomenology" (see **glossary**; Berkovich-Ohana et al., 2020a) and employ "radical neurophenomenology" (Berkovich-Ohana, 2017), which integrates subjective experience into the scientific study of consciousness as a source of insight, inspiration, and a means for expanding the discourse. This approach has the potential to expand the science of consciousness from a narrow focus on cognitive functions or the search for neural correlates to include nuanced states of expanded consciousness and explore consciousness as multidimensional and dynamic (Varela, 2002).

This reminds me of the words of Professor Kenneth Pargament, a renowned researcher and pioneer in integrating spirituality into psychotherapy. When Pargament visited the University of Haifa a couple of years ago, he gave a keynote lecture at a conference. Sharing his four-decade career path, he described how spirituality and related altered states of consciousness emerged as a legitimate pathway within psychotherapy, concluding his lecture by saying, "We don't need to whisper anymore." In comparison, the integration of such experiences into the scientific study of consciousness is still in its infancy. I would like to end by asking: Why are we whispering?

Glossary

Altered states of consciousness: This can be defined as alternate patterns or configurations of experience, which differ qualitatively from a regular state. While altered states can also include cognitive and psychological impairments, the social change of the 1960s has decisively led to tying this term with the effects of psychoactive drugs (chemical substances that cause changes in awareness, consciousness, and behavior). In this widely used context, scholars focused on seemingly beneficial aspects of altered states, such as their potential to promote creativity or treat addiction (Pahnke & Richards, 1972).

Enactive approach: This is focused on the purposeful actions and reciprocal dynamic interactions with the environment by bringing together the mind/brain, mind/body, and body/environment.

Mystical experience: This can be defined as "a belief in direct subjective experience of transcendent reality, beyond perceptual or intellectual apprehension, especially induced by means of contemplation instead of rational thought."[2]

Neurophenomenology—mild versus radical: Mild neurophenomenology aims at building correlations between first-person descriptions and neural recordings, and tries to evaluate the validity of first-person descriptions through objective criteria; in contrast, radical neurophenomenology aims to investigate the process of co-constitution of the subjective and the objective poles, within lived experience.

Notes

1. A cyclicality of all life, matter and existence, a fundamental belief of most Indian religions, which is linked to the karma theory.
2. https://www.thefreedictionary.com/Mystical+experience.

References

Ataria, Y., Dor-Ziderman, Y., & Berkovich-Ohana, A. (2015). How does it feel to lack a sense of boundaries? A case study of a long-term mindfulness meditator. *Consciousness and Cognition*, *37*, 133–147.

Berkovich-Ohana, A. (2017). Radical neurophenomenology: We cannot solve the problems using the same kind of thinking we used when we created them. *Constructivist Foundations*, *12*(2), 156–159.

Berkovich-Ohana, A., Brown, K. W., Gallagher, S., Barendregt, H., Bauer, P., Giommi, F., Nyklíček, I., Ostafin, B., Raffone, A., Slagter, H. A., Trautwein, F-M., Vago, D. & Amaro., A. (2024). From self to selflessness: How meditation transforms the self-pattern. Mindfulness *15*, 2114–2140. https://doi.org/10.1007/s12671-024-02418-2.

Berkovich-Ohana, A., Dor-Ziderman, Y., Trautwein, F.-M., Schweitzer, Y., Nave, O., Fulder, S., & Ataria, Y. (2020a). The Hitchhiker's guide to neurophenomenology—The case of studying self boundaries with meditators. *Frontiers in Psychology*, *11*. https://doi.org/10.3389/fpsyg.2020.01680

Berkovich-Ohana, A., Furman-Haran, E., Malach, R., Arieli, A., Harel, M., & Gilaie-Dotan, S. (2020b). Studying the precuneus reveals structure–function–affect correlation in long-term meditators. *Social Cognitive and Affective Neuroscience*, 1–14. https://doi.org/doi:10.1093/scan/nsaa137

Berkovich-Ohana, A., Harel, M., Hahamy, A., Arieli, A., & Malach, R. (2016). Alterations in Task-induced activity and resting-state fluctuations in visual and DMN areas revealed in long-term meditators. *NeuroImage*, *135*, 125–134.

Brewer, J. A., & Garrison, K. A. (2014). The posterior cingulate cortex as a plausible mechanistic target of meditation: Findings from neuroimaging. *Annals of the New York Academy of Sciences*, *1307*(1), 19–27.

Cacioppo, J. T., Chen, H. Y., & Cacioppo, S. (2017). Reciprocal influences between loneliness and self-centeredness: A cross-lagged panel analysis in a population-based sample of African American, Hispanic, and Caucasian adults. *Personality and Social Psychology Bulletin*, *43*(8), 1125–1135. https://doi.org/10.1177/0146167217705120

Christoff, K., Cosmelli, D., Legrand, D., & Thompson, E. (2011). Specifying the self for cognitive neuroscience. *Trends in Cognitive Sciences*, *15*(3), 104–112.

Dahl, C. J., Lutz, A., & Davidson, R. J. (2015). Reconstructing and deconstructing the self: Cognitive mechanisms in meditation practice. *Trends in Cognitive Sciences*, *19*(9), 515–523. https://doi.org/10.1016/j.tics.2015.07.001

Dambrun, M. (2016). When the dissolution of perceived body boundaries elicits happiness: The effect of selflessness induced by a body scan meditation. *Consciousness and Cognition*, *46*, 89–98.

Dambrun, M., & Ricard, M. (2011). Self-centeredness and selflessness: A theory of self-based psychological functioning and its consequences for happiness. *Review of General Psychology*, *15*(2), 138–157.

Dor-Ziderman, Y., Ataria, Y., Fulder, S., Goldstein, A., & Berkovich-Ohana, A. (2016). Self-specific processing in the meditating brain: A MEG neurophenomenological study. *Neuroscience of Consciousness*, *1*, 1–13. https://doi.org/10.1093/nc/niw019

Dor-Ziderman, Y., Berkovich-Ohana, A., Glicksohn, J., & Goldstein, A. (2013). Mindfulness-induced selflessness: A MEG neurophenomenological study. *Frontiers in Human Neuroscience*, *7*, 582. https://doi.org/10.3389/fnhum.2013.00582.

Fox, K. C., Nijeboer, S., Dixon, M. L., Floman, J. L., Ellamil, M., Rumak, S. P., Sedlmeier, P., & Christoff, K. (2014). Is meditation associated with altered brain structure? A systematic review and meta-analysis of morphometric neuroimaging in meditation practitioners. *Neuroscience & Biobehavioral Reviews*, *43*, 48–73.

Fox, K. C. R., Dixon, M. L., Nijeboer, S., Girn, M., Floman, J. L., Lifshitz, M., Ellamil, M., Sedlmeier, P., & Christoff, K. (2016). Functional neuroanatomy of meditation: A review and meta-analysis of 78 functional neuroimaging investigations. *Neuroscience & Biobehavioral Reviews*, *65*, 208–228. https://doi.org/10.1016/j.neubiorev.2016.03.021

Gallagher, S. (2013). A pattern theory of self. *Frontiers in Human Neuroscience*, *7*. https://doi.org/10.3389/fnhum.2013.00443

Gallagher, S. (2000). Philosophical conceptions of the self: Implications for cognitive science. *Trends in Cognitive Sciences*, *4*(1), 14–21.

Gallagher, S., Raffone, A., Berkovich-Ohana, A., Barendregt, H. P., Bauer, P. R., Brown, K. W., Giommi, F., Nyklíček, I., Ostafin, B. D., Slagter, H., Trautwein, F.-M., & Vago, D. R. (2023). The self-pattern and Buddhist psychology. *Mindfulness*, *15*, 795–803. https://doi.org/10.1007/s12671-023-02118-3

Garrison, K. A., Santoyo, J. F., Davis, J. H., Thornhill IV, T. A., Kerr, C. E., & Brewer, J. A. (2013). Effortless awareness: Using real-time neurofeedback to investigate correlates of posterior cingulate cortex activity in meditators' self-report. *Frontiers in Human Neuroscience*, *7*(440), 1–9. https://doi.org/10.3389/fnhum.2013.00440

Giommi, F., Bauer, P. R., Berkovich-Ohana, A., Barendregt, H., Brown, K. W., Gallagher, S., Nyklíček, I., Ostafin, B., Raffone, A., & Slagter, H. A. (2023). The (in) flexible self: Psychopathology, mindfulness, and neuroscience. *International Journal of Clinical and Health Psychology*, *23*(4), 100381.

Goleman, D. (1988). *The meditative mind—The varieties of meditative experience*. GP Putnam & Sons.

Hanley, A. W., Baker, A. K., & Garland, E. L. (2017). Self-interest may not be entirely in the interest of the self: Association between selflessness, dispositional mindfulness and psychological well-being. *Personality and Individual Differences*, *117*, 166–171.

Hartman, D., & Zimberoff, D. (2008). Higher stages of human development. *Journal of Heart-Centered Therapies*, *11*(2), 3–95.

James, W. (1958). *The varieties of religious experience*. Modern Library.

Laukkonen, R. E., Sacchet, M. D., Barendregt, H., Devaney, K. J., Chowdhury, A., & Slagter, H. A. (2023). Cessations of consciousness in meditation: Advancing a scientific understanding of nirodha samāpatti. In *Progress in Brain Research*, *280*, 61–87. https://doi.org/10.1016/bs.pbr.2022.12.007

Lehmann, D., Faber, P. L., Achermenn, P., Jeanmonod, D., Gianotti, L. R. R., & Pizzagalli, D. (2001). Brain sources of EEG gamma frequency during volitionally meditation-induced, altered states of consciousness, and experience of the self. *Psychiatry Research: Neuroimaging*, *108*(2), 111–121.

Limanowski, J., & Friston, K. (2020). Attenuating oneself: An active inference perspective on "selfless" experiences. *Philosophy and the Mind Sciences*, *1*(I), 6–6. https://doi.org/10.33735/phimisci.2020.I.35

Loy, L. S., Clemens, A., & Reese, G. (2022). Mind–body practice is related to pro-environmental engagement through self-compassion and global identity rather than to self-enhancement. *Mindfulness*, *13*(3), 660–673. https://doi.org/10.1007/s12671-021-01823-1

Maslow, A. H. (1971). *The farther reaches of human nature*. Viking.

Millière, R., Carhart-Harris, R., Roseman, L., Trautwein, F. M., & Berkovich-Ohana, A. (2018). Psychedelics, meditation and self-consciousness. *Frontiers in Psychology*, *9*, 1–29. https://doi.org/10.3389/fpsyg.2018.01475.

Nave, O., Trautwein, F.-M., Ataria, Y., Dor-Ziderman, Y., Schweitzer, Y., Fulder, S., & Berkovich-Ohana, A. (2021). Self-boundary dissolution in meditation: A phenomenological investigation. *Brain Sciences*, *11*(6), 819. https://doi.org/10.3390/brainsci11060819

Pahnke, W. N., & Richards, W. A. (1972). Implications of LSD and experimental mysticism. In C. T. Tart (Ed.), *Altered states of consciousness* (pp. 409–439). Anchor.

Park, H.-D., & Blanke, O. (2019). Coupling Inner and Outer Body for Self-Consciousness. Trends in Cognitive Sciences, 23(5), 377–388. https://doi.org/10.1016/j.tics.2019.02.002

Raichle, M. E., MacLeod, A. M., Snyder, A. Z., Powers, W. J., Gusnard, D. A., & Shulman, G. L. (2001). A default mode of brain function. *Proceedings of the National Academy of Sciences of the United States of America*, *98*(2), 676–682.

Siderits, M., Thompson, E., & Zahavi, D. (Eds.). (2011). *Self, no-self: Perspectives from analytical, phenomenological, and Indian traditions*. Oxford University Press.

Stace, W. T. (1960). *Mysticism and philosophy*. Jeremy P. Tarcher.

Tang, Y.-Y., Hölzel, B. K., & Posner, M. I. (2015). The neuroscience of mindfulness meditation. *Nature Reviews Neuroscience*, *16*(4), 213–225.

Trautwein, F.-M., Schweitzer, Y., Dor-Ziderman, Y., Nave, O., Ataria, Y., Fulder, S., & Berkovich-Ohana, A. (2024). Suspending the embodied self in meditation attenuates beta oscillations in posterior medial cortex. *Journal of Neuroscience*, *44*(26). https://doi.org/10.1523/JNEUROSCI.1182-23.2024

Varela, F. J. (1996). Neurophenomenology: A methodological remedy for the hard problem. *Journal of Consciousness Studies*, *3*(4), 330–349.

Varela, F. J. (2002). Upwards and downwards causation in the brain. *No Matter, Never Mind: Proceedings of Toward a Science of Consciousness: Fundamental Approaches (Tokyo'99)*, 95.

Varela, F. J., Thompson, E., & Rosch, E. (1991). *The embodied mind: Cognitive science and human experience*. MIT Press.

Yang, W. F. Z., Chowdhury, A., Bianciardi, M., van Lutterveld, R., Sparby, T., & Sacchet, M. D. (2023). Intensive whole-brain 7T MRI case study of volitional control of brain activity in deep absorptive meditation states. *Cerebral Cortex*, *34*(1). https://doi.org/10.1093/cercor/bhad408

3 Horizons of Examining Experience

Ema Demšar and Urban Kordeš

Personal Reflections

Ema Demšar: *It was a late summer afternoon in 2014, and I was about to start my first serious attempt at meditation. I closed my eyes. I inhaled, and exhaled. I opened a non-judgmental (or so I tried) attention to whatever was there. What did I notice? A colorful kaleidoscope of bodily sensations, brief images, half-finished spoken sentences; a series of hasty shifts in my focus; the feeling of the inhale in my belly; a background feeling of being at home in the heavy warmth of the summer air. There was also a sense of surprise at what I saw—I hadn't expected to find experiences this numerous, varied, and rich while simply sitting there, doing nothing, being me. How could I have not already been familiar with so many of the things I was now noticing? Hadn't this experience—the experience that remains as I am just existing, breathing in, breathing out—been at the center of what it was like to be me all along?*

It is this potential for discovering something so intimate, yet perhaps previously unnoticed, that sits at the core of my personal interest in studying experience. People's experiential lives are magnificently diverse and cannot be reliably predicted from the outside. Behind every seemingly simple act—having a chat, letting one's mind wander while reading this text, participating in a neuroscientific experiment—there is a myriad of possible experiences, which one (including the experiencing person themself) might not become aware of without attending to them. Besides being fascinating on a day-to-day level, I believe that charting the diversity and complexity of people's experiences is critical for future advances in understanding human cognition. While mind sciences today collect unprecedented amounts of neurophysiological and behavioral data associated with different phenomena of consciousness, participants' actual lived experience of these phenomena—and thereby, the bulk of what even the third-person approaches are trying to get at—often remains undifferentiated or even unknown. In my view, developing and employing rigorous phenomenological methods is an important contribution to the multi-method toolkit of contemporary consciousness research and cognitive science.

I am interested in first-person research because of—rather than despite—its thorny epistemological and methodological challenges. With its dynamic, transformative, and self-referential character, scientific research into experience requires not only developing and implementing first-person methods, but also a reexamination and transformation of the scientific culture, including the image of how science is and can be done. This presents an exciting opportunity for thinking about science itself. Compared to the third-person side—for which the conceptual frameworks and standards of validity have been refined through centuries-long self-correcting advancements carried out within the scientific community—the epistemological and methodological ground of researching experience is just beginning to be developed. My own motivation for the study of experience is incited by a wish to contribute to this development, and thus aims as much at investigating experiential phenomena as it does at examining the constraints and possibilities of such investigation. Starting from its uncharted terrain, I see the study of experience as necessarily humble and open-ended, but at the same time also an excitingly promising scientific endeavor.

Reductionist from Monday to Thursday

Urban Kordeš: *Studying consciousness as it presents itself to us—through subjective experience—can be confusing, paradoxical, and sometimes nearly impossible. Despite these challenges, or perhaps because of them, I feel extraordinarily fortunate that this is part of my job.*

Originally trained as a physicist, I carried the mindset of my scientific training into cognitive science: there is a physical world out there that needs to be studied. This mindset implicitly assumes that everything worth studying—elementary particles, biological organisms, human behavior, as well as human experience—lies within this domain. However, I eventually realized that, no matter how much time we spend observing neuronal activity or measuring reaction times (or conducting quantum physics experiments), we cannot seem to find lived experience. Cognitive neuroscience is becoming adept at measuring patterns of brain activation, yet it cannot even imagine directly detecting the supposed result of this activity: experience itself.

So, the big questions I had to face early on my research path were "Why is consciousness so elusive? Where is it hiding?" The answer, provided by Husserl's Crisis *(1970), was quite devastating in its simplicity and finality: within the domain studied by natural sciences there is no such thing as lived experience. Why? Because experience has been eliminated from natural science by design.*

Pioneers of Western science discovered that only those aspects of the world that can be treated as experience-free are amenable to intersubjective comparison and quantitative measurement. As natural scientists, we tend to forget that our field of research developed by selecting a specific part of the experienced world as being eligible for research—and ignoring everything that bears any sign of the existence of the subjective. Armed with this amnesia, the physical world appears to be all there is. Natural science not only accepts the everyday (or "natural") intuition—an intuition striving at all times to conceal the medium in which the world is emerging. It is built on it.

In the rare moments when we manage (or are forced by circumstances) to evade the lure of the natural attitude, it becomes clear that lived experience is everywhere. I see (i.e., experience) this computer, I hear (i.e., experience) the noise that I believe (i.e., experience) indicates a passing car, I observe recent functional magnetic resonance imaging (fMRI) data and think about it (experience again and again). Noticing that the natural attitude is only a way of interpreting experience allows for a surprising insight: consciousness is hiding in plain sight. It is everywhere. It is epistemically primary.

In this way, I know where to look—or better, how to look—to find consciousness: I have to bracket my interest in things and instead notice that I have been all the while bathing in consciousness. Such an attitude (sometimes called phenomenological) enables me to notice and research lived experience—not as if lived experience was a product of the physical world, but as a medium within which this world (and with it the self) constantly comes into being. The world and nature as they present themselves are manifestations of experience.[1] *A big advantage of my area of research is that the lab is always at hand.*

The discovery of consciousness, as earth shattering (for a natural scientist) as it is, is also very problematic. When one realizes that lived experience is all that is epistemically available and the data on lived experience is therefore the only kind of epistemically reliable data, the trouble only begins. Adopting the natural science research framework, one can simply study without needing to convince fellow scientists of the epistemological correctness of one's methods, even when the domain of interest is patently inaccessible within the chosen framework. On the other hand, the phenomenological perspective enables full contact with the domain of lived experience, but here, quite different rules apply—rules that are mostly unfriendly to what we understand as "research." For example, when we try to examine experience, this very examination becomes part of the experience. This means we cannot talk about experiential data as being collected or extracted; instead, it is co-constructed through the research process. What to do?

This is the dilemma I encounter time and again—not only in my scientific work but also in everyday life. When I meditate, I often come to understand that consciousness is all I have. These moments seem more real than the everyday focus on things. But at the same time, living with this understanding would make it difficult to function in the everyday world. Additionally, the gaze focused on things, content, and ego offers much more fun and excitement.

The two perspectives—reductionist (following an intuition of the primacy of the physical and physiological) and phenomenologist (I bracket the everyday natural attitude and discover that lived experience is all I have got to work with)—are complementary. It seems that a balancing act between the two positions is needed to build a full-fledged science of consciousness. Perhaps one could oscillate between the perspectives: adopting reductionism from Sunday to Thursday (when chances of encountering fellow neuroscientists are highest), followed by three days of a phenomenological attitude.

Overview

Lived experience has been proving to be an elusive object of examination throughout the history of the scientific study of consciousness. Neuroscientists and psychologists have invented a myriad of methods for tapping into consciousness "from the outside"—that is, by identifying and measuring its physiological and behavioral correlates. However, such third-person examination remains largely blind to the way consciousness appears "from within"—that is, from the first-person point of view of the experiencing subject. But first-person investigation of consciousness is facing a variety of epistemological and methodological challenges. Different first-person approaches produce divergent and sometimes even opposing results, and the field of first-person research has yet to develop a shared framework that would enable researchers to compare, integrate, and assess the validity of the various methods and their findings. At the heart of this challenge lies a peculiar, but unavoidable feature of examining experience: the way in which we attend to experience importantly shapes how this experience appears to us.

This chapter focuses on investigating this constructive and self-referential research feature and—starting from there—summarizes our main suggestions for the epistemology of first-person research. Grounded in our past empirical and theoretical work on studying experience in the context of cognitive science, we argue that the epistemic acts involved in observing and reporting experience do not merely retrieve information from an observer-independent stream of consciousness, but rather necessarily constructively contribute to what is acquired as phenomenological data. In this line, we describe examining experience as a process of enaction, and the experience under examination as co-defined by the way it is being investigated, arguing that. We introduce the notion of the *horizon of attending to experience* (i.e., the way in which experience is being examined within the reflective act) as an inherent factor shaping what ends up being observed and reported. We conclude by putting forward some lines of inquiry that could allow researchers to identify and articulate the horizons of particular methods. The proposed non-objectivist epistemological framework, in which results of examining experience are regarded as inseparable from the acts of examination, could present the foundation for constructing an overview of complementary methods and results, opening the way toward joining findings of different first-person studies into a more integrative map of the phenomena of lived experience.

To speak of experience as being standard, raw, or pure generally makes no sense. All we have is experience at its own level of examination, and depending on the kinds of effort and methods brought into play. (Varela & Shear, 1999, p. 14)

Introduction

What happens when you direct your attention toward an aspect of your experiential field and observe it? You can try it for yourself: Pay attention to the felt contact of your body against the surface of your seat, the feeling of your eyes focusing on reading these sentences, the sensation of the breath in your nostrils, the feeling of your left foot. After you've noticed what it's like to feel any one of these aspects of your experience, ask yourself: Was this aspect present already before I directed my attention toward it? Or did my particular way of observing it enhance it, transform it, or even create it anew?

It might appear that you are merely noticing what had been there all along—that by paying attention to the prickling in your eyes, for instance, you have *detected* it, rather than created it from scratch. But even if that's how it *feels,* how can you really *know* that the act of attending hasn't transformed or even constructed something in your experience?

Furthermore, after having turned your attention toward a particular aspect of your experience, you might be able to continue observing it in different ways, or even notice a change in what you observe in relation to your way of observation. If you pay attention to the feeling of the prickling in your eyes with aversion ("Ah, this is uncomfortable—I don't want it!"), concern ("Is my eyesight starting to deteriorate?"), or with a mindful attitude, attending to your experience might result in *different ways of experiencing*, rather than merely observing the "same" object (i.e., the "same" experience) in different ways.

The dilemmas raised by this simple experiential exercise point to a central question for the field of first-person research: Does the reflective act (see **glossary**)[2] merely bring forward what *is* (or *was*) present in experience, in precisely the way in which it is (or was) present, or does it instead play a more active role in constructing what is noticed?

The question of the epistemic status of reflection is as old as the history of first-person research itself. More than a century ago, William James—otherwise an avid proponent of self-observation—compared observing certain aspects of experience to attempts at catching a snowflake: in the same way a snowflake crystal turns into a drop when caught in a warm hand, so the continuous, ever-changing stream of consciousness is transformed into something else when arrested in an attempt at observation (James, 1890). More recent authors have discussed the reflective act in relation to the "refrigerator light" fallacy (Schear, 2009) and the "excavation" fallacy (Varela & Shear, 1999): When trying to quickly open our fridge to check if the refrigerator light is on, or to excavate an archaeological artifact that will get contaminated by the influx of air, how can we know what was there *before* we decided to look and unavoidably affect our object of observation? In attending to our experience, how can we disentangle experience from the acts of its examination, and get to what was there in the absence of looking?

Compared to the typical research objects of the natural sciences—a planet, a tree, a neuron—experience is a peculiar object of observation; the way in which it is examined appears to co-determine the way it manifests in observation. In this chapter, we will describe this circumstance not as a flaw, but as a central and unavoidable feature of the study of experience, and discuss the implications of this view for the epistemology of first-person research. While the experiential exercise and the examples used above referred to a singular reflective act, we will be primarily interested in how the constructive dimension of examining experience is present at the level of different first-person methods and their findings. We will first look at how different methods might yield diverging accounts of purportedly similar experiential phenomena through examining selected examples from studies of the experience of thinking. We will then consider cases when cultivating a particular kind of observational attitude involved in noticing and describing experience can be related to a more long-term transformation in how one relates to their experiential field. Finally, we will put forward some central concepts of a non-objectivist epistemology of first-person research, which we developed in more detail in our previous publications.[3] In particular, we will suggest regarding examining experience as a process of enaction and construing the examined experience as co-defined by the acts of examination; furthermore, we will explore the notion of horizons of examining experience, and propose several lines of inquiry that would enable researchers to identify and articulate horizons of particular first-person methods.

Our hope is that approaching the charting of experiential phenomena enacted by different first-person methods in a more systematic way could offer a way of beginning to join results of different methods into a more integrative understanding of lived experience. This endeavor, we believe, is relevant both for first-person research as an independent scientific field, and for the research program of neurophenomenology (Varela, 1996), which aims to establish correlations and mutual constraints between first-person accounts of experience and third-person accounts of corresponding neurophysiological and/or behavioral activity.

Do Different Methods Lead to Different Results? The Example of the Experience of Thinking

While the process of thinking has historically presented an important research interest for cognitive and consciousness science, different first-person approaches have typically described it in surprisingly divergent ways. In studies using *think-aloud protocols* (e.g., Ericsson & Simon, 1993), for instance, the experience of thinking tends to be conceived of—and reported—as a linear succession of verbal or verbal-like thoughts that are "passing through [participants'] attention" (see table 3.1, part A for reports of thinking during solving a math task (task 1) and a long-distance run (task 2)). A similar conception can be found in research by Benny Shanon, who examined thinking processes by documenting spontaneously emerging *thought sequences*, described as "ordered series of verbal-like expressions that pass spontaneously through one's mind and which are characterized by well-demarcated beginnings and ends" (Shanon, 1989, p. 129; see table 3.1, part B).

In contrast to viewing the experience of thinking as a linear sequence of disembodied states, other methods put more emphasis on bodily and prediscursive aspects that might accompany the emergence of thought. Research using the *micro-phenomenological interview* method (Petitmengin, 2006), for instance, points to a dimension of experience "situated at the source of our thoughts" (Petitmengin, 2007, p. 55) that, although continuously present in the depths of experience, usually remains unnoticed and is only brought to awareness under particular circumstances or by using particular tools. Micro-phenomenology studies suggest that the emergence of thought is often accompanied by a *felt meaning* (cf. Gendlin, 1962): an embodied feeling that is experienced as blurred and fuzzy, yet specific with regard to the meaning of the thought, and cannot be assigned to any particular sensory modality, but has "sensory submodalities such as shape, intensity, rhythm and movement . . . which [are] transposable from one sense to another" (Petitmengin, 2016, p. 41; see table 3.1, part C).

The *descriptive experience sampling* (DES) method (Hurlburt, 2011)—an approach that emerged from its proponent's ultimately abandoned quest for establishing a method for "thought sampling"—offers a quite different (but not necessarily opposing) insight into the "experience of thinking"—namely that there is no such thing as a general "experience of thinking" at all! Instead, DES studies show that random samples of participants' naturally occurring experience, which participants themselves initially describe as instances of "thinking," can fall into a variety of different experiential modalities—from *inner speech, inner seeing*, and *unsymbolized thinking* to *feelings* and *sensory awareness*—thereby lacking a single common experiential denominator (Heavey and Hurlburt, 2008; see table 3.1, part D).

Our aim in pointing to these—admittedly selective—examples is not to provide an exhaustive overview of research on the experience of thinking, but to demonstrate that different

Table 3.1
Examples of descriptions of thinking from studies using four different techniques.

A	B
Task 1: (1) "OK, 36 times 2 4, um, 4 times 6 is 2 4, 4, carry the 2, 4 times 3 is 12, 14, 144, 0, 2 times 6 is 12, 2, carry the 1, 2 times 3 is 6, 7, 72 0, 72 0, 144 plus 72 0, so it would be 4, 6, 864." (2) "36 times 2 4, 4, carry the—no wait, 4, carry the 2, 14, 144, 0, 36 times 2 is, 12, 6, 72, 72 0 plus 144, 4, uh, uh, 6, 8, uh, 864." Task 2: "I'm a fast runner but there's a certain speed that I want to do it and I really don't know if I can run that fast" (Emily) "my right foot is getting a little numb from going back up the hill. It doesn't feel very comfortable" (Jenny) "a little bit of pain in my left hip. Ouch. It'll get better . . . loosen up . . . it's probably a little cold" (Enzo) "I am so lucky and so blessed I can come out here and run for an hour and a half . . . it's beautiful. It's amazing; we live in a beautiful place" (Henry)	Example 14, Shanon (1989) 0. Hearing someone outside the room sneeze 1. (thinking) Bless you Example 18, Shanon (1989) 0. Eating soup with noodles, the noodles swim away 1. One should invent an eating utensil that has holes for the soup to drop off so that the noodles would stay on it 2. But there is a fork Example 33, Shanon (1989) 0. Drinking coffee with more sugar in it than usual 1. Tasty 2. It's the sugar 3. The sugar with the coffee tastes good

C	D
"Beneath the word, there is the idea, which is a particular inner space, with a specific density and texture, an orientation, a direction, tensions, inner movements . . . The space of this idea is quite different from the space of this other one . . . There's a rhythm that emerges, sometimes through only two or three words, which carry a specific rhythm, and this rhythm carries writing. It's an inner rhythm, specific to this idea that I want to express. When I feel this rhythm, I am filled with it, inhabited by it. I know that what I'm going to write will be alive, vibrant. But this rhythm comes or does not come, you cannot force it. . . . I become impregnated with it, and I feel it resonate inside me. Then I can write."	"RD was saying to himself in his inner voice, in a quiet yet surprised voice, 'Wow! Those are nice shoes!' It was a light thought, but it had started off as a solid thought deep inside his head, and then came up through his head getting progressively lighter the closer it got to the surface. As this thought made [its] way up from the depths of his mind to the surface it transformed into a medium thought, and then into a light thought—at which point he grabbed it. Thus this thought originated from within the depths of his mind and floated forward, apparently reducing in density along the way."

(A) *Think-aloud protocols*: task 1 reports were acquired by asking a participant to think aloud while solving a math task on two occasions (Ericsson, 2003, p. 10); task 2 reports were acquired by asking participants to think aloud and "verbalize everything that passes through [their] head" during a long-distance run (Samson et al., 2017, pp. 180–184). (B) *Recording thought sequences*: selected examples of "thoughts that spontaneously appeared and spontaneously ended in [participants'] minds" recorded immediately after their termination (examples 14, 18, and 33 in Shanon [1989]). (C) *Micro-phenomenology*: example in which participant "Lise" describes her experience of developing a philosophical idea through writing (Petitmengin, 2016, pp. 41–42). (D) *Descriptive experience sampling*: example of randomly sampled naturally occurring experience described (by the participant) as an experience of a thought, provided by participant "RD" (Hurlburt, 2011, p. 402, sample 5.3).

methods can generate divergent accounts of what is taken to be a certain type of experience. How are we to make sense of this divergence? Does the multiplicity of these accounts—echoing the "imageless thought" controversy that added the final blow to the project of introspectionist psychology a century ago (cf. Monson & Hurlburt, 1993)—perhaps testify to the unreliability of the field of first-person research as a whole? If not, is there a way to determine which account is correct, or at least more accurate than the others? Or might we instead say that different accounts focus on different aspects or different levels of experience, leading to different—but perhaps complementary—phenomenological descriptions that can be added together? If so, are these differences in focus primarily pragmatic, or do they reflect the method's reliance on different conceptions of consciousness?

Starting from the position of methodological pluralism, the field of first-person research is increasingly faced with the challenge of how to join the accumulating results, yielded by different methods, into coherent accounts of experiential phenomena. This calls for a development of epistemological and methodological frameworks that would enable researchers to compare, combine, and evaluate different first-person methods, as well as to begin integrating their findings into a more comprehensive body of knowledge about the phenomena of lived experience.

Becoming Aware of Experience: When Discovery Relies on Transformation

The transformative dimension of first-person research can be observed not only within a singular reflective act (in the sense of transforming the experiential episode under examination), but also on a longer-term scale (in the sense of transforming how the experiencing subject typically relates to their experience—and therefore, how they tend to experience). Compared to the field of scientific first-person research, the long-term transformation resulting from cultivating one's way of attending to experience is perhaps more explicitly acknowledged in two of its neighboring fields: phenomenology and contemplative research.[4] In the case of contemplative practices, the notion that developing particular ways of attending to experience during meditation sessions can reshape one's general way of experiencing seems especially uncontroversial: after all, this transformative potential is precisely why many practitioners take up meditation in the first place! Long-term experiential (and existential) transformation can also be seen as integral to the method of the phenomenological tradition, with Edmund Husserl emphasizing that a genuine adoption of the phenomenological attitude, while enabling the only suitable way of attaining knowledge about experience, also entails "a complete personal transformation, comparable in the beginning to a religious conversion" (Husserl, 1970, p. 137).

The idea that research practice may shape how participants relate to—and therefore, experience—their experiential landscapes in the long run is also partially acknowledged in scientific first-person research, where some authors have suggested that acquiring valid and high-quality phenomenological data demands systematic participant training. In this vein, many early introspectionists only admitted introspective reports of participants with extensive practice in observing and reporting experience (Humphrey, 1951). Among contemporary approaches, the demand for participant training is most explicitly voiced in DES (Hurlburt, 2011), where phenomenological data is acquired through a longitudinal and iterative process: a series of sampling days (i.e., days in which participants randomly sample their everyday experience using beepers), each of which is followed by an interview clarifying and expanding on the acquired samples. Through this process, participants gradually get better at observing and reporting their ongoing experience at the moment of the sampling and are likely to generate increasingly more "faithful" reconstructions of experience.

Across different DES studies, Hurlburt reports a telling pattern: after a few days of experience sampling, most participants are surprised to discover that their initial beliefs about their experience have turned out to be largely mistaken. In a study with women with bulimia nervosa, for instance, all participants eventually began to notice that most of their sampled experiences were "fragmentedly multiple"; however, none of them had been aware of this characteristic of their experiential lives, which seems to be infrequent in general population, prior to the sampling (see Hurlburt, 2011, pp. 34–35). Similar dynamics of participants becoming sensitized to specific aspects of their experience has been reported for methods that do not specifically demand participant training (see for instance Petitmengin et al. 2007 for examples in which participants learned to recognize previously unnoticed signs of the approaching epileptic seizure by becoming aware of them in a micro-phenomenological interview).

First-person research, as well as contemplative and phenomenological research all seem to share a tension between the realism implied in their aim to "become aware of" (Depraz et al., 2003), to "come into contact with," or to "unveil" (Petitmengin, 2021) lived experience—that is, to notice lived experience precisely as it is (or was) given in consciousness—and the transformation of the experiencing subject that they often acknowledge as a consequence of, or even a condition for their epistemic gains. Treating this tension with a dichotomizing gaze can give rise to metaphysical dilemmas paralleling those that we asked after the experiential exercise at the beginning of this chapter: Does encountering and/or training new ways of examining experience enable research participants (or meditators, or phenomenologists) to *better access* their experience—or does it instead lead them, this time in the long run, to *experience differently*?

In our view, this dilemma between discovery and transformation is part of a metaphysical discussion that cannot be resolved by empirical evidence. Instead of engaging in it, we suggest that understanding first-person research could benefit from "bracketing the metaphysical attitude" (Kordeš, 2021) and starting from a pragmatic focus on what is empirically available.

Epistemological Framework for First-Person Research: Reflective Act as a Process of Enaction, Shaped by the Horizons of Examining Experience

In the previous two sections, we used various examples to illustrate that the way in which we investigate experience appears to play a constructive role in shaping what ends up being observed and reported. How you *look at* your experience appears to co-determine what you (can) *see*—and, by extension, what phenomenological data you (can) provide as a participant in a first-person study. In our view, this constructive and self-referential nature of examination is an inherent feature of examining experience (within and outside the scientific context), and should thus be accepted as one of the constitutive constraints of the field of first-person research. In what follows, we present some central concepts of an alternative, non-objectivist epistemological framework for first-person research, grounded in conceiving of the process of examining experience as a process of enaction.

Observing and Reporting on Experience Is a Process of Enaction

We suggest that any method for examining lived experience must acknowledge that the act of observing (and reporting) experience does not merely retrieve information from an independent, undisturbed stream of consciousness, but constructively contributes to the generated phenomenological data. In line with these considerations, the act of acquiring data on and forming knowledge about experience can be best described as a process of *enaction.* In *The Embodied Mind*, Varela, Thompson, and Rosch introduced the concept of enaction as a means to "negotiate a middle path" between "cognition as the recovery of a pregiven outer world" and "cognition as the projection of a pregiven inner world" (Varela et al., 1991, p. 172), to stress the inseparability of action and perception, and to emphasize the mutual specification of the knower (the cognizing mind) and the known (the cognized world). Starting from their treatment of cognitive and perceptual processes, we have suggested (see especially Kordeš & Demšar 2018) that an analogous "middle path" can be identified in the investigation of experience. We propose that experience, manifested in the process of attending to it within the reflective act, can be described as *enacted*—brought forth as an outcome of an unresolvable intertwinement of the examined object and the act of examination.

This view offers a way of overcoming the dichotomy in understanding the results of observing experience either as a replica of the original experience or an arbitrary construction. In our view, neither of the two options provides a suitable answer, as the question itself starts from an untenable objectivist premise. According to the idea of examining experience as a process of enaction, the puzzle of what was *really* present in experience—and what wasn't—cannot be disentangled empirically: all we can know is what was brought forth in the process of reflection. Consequently, the final outcomes of observation (i.e., phenomenological data) are no longer conceived as descriptions of a domain existing independently of the acts of examination, but as descriptions of this domain's reactions to particular ways of probing (with the caveat that the act of probing is itself part of the domain).

Experience Under Investigation Is A Phenomenon, Co-Defined by the Way of Investigating

In discussing the epistemological situation in the field of first-person research, our past work (see especially Kordeš and Demšar 2019) has borrowed from Niels Bohr's philosophy of quantum mechanics—specifically his notion of the *phenomenon*, referring to "the observations obtained under specified circumstances, including an account of the whole experimental arrangement" (Bohr, 1949, p. 238). In quantum observation, the experimental arrangement—encompassing the totality of observational characteristics—appears to shape phenomena that end up being observed: if we are interested in particle-like attributes of a quantum object (i.e., by measuring its position), it will behave as a particle; if we want to find out its wave-like attributes (i.e., by measuring its wavelength), it will behave as a wave. Related to this and similar observational paradoxes, Bohr suggested understanding quantum phenomena (i.e., the experientially and epistemically accessible *manifestations* of quantum systems) as *co-defined* by the selected experimental arrangement. He further argued that accounting for the experimental arrangement is necessary in accounting for any phenomenon that ends up being observed (Bohr, 1949; cf. Bitbol, 2001). Similarly, we suggest that experience being observed in the reflective act can be regarded as a phenomenon, co-defined by the way in which it is being attended to and examined.

Horizons Shape the Way of Attending to and Examining Experience

Taking the above two suggestions seriously implies that understanding phenomenological data requires analyzing the characteristics of the way in which this data was acquired. To refer to the totality of these characteristics with a unitary concept, we have borrowed the Husserlian notion of the "horizon" (see especially Kordeš & Demšar, 2021b) and suggested that horizons of examining experience can be observed on two levels: that of the *horizon of attending to experience* (shaping how experience is attended to and examined within a particular reflective act), and more broadly, that of the *horizon of the method* (shaping the propensity of a particular first-person method to generate particular types of results; see **glossary**).

Horizons may be related to a variety of factors: the initial attitude with which one first directs their attention toward a specific aspect of their experience; their beliefs and expectations about what they would (or should) notice in reflection; their overall aim in attending to experience; the ongoing evolution of the reflective act; and the accessibility of concepts and terms for articulating the experience under examination. In the context of a scientific study, all these factors might intertwine with a wider framework shaped by the study's research questions and goals, the theoretical background that contributed to its design, and the social dimension of the research environment, including the demand characteristics (e.g., Orne, 1962) and the relational dynamics between the participant and the researcher (e.g., Weber & Cook, 1972). The horizon cannot be avoided: even with an impeccable phenomenological attitude (whose precise nature is still to be described), attending to one's experience is still carried out with—precisely—a particular attitude.

Horizons of Different First-Person Methods Can Be Investigated by Articulating Aspects of These Methods

The proposed epistemology, grounded in the notion of enaction, allows for a pluralist understanding of first-person research: different first-person methods employ specific, characteristic ways of probing, which lead to different types of responses of the probed domain. Thus, in order to understand the answers (i.e., the emerging phenomena, phenomenological data, and final results) acquired with a particular method, we must understand its way of questioning (i.e., the horizon of the particular method). By analogy to quantum observation, each particular horizon of the method will allow us to inquire into some aspects of experience, and prevent us from accessing others.

We suggest that investigating and mapping the horizons of different methods could involve three intertwined axes of analysis:

1. *Examining the method's underlying theoretical views, assumptions, and expectations*—for instance, whether a particular method subscribes to a "deep" or a "shallow" view of consciousness (identified as respectively adopted by DES and microphenomenology in Froese et al., 2011), or whether it focuses on the instantaneous (synchronic) or the temporarily extended (diachronic) structure of experience (cf. Petitmengin, 2006). (A good

example of such examinations can be found in Hurlburt's continuous engagement with philosophical challenges to DES; see, for instance, Hurlburt and Schwitzgebel, 2007.)

2. *Examining specific techniques employed in the methodological framework* of the method—for instance whether (and how) the method uses retrospection or real-time observation of concurrent experience; whether (and how) it employs sampling, interview, or diary techniques, and what are their technical specificities; whether the method requires training of the participants, and the type of this training; the method's relationship to phenomenological attitude; the techniques involved in managing and analyzing acquired descriptions of experience and presenting the results.

3. *Examining the method's results*, specifically with regard to the enacted experiential phenomena and aspects of experience they tend to describe—for instance comparing the results acquired with a particular method with those acquired with other methods that have been applied to investigating similar experiential phenomena (e.g., a systematic metastudy of first-person studies of the experience of thinking, carried out along the lines sketched at the beginning of this chapter).

The project of examining horizons of different methods would aim at constructing a space in which results of different first-person methods would not be regarded as opposing (or competing), but recognized as complementary (insofar as they satisfy validity criteria—a complex issue that exceeds the scope of this chapter; but see Kordeš & Demšar, 2021b). This could—or so we hope—eventually enable consciousness researchers to start charting different experiential phenomena, enacted and described by a variety of first-person methods, within an integrative "map" representing not only the phenomena, but also the horizons through which they were acquired. Along with presenting an epistemologically and methodologically improved foundation for future research, this approach might also enable an integration of already existing findings, acquired in past first-person (or neurophenomenological) studies. At the same time, an integrative system of mapping different types of experiential phenomena could eventually help to identify current gaps in understanding lived experience and guide the development of new methods that would allow for acquiring data pertaining to these gaps.

Concluding Reflections: Good Fences Make Good Neighbors

The notion that experience can be investigated in the same way as the elements of the natural world has led to many confusions along the path of integrating its study into cognitive science, chiefly contributing to the continuous failures of the field of first-person research. When investigating experience, we cannot ignore—at least not without ending up in inconsistencies—that the act of investigating co-constructed the results. Rather, to understand what we have observed, we must consider not only the acquired phenomenological data, but also the characteristics of the epistemic acts that brought this data to the fore.

Questioning the idea that consciousness, as it is given in the first-person, is already known is relevant not only for first-person research as an independent scientific field, but also for third-person inquiry. For understanding any potentially measured physiological or behavioral correlates of conscious processes or states, we must strive for a phenomenological understanding of the experiential phenomenon that they are correlated with. There is a prevalent intuition in cognitive science that we already know our experience, and that the still unknown research object—one whose future study promises to bring most findings about consciousness—is the brain. However, systematic investigation of experience shows that this intuition is illusory: when we turn the lens of scientific curiosity onto experience, we discover that very little is known about the basic structures of experiencing, the nature of experiential gestures, or the intersubjective differences in people's experiential fields.

In this line, we believe that the scientific study of consciousness as a whole would greatly benefit from encouraging the very "elementary" exploration of lived experience—"elementary" in the sense of persistently and patiently gathering samples of experience, even if those don't immediately lead to intersubjectively invariant and/or statistically significant results. This will require enabling an ethnography of experiencing, inviting a precise, systematic, qualitative, and in-depth familiarization with the field that at a first glance appears to be familiar and obvious.

We must acknowledge the distinct nature of the domain of first-person research, and accept its peculiarities as features, rather than bugs. Instead of evaluating the first-person side against the yardstick of the third-person side (or the other way around), we must cultivate the particularity of both sides—striving for neurophenomenological connections that naturally arise in the form of correlations or mutual constraints (Varela, 1996).

Acknowledgments

We would like to thank Yochai Ataria, Natalie Depraz, Yann Harel, Ole Höffken, Russ Hurlburt, Ohad Nave, and Aleš Oblak for their comments on earlier versions of this chapter.

Glossary

Enaction: A notion adapted from Varela et al. (1991), used here to describe the epistemological nature of the acts of knowledge generation in examining experience. By suggesting that in reflection, experiential phenomena (and further on, phenomenological data acquired on them) are *enacted*, we aim to emphasize that they are neither arbitrarily invented nor simply "recovered" from an independent domain of observation: instead, they are brought forth through the mutual specification between the examined experience and the act of examination.

First-person method: Any research method aimed at acquiring first-person or phenomenological data (this includes interview-based approaches sometimes called "second-person methods").

Horizon of attending to experience: The totality of characteristics involved in relating to one's experience within the reflective act. We argue that every attending to experience (directed either at the concurrent experience from the present or an already transpired experience from the past; phenomenological or not) entails a horizon. Horizons necessarily co-determine how experience is made manifest in the reflective act, and can be at the same time themselves recognized as elements of experience.

Horizon of the method: The inherent perspectivity of the entire process of first-person inquiry. It is the sum of "horizons" adopted not only in attending to experience, but also in further processes involved in generating phenomenological data and final results of first-person studies—acts of (verbal) articulation, intersubjective co-construction, analysis, and assessment.

Phenomenological data: Data on lived experience (cf. Varela & Shear, 1999), usually taking the form of verbal descriptions of a particular experiential episode, or certain aspects thereof.

Phenomenon: The object of observation *as it is manifested* in the act of observation (in this case, lived experience as it appears to the experiential subject as she is attending to it). By analogy to perception, experience-as-phenomenon refers to what is "seen" when "looking at" experience.

Reflection/reflective act: The process of becoming reflectively aware of one's ongoing or past experience, most crucially consisting of attending to (or observing) experience (what Depraz et al. [2003] refer to as the "gesture of becoming aware"), but potentially also involving verbal or nonverbal articulation of experience. We are primarily interested in *phenomenological* reflection (i.e., reflection broadly carried out in line with the general methodological guidelines of the phenomenological approach) that is aimed at yielding phenomenological data.

Notes

1. This does not mean, however, that some of these manifestations do not reflect a broader, "external" reality. There might exist a very clever metaphysical argument why they do, but this does not in any way change the epistemic primacy of lived experience.

2. See **glossary** for clarification of our use of the following terms: first-person method, phenomenological data, reflection (or reflective act), phenomenon, enaction, horizon of attending to experience, and horizon of the method.

3. Summarizing elements of our previous work, parts of this chapter are similar to the content presented in some of our other recent publications (Kordeš & Demšar, 2018, 2019, 2021a, b; Kordeš et al., 2019).

4. Rather than presenting entirely separate fields of inquiry, phenomenology, certain contemplative practices, and first-person research share several commonalities (cf. Depraz, 2019; Varela & Shear, 1999), starting with their shared striving for "becoming aware" of lived experience precisely as it is (or was) given in consciousness (Depraz et al., 2003). In this way, meditative practice has been said to assist in "lowering the threshold" of consciousness (Sparby, 2015) and even described as a "telescope" for consciousness research (Wallace, 1999, p. 177). Furthermore, both phenomenological and contemplative insights have been at the core of designing first-person approaches such as microphenomenology (Petitmengin, 2006).

References

Bitbol, M. (2001). Non-representationalist theories of knowledge and quantum mechanics. *SATS*, *2*(1), 37–61.

Bohr, N. (1949). Discussion with Einstein on epistemological problems in atomic physics. In P. A. Schilpp (Ed.), *Albert Einstein: Philosopher-scientist* (pp. 200–241). The Library of Living Philosophers.

Depraz, N. (2019). Epoché in light of samatha-vipassanā meditation. *Journal of Consciousness Studies*, *26*(7–8), 49–69.

Depraz, N., Varela, F. J., & Vermersch, P. (2003). *On becoming aware: A pragmatics of experiencing.* John Benjamins.

Ericsson, A. (2003). Valid and non-reactive verbalization of thoughts during performance of tasks towards a solution to the central problems of introspection as a source of scientific data. *Journal of Consciousness Studies*, *10*(9-10), 1–18.

Ericsson, K. A., & Simon, H. A. (1993). *Protocol analysis: Verbal reports as data.* MIT Press.

Froese, T., Gould, C., & Seth, A. K. (2011). Validating and calibrating first- and second-person methods in the science of consciousness. *Journal of Consciousness Studies*, *18*(2), 38.

Gendlin, E. (1962/1997). *Experiencing and the creation of meaning.* Northwestern University Press.

Heavey, C. L., & Hurlburt, R. T. (2008). The phenomena of inner experience. *Consciousness and Cognition*, *17*(3), 798–810.

Humphrey, G. (1951). *Thinking: An introduction to its experimental psychology.* Methuen.

Hurlburt, R. T. (2011). *Investigating pristine inner experience: Moments of truth.* Cambridge University Press.

Hurlburt, R. T., & Schwitzgebel, E. (2007). *Describing inner experience? Proponent meets skeptic.* MIT Press.

Husserl, E. (1970). *The crisis of European sciences and transcendental phenomenology: An introduction to phenomenological philosophy* (D. Carr, Trans.). Northwestern University Press.

James, W. (1890). The principles of psychology. New York : Holt.

Kordeš, U. (2021). Bracketing the metaphysical attitude. *Constructivist Foundations*, *16*(2), 156–159.

Kordeš, U., & Demšar, E. (2018). Excavating belief about past experience: Experiential dynamics of the reflective act. *Constructivist Foundations*, *13*(2), 219–229.

Kordeš, U., & Demšar, E. (2019). Towards the epistemology of the non-trivial: Research characteristics connecting quantum mechanics and first-person inquiry. *Foundations of Science*.

Kordeš, U., & Demšar, E. (2021a). Being there when it happens: A novel approach to sampling reflectively observed experience. *New Ideas in Psychology*, *60*, 100821.

Kordeš, U., & Demšar, E. (2021b). Horizons of becoming aware: Constructing a pragmatic-epistemological framework for empirical first-person research. *Phenomenology and the Cognitive Sciences*, 1–29.

Kordeš, U., Oblak, A., Smrdu, M., & Demšar, E. (2019). Ethnography of meditation: An account of pursuing meditative practice as a tool for researching consciousness. *Journal of Consciousness Studies*, *26*(7–8), 184–237.

Monson, C. K. & Hurlburt, R. T. (1993). A comment to suspend the introspection controversy: Introspecting subjects did agree about imageless thought. In R. T. Hurlburt (Ed.), *Sampling inner experience in disturbed affect*. Plenum.

Orne, M.T. (1962). On the social psychology of the psychological experiment: With particular reference to demand characteristics and their implications. *American Psychologist*, *17*(11), 776–783.

Petitmengin, C. (2006). Describing one's subjective experience in the second person: An interview method for a science of consciousness. *Phenomenology and the Cognitive Sciences*, *5*(3), 229–269.

Petitmengin, C. (2007). Towards the source of thoughts: The gestural and transmodal dimension of lived experience. *Journal of Consciousness Studies*, *14*(3), 54–82.

Petitmengin, C. (2016). The scientist's body at the source of meaning. In Schoeller D. & Saller V. (Eds.), *Thinking thinking. Practicing radical reflection* (pp. 28–49). Verlag Karl Alber.

Petitmengin, C. (2021). On the veiling and unveiling of experience: A comparison between the micro-phenomenological method and the practice of meditation. *Journal of Phenomenological Psychology*, *52*(1), 36–77.

Petitmengin, C., Navarro, V., & Le Van Quyen, M. (2007). Anticipating seizure: Pre-reflective experience at the center of neuro-phenomenology. *Consciousness and Cognition*, *16*(3), 746–764.

Samson, A., Simpson, D., Kamphoff, C., & Langlier, A. (2017). Think aloud: An examination of distance runners' thought processes. *International Journal of Sport and Exercise Psychology*, *15*(2), 176–189.

Schear, J. K. (2009). Experience and self-consciousness. *Philosophical Studies*, *144*(1), 95–105.

Shanon, B. (1989). Thought sequences. *European Journal of Cognitive Psychology*, *1*(2), 129–159.

Sparby, T. (2015). Investigating the depths of consciousness through meditation. *Mind and Matter*, *13*(2), 213–240.

Varela, F. J. (1996). Neurophenomenology: A methodological remedy for the hard problem. *Journal of Consciousness Studies*, *3*(4), 330–349.

Varela, F. J., & Shear, J. (1999). First-person methodologies: What, why, how. *Journal of Consciousness Studies*, *6*(2–3), 1–14.

Varela, F. J., Thompson, E., & Rosch, E. (1991). *The embodied mind: Cognitive science and human experience*. MIT Press.

Wallace, B. A. (1999). The Buddhist tradition of samatha: Methods for refining and examining consciousness. *Journal of Consciousness Studies*, *6*(2–3), 175–187.

Weber, S. J., & Cook, T. D. (1972). Subject effects in laboratory research: An examination of subject roles, demand characteristics, and valid inference. *Psychological Bulletin*, *77*(4), 273.

4 Disentangling Consciousness from the Self: A Buddhist Contribution to the Mind Sciences

Georges Dreyfus

Personal Reflections

I come to the study of consciousness and the self through my training as a Buddhist monk and scholar, who has learned about the mind and the self from Tibetan teachers, the meditations they taught, and the voluminous literature of this tradition. It was a welcome surprise to realize that the insights that I gained through my studies and practice could be of use in the scientific study of consciousness, not just as objects to be investigated but as sources of insights and philosophical analysis that can inform scientific research.

Overview

In recent years, a lot of attention has been paid to the possible contribution of contemplative practices to the understanding of consciousness. My claim here is that Buddhist traditions can not only provide outstanding practitioners of meditation but also help researchers to understand conceptually what they are trying to study. One of the areas in which Buddhism can make a clear conceptual contribution is the question of the self. In many writings concerning consciousness, one often notices a conflation between consciousness and the self. This conflation, I believe, is very unfortunate as it leads to confusing some of the central issues surrounding consciousness and the self. My belief is that Buddhism can help clear away this confusion by providing an analysis of the self as a complex construction that needs to be disentangled and analyzed carefully. Buddhism has a long tradition of such analyses, and contemporary mind-science discussions can greatly benefit from learning about this tradition.

To make my case,[1] I first explore, from a slightly unusual angle, the question of the self from a Buddhist perspective. Instead of adopting a metaphysical approach and analyzing whether there is a self as it is often traditionally understood, I take a phenomenological approach to explore what our senses of self are. In doing so, I offer a simple three-fold typology that can be used easily and yields important conclusions. Distinguishing minimal, core, and extended senses of self provides an understanding of what Buddhists mean when they speak of no-self and underlines the complexities of self-construction.

Three Levels of Self-Construction

The view about the self usually associated with Buddhism is the doctrine of no-self (*anātman, anatta*), which is widely (though not universally) shared among Buddhist traditions.[2] This view argues that the self does not exist—that is, it is not part of the reality of what a person is. It is an illusory construction superimposed on the diversity of the mind-body complex.

In the Buddhist tradition, this view is often understood from a metaphysical point of view, as a response to questions such as "Is there really a self?" and "What could this self be?" This understanding of the question of the self as an ontological question has been important traditionally (Siderits, 2003) but it can be difficult to connect to the empirical concerns of thinkers involved in the modern mind sciences. Hence, instead of taking such an approach, I think it is more fruitful to look at how Buddhist traditions approach this question phenomenologically in terms of how the self appears to the ordinary mind. That is, rather than discuss whether there is a self in reality or not, my analysis focuses on our experience or sense of self. The question then is less whether there is a self in reality but what are our senses of self and whether those are realistic or, as Susan Blackmore aptly suggests, illusory (Baggini, 2011, p. 149).

In approaching the self phenomenologically, I look at how the self is given in our ordinary experiences. What we notice is that the self appears in many different ways. This has been noticed by many thinkers, including a number of traditional Buddhists. For example, the twentieth-century Tibetan master Papongka says, "In this process (of examining the self), you are looking into exactly how the 'I' instinctively appears to you. The 'I' may present itself in a variety of ways. Sometimes the 'I' may appear as something imputed on the body, sometimes on the mind" (p. 687).

When we look at how the self is given to us ordinarily, we find that it appears in different ways. Papongka mentions two of these ways—the self appearing as the body and the self appearing as the mind—but there are many more ways in which the self can appear in our ordinary experiences. This multiplicity is well documented in the contemporary cognitive science literature in which a number of types of self are listed. For example, Gallagher and Zahavi list twenty-five different selves such as the material self, social self, ecological self, and so on (Gallagher and Zahavi, 2008, p. 197).

This multiplicity in our sense of self makes a great deal of sense within the context of the Buddhist view that the self is not an enduring real entity but is constructed in experience by imposing a useful, but misguided, sense of unity on the diversity of the body-mind complex. Given that the self is not given in reality but only constructed in experience, it makes sense that it appears in a multiplicity of ways depending on the variety of experiences. This is why our sense of self is multiple. Its object, the self, is a constant re-creation of unity imposed on various aspects of the body-mind complex dependent on various contexts. This enduring unity is illusory: the self is constructed in different ways depending on the context but it is an illusion that is constant.

I think this point clarifies only the degree to which Buddhist philosophy can help us understand the constructed nature of the self. This is an important point but it is still not sufficient to understand the relation between consciousness and the self. To do so, we need to explore the different types of self-construction one can observe and see how they relate to consciousness. To do so, the list of twenty-five forms of self (and counting) mentioned above may not be the most useful. To get a handle on the kind of phenomenology of the self that is relevant to the Buddhist tradition, I suggest that we use a simpler typology. This will help us to understand how Buddhist thinkers conceptualize the self and differentiate it from consciousness and what it means for the sense of self to be illusory. This simpler typology consists of the three-fold distinction between the *minimal self*, the *core self*, and the *extended self*.[3]

The *minimal self* is constituted by the continuum of momentary mental states with their first-personal self-givenness. That is, rather than being given anonymously and appropriated retrospectively as mine, mental states are given immediately in my experience as self-manifesting and self-owned—that is, as happening to me. When I undergo an experience, this experience is already given (phenomenologically) as happening to an *I*, without any possibility of mistake or doubt (immunity from misidentification). This minimal level of subjectivity is what I identify as the *minimal sense of self*. The psychologist Daniel Stern calls this sense of self the *emergent self* and argues that it forms at around two months, when experiences receive a minimal level of coherence. This level of psychic unity provides a kind of sense of self in that it involves an element of reflexivity, an *I*-ness that is perhaps intrinsic to having a subjectivity.

Buddhism does not recognize this sense of I-ness as being a sense of self. It is too preoccupied by the idea of deconstructing the sense of self to do so, but some of its schools recognize that such a sense of I-ness exists and that it has a certain validity. This recognition is particularly well captured by the doctrine of self-cognition (*svasamvitti*) or self-experience (*svasaṃvedana*) first articulated by Dignāga and further developed by Dharmakīrti (Dreyfus, 1997, ch. 25; Dreyfus & Thompson, 2007). In their view, consciousness implies a kind of apperception, the sense that we have of knowing our mental states are our own. This self-awareness is neither introspective nor reflective, for it does not take inner mental states as its objects. Rather, it is the self-specifying function of every mental episode that brings about a non-thematic awareness of mental states as our own so that the person automatically knows whose experiences he or she is experiencing. This self-awareness is prereflective, providing the basis for introspection, the paying of attention to some of our mental states. Regardless of whether we call this prereflective self-awareness a *self* (as some phenomenologists and I do) or not, it is important to understand that this minimal sense of self is not the self that basic Buddhist practice seeks to undermine because it is not the bounded entity with which we identify and that serves as the seat of our deepest affects such as desire, fear, aversion, and so on. It is also important to understand this minimal self to understand the relation between self and consciousness that I will explore below.

To understand what Buddhist practice seeks to expose as illusory we need to look at the second sense of self, the *core self*, which is quite different from this minimal sense of self. Core self is an illusory reification of subjectivity as being a bounded agent enduring through time rather than a complex flow of fleeting self-specified experiences. From a Buddhist perspective, this illusory entity is based on a mechanism of identification through which the subject creates a boundary, identifies itself with this bounded entity and appropriates elements of the mind-body complex (the five aggregates of Buddhism; see Gethin, 1998, ch. 6) thus separating itself from everything else and creating a sense of being in control of the mind-body complex. For example, when I feel really healthy, I identify with the body, I am that healthy body at the same time that I have the sense that I control it, that I can do whatever I want with it. It is as if the experiencing subject becomes fused with the body felt as healthy and strongly distinct from everything else. In this way, an entity is constructed to which various qualities are attributed, a phenomenal avatar (Metzinger, 2005) that is imagined as an enduring, bounded entity through identification with the various parts of the body-mind complex and thus sharply separated from the rest of the world. This *core sense of self* arises transiently, being constantly regenerated anew when we engage in affectively valenced activities.[4] It does not depend on conventional memory, language, or reasoning and hence is not exclusively human. For Stern, this core sense of self develops around six months when the infant is able to experience a sense of agency based on differentiation between movements that are self-initiated from those induced by others.

Finally, the core self is extended through its being represented by memory and language in relation to others (*extended self*). Coan and Sbarra (2015) has argued that an important aspect of the self-construction process consists of the incorporation of relational patterns in the representation of the self.[5] We are constantly assessing how we are doing in relation to others, how much we can rely on them to deal with potential risks and share burdens. This constant preoccupation is not just a central concern of ours but is a vital aspect of the process of self-construction that develops from a very young age. Tomasello (2010) has argued that this construction of an *expanded sense of self* is rooted in the process of joint attention, through which the very young infant learns how to play and cooperate with others. This expanded sense of self further develops through the construction of oneself as an agent able to play varied social roles based on the expectation of having access to relationships based on reciprocity and shared goals. It is also this expanded sense of self that is at work in the often destructive distinctions that we make between our group and others, with all the affective and cognitive biases that this distinction entails. This extension of the self develops further through the narrative ability to understand ourselves over longer periods of time. It is the sense of self that is closely connected to our sense of identity. For example, I feel that I am a Swiss, a professor, a Buddhist. This sense of self requires highly developed symbolic abilities to represent oneself over an extended period of time and therefore is perhaps uniquely human. Although we often think of this sense of self as being our own ("it is me who is Swiss") and as being rooted exclusively in our own narrative abilities, it is in fact born out of our interactions with others and hence is based on the expanded self, created through social interactions. As MacIntyre puts it, it is not our exclusive creation, for "we are never more (and sometimes less) than the coauthors of our own narratives" (1981, 1984, p. 213).

The fact that this extended sense of self is based on different affective and cognitive functions, which are developed at different times in an infant's life, shows the degree to which this three-fold typology of the self is heuristic and reductive of the actual complexity of the process of self-construction. For example, Stern differentiates the *subjective self,* which develops between seven to fifteen months and forms the basis of the infant's ability to interact with persons other than her caretaker, from the *verbal self,* which develops significantly later (three to four years) on the basis of the acquisition of language and involves the infant's ability to perform a role. Nevertheless, this three-fold typology is useful in so far as it gives us ways to understand how Buddhist traditions understand the nature and problems of self-construction.

The Buddhist Deconstruction of the Self

As I argued above, I do not believe that the minimal self is the target of basic Buddhist practice. It is not the self of no-self as it is not a bounded entity. Whether such a sense of self is also illusory or not is a question I will briefly broach below, but in any case it is clear that when Buddhists attempt to expose the illusory nature of the self they are not referring to this minimal sense of self (Dreyfus, 2011).

I also do not believe that the extended self is the self that Buddhist traditions mainly seek to undermine. Siderits (2003) has argued that it is the extended self that is the target of the Buddhist deconstruction of the self. For him, it is this self that gives rise to existential anxiety, which is born from understanding one's life as being that of a self who seeks to make sense of life but is confronted by disease, old age, and mortality. For him, the goal of Buddhist practice is to understand that this meaning-seeking self does not exist. There is no fixed entity in relation to which it would make sense to talk about the meaningfulness of life except as a useful convention to speak about the satisfactions that one can reach in life through various actions and practices. It is also this extended self that is the seat of social emotions such as pride, shame, remorse, and as such is a source of suffering and dissatisfaction.

Although this view is not without merit, I don't think that the target of Buddhist practice can be limited to the extended self. It is not the main target, inasmuch as it is not the seat of the most basic affects that Buddhism seeks to eliminate (desire, aversion, fear). To be free from suffering, it is not enough to be free from the compulsion to understand one's life as that of a single self. One must also be able to remain unaffected by the greatest dangers, the greatest ordeals, and hence be free from the deepest affects such as attachment to oneself, fear, aversion, and so on associated with the core self. This is the state of the Arhat (the fully liberated person), who is described as being not just free from desire and aversion but also from fear and hence unmoved by death.

This freedom from some of the deepest human emotions does not just concern the extended self but goes deeper and concerns the core self, the main target of Buddhist practice. This view is shared by Tibetan thinkers, who distinguish two forms of self-grasping. There is the grasping to a conceptually imputed sense of self, which is linguistic and is based on a deeper innate sense of self (Dreyfus, 2011). It is this sense of self that is the seat of deep affects such as fear, desire, and aversion and that lead us to morally valenced egoic actions (*karma*).

It is also this core sense of self that is experienced as the seat of agency, the boss, or CEO in charge of leading the mind-body complex. It is this sense of self that is involved in the more morally valenced actions induced by various affective states. There we feel that we are in command. We freely decide to act and initiate the action, which we try to bring to a successful conclusion. Obviously the result of our action is not in our control but the action itself is, or so we think. The reality is quite different. Although we have some control over some of our actions, most of them are the result of habits, automatisms, and unconscious or partly conscious affects. Moreover, as we will suggest later, the various functions that the self is experienced as performing do not depend on any central core of our subjectivity. They are largely distributed throughout the cognitive and affective systems that make our psyche and hence can happen without any reference to the self. Hence, our sense of being a self in control is largely illusory—that is, it is misleading, cognitively and affectively. This is not the way we actually function, but merely the way we experience ourselves functioning. It is this sense of being a bounded self in charge of the body-mind complex that is the seat of our deepest emotions, desires, and fears, at least as these are consciously experienced. It is also this sense of self that is involved in our instinct for survival. Hence, it is this sense of self that needs to be taken apart by Buddhist practice if we want to free ourselves from the deep conditionings that bind us as sentient beings.

Does Consciousness Require Having Experiences?

What can we learn from this discussion of the self? First we notice the complexity of this question. There are many senses of self and we should try to be clear when we speak about the self and to which sense of self we are referring. This is important for Buddhists as many thinkers have recognized. For example, the fifteenth century Tibetan thinker Tsong-khapa writes: "In order to be certain of the meaning of 'selflessness' . . . , you must carefully identify the self . . . that does not exist. For, if you do not have a clear concept of the object to be negated, you will also not have accurate knowledge of its negation." (Tsong-kha-pa, p. 126) But the importance of identifying correctly the various senses of self is not limited to Buddhist thinkers and should also concern mind-science researchers. They should not assume that when they talk about the self they are referring to a self-evident entity that can be defined in abstraction from a careful exploration of the different modes of self-construction.

Second, we can now differentiate between consciousness and the self. Often in contemporary mind-science literature this distinction is ignored and a conflation of consciousness and the self is assumed. I believe that the three-fold typology mentioned here helps to clear the confusion. It is clear that consciousness is not identical to the core or the extended senses of self, which are born from a division of the field of awareness, a division that creates a boundary between the organism and others. This sense of being a bounded entity is also evident in the extended self, a sense of self created by identifying oneself symbolically or conceptually. The core sense of self may be more primitive and less dependent on symbolization but it is also born from the creation of a boundary created between me, the focus of my emotional concerns, and others. These senses of self tend to impose limitations on the field of awareness, which is by itself unbounded in this regard.

But what about the minimal self? This type of self does not involve any strong boundary between me and others.[6] It merely consists of locating myself as a subject of experience. Is this minimal sense of self identical to consciousness? To answer I need to recognize the difficulty of defining *consciousness*. The broadest definition of this term (from the first-person perspective) in the contemporary Western philosophical literature is the awareness of what it means to undergo an experience, as articulated by Thomas Nagel (1974) in his famous essay. This meaning of consciousness as phenomenality is not the standard way Buddhist traditions define *vijñāna,* the Buddhist term usually translated as *consciousness,*[7] but it is well-known among Buddhist thinkers and meditators. When asked this question, a Tibetan Buddhist master gave this definition: consciousness is the "knowing of sensations" (*tshor ba shes pa*).[8] This description points to consciousness understood as the seat of phenomenality, how it is for the subject to undergo an experience. This is also the most minimal sense of being a self and the question is then whether this knowing of how it feels—that is, the minimal self, exhausts the meaning of consciousness?

This is a question that is of interest not just for the mind sciences but also for Buddhism as there are practices in that tradition that bring the person into states that seem to go beyond this minimal sense of self, in which there is no sense of self whatsoever. For example, there is the attainment of the state of cessation (*nirodha samāpatti*) in which the meditator is described as not experiencing anything, the complete cessation of any affective or cognitive activity. This state is described at times by some Theravāda commentators as involving the complete cessation of consciousness. This would not raise a problem for our definition of consciousness as phenomenality and allows us to maintain the identification of consciousness with the minimal self. Nevertheless, even in that tradition, there is the idea that in such a state there is the *bhavaṅga*, a kind of preconscious default state (Ganeri, 2017). Other schools such as the Yogācāra, however, argue that even in this state there is still consciousness thereby suggesting that consciousness can exist even without any awareness of experience (Griffiths, 1987).

There are also advanced states of non-duality in which every sense of a self seems to disappear. In such a state, there is no object of experience and hence no experiencing subject. Since there is neither subject nor object, such a state cannot even be remembered. What practitioners remember are the border mental states that happen before entering or after exiting such a deep state but they cannot remember the state itself because there is no object to remember. Can we then still speak about an experience? And if we cannot, can we still say that there is consciousness in such a state? And if so, what would be such a consciousness?

There is also the question of how seriously we take these questions? Should we decide that consciousness cannot even be understood as necessitating this minimal sense of being a subject of experience because of these limit states of complete absence of any sense of self? This would imply that even the minimal sense of self is illusory, that consciousness is, in its deepest nature, completely free from even the most minimal level of self-reference or self-ownership. Or do we decide that these experiences are intriguing but are not really important for understanding consciousness given that they concern only very few limit states? I am not sure how to answer these questions. I am merely offering them for further reflection.

It is clear that in such a state the practitioner does not have the subjective impression of having an experience and yet most Buddhist traditions hold that there is still consciousness in such a state. For example, the Theravāda tradition argues that there is no *vijñāna*, the term usually translated as consciousness, but holds that there is a default state (*bhavaṅga*), a kind of preconscious state. What this state is may not be easily explainable. Perhaps, it is a state of pure reflexivity, the state described as *clarity* (*gsal*) in the Mahāmudrā literature, a state in which there is no content whatsoever and just the presence of consciousness to itself (Thrangu Rinpoche, 2003). In such a state, consciousness is described as being aware of itself in a non-objectified way rather than the usual thetic way (having an object). But because there is no object to be aware of, consciousness cannot retain any impression of such an experience.

These are mere suggestions of how we may think about these interesting but limit states. In any case, the analysis of consciousness as being an awareness of how it feels to undergo experience remains valid for most cases. In fact, I find it hard to imagine to which degree a person deprived of such a minimal self could still be a subjective being. Can we say of such a person that she is undergoing an experience? I really doubt it. Hence, the description of consciousness as co-extensive with the minimal self is good enough to understand consciousness in most situations. It also allows us to separate consciousness from most forms of self-construction given that in most cases what we mean by self is not the minimal self who undergoes the experience but a more restricted sense of self such as the core self, a sense of self in which there is a separation from others. It is this sense of a separate self that Buddhist practices seek to undermine. I would also venture to suggest that this deconstruction of the self is not just the goal of Buddhist practices but that of the many spiritual traditions that see what is usually called the *ego* as the main obstacle to the good life. The contribution of Buddhism is perhaps a special emphasis on this goal and a clarity on the difference between self and consciousness, a difference that I believe should be of interest not just to Buddhists but to also to those interested in the contemporary mind sciences.

Notes

1. Thanks to David Germano and Evan Thompson for providing helpful insights and sources.
2. The best known exception is the Pudgalavadins, who argue for the existence of inexpressible self that would be neither identical nor completely different from the mind-body complex (the aggregates). See Kapstein (2001, pp. 347–391). For a good summary of the no-self view from a philosophical perspective, see Siderits (2007, chap 3).
3. Dreyfus (2011).
4. I am appropriating here Damasio's idea for my own purpose, which is quite different from the way he uses this concept (see Damasio, 1999, pp. 108, 112–113, 196).
5. Coan (2015).
6. I am distinguishing having an affectively charged difference between self and others (strong sense of boundary) from merely distinguishing what is relevant to self and what is not (weak sense of boundary).
7. The standard definition is given by the various Abhidharmas is the cognizing of an object in Bodhi (1993, pp. 27–32).
8. Tshur-pu, oral interview with the author in Tibet, 1993.

References

Baggini, J. (2011). *The ego trick.* Granta.

Bodhi, B. (Ed.). (1993). *A comprehensive manual of abhidharma.* Buddhist Publication Society.

Coan, J., & Sbarra, D. (2015). Social baseline theory: The social regulation of risk and effort. *Current Opinion in Psychology, 1,* 87–91.

Damasio, A. (1999). *The feeling of what happens.* Harcourt.

Dreyfus, G. (1997). *Recognizing Reality: Dharmakīrti and his Tibetan interpreters.* SUNY Press.

Dreyfus, G. (2012). Self and Subjectivity: A middle way approach. In M. Siderits & al, *Self, No-Self: perspectives from analytical phenomenology, and Indian traditions.* Oxford University Press.

Dreyfus, G., & Thompson, E. (2007). Indian theories of mind. In M. Zelazo, Moscovitch, & Thompson (Eds.), *Cambridge Handbook of Consciousness.* Cambridge University Press.

Gallagher, S., and Zahavi, D. (2008). *The phenomenological mind: An introduction to philosophy of mind and cognitive science*. Routledge.

Ganeri, J. (2017). *Attention, not self.* Oxford University Press.

Gethin, R. (1998). *The foundations of Buddhism.* Oxford University Press.

Griffiths, P. (1987). *On being mindless: Buddhist meditation and the mind-body problem*. Open Court Publications.

Kapstein, M. (2001). *Reason traces.* Wisdom.

MacIntyre, A. (1985). *After virtue: A study in moral theory.* Duckworth.

Metzinger, T. (2005). Being no one. *Psyche, 11*, 1–35.

Nagel. (1974). What it is to be a bat. *Philosophical Review, 83*, 435–450.

Papongka Rinpoche. (1991, 1997). *Liberation in the palm of your hand; a concise discourse on the path to enlightenment.* Wisdom.

Siderits, M. (2003). *Personal identity and Buddhist philosophy.* Ashgate.

Siderits, M. (2007). *Buddhism as philosophy.* Hackett.

Stern, D. (1985, 2000). *The interpersonal world of the infant.* Basic Books.

Tomasello. (2010). *Origins of human cognition.* Bradfort Book

Thrangu, Khenchen Rinpoche. (2003). *Pointing out the dharmakay: Teachings on the ninth Karmapa's text.* Snow Lion.

Tsong-kha-pa. (2002). *The great treatise on the stages of the path.* Snow Lion. Vol III.

5 Exploring Consciousness from the Inside

Stephen Fulder

Consciousness arises dependent on sense data and stimulation of the senses. Therefore it is dependent and has no self-existence.
—Anguttara Nikaya 41:7 (Bodhi, 2012)

Personal Reflections

I have been interested in spiritual practice and inner exploration for most of my life. I can remember as a child, realizing that, like Mowgli in The Jungle Book, *I am a particular being because of conditioning, and so all reality can only be assumed and co-created. This led to years in India and a fascination with the huge possibilities available in Buddhist inspired practices, both meditative and in relation to a wise and wholesome, caring life. Over the years I have been exploring what a liberated awareness is like and what are its sources and how it emerges in daily life experience, whether teaching retreats (up to now hundreds of retreats!) in peace work with Palestinians and Israelis, within the family and grandfathering, or in digging the vegetable garden. For me it has not been an ecstatic roller-coaster, but rather a continuous and steady sense of merging with the totality and letting the universe see itself through my eyes.*

Overview

Buddhist teachings are a vast resource for studying, contemplating and exploring the nature of consciousness, since consciousness or awareness is one of the key engines that drive the journey to awakening. Consciousness is regarded as one of a group of transient constituents of an experience. It is the element that processes raw materials arriving from the senses, including thoughts and memories, and constructs recognizable entities. It has no definable qualities but as a potential for knowing, it co-arises when the senses are presented, at which point it wraps itself around the sense data, like a hand taking the shape of an apple, and there is a recognizable apple experience. There is an endless historical debate over whether consciousness, or awareness, can exist at all in and of itself in the absence of an object to meet and land on. In any event, states of awareness in which consciousness meets subtle objects, such as space, are known through meditation and are helpful to the project of liberation, which itself is not an event in consciousness. Loyal to this practical Buddhist tradition I have written some guidance on how to meditate on consciousness.

Introduction

When we try and use consciousness to explore the nature of consciousness, whether or not we use equipment and intellectual rigor, the mindset that we use will inevitably color the questions, the observations, and the interpretations. In particular, conceptual thinking about consciousness results in the mindset that consciousness is a concept. This reification bears a cost, including the sense that consciousness is basically unknowable from any "external" position, though some of its properties can be defined. This would invite a greatly broadened and non-materialistic science, which has an epistemological and metaphysical basis that allows exploration into an indescribable consciousness. Or observation from within. But just as wise observation from outside needs some expertise, so also from the inside. Aldous Huxley once said, "Knowledge is a function of being. When there is a change in the being of the knower, there is a corresponding change in the nature and amount of knowing" (Huxley, 1946, p. 1). Buddhist teachings over millennia have developed many sophisticated methods to explore the nature of consciousness from within, especially meditation.

This exploration does have an agenda, which is liberation, which can be described within this context as freedom from the pain of living automatically, controlled by fixations, views, needs, and aversions. Meditation is a way of seeing with penetrating insight and kindness, how we make experience, and what is beyond experience. In relation to consciousness, it is a way of exploring the nature of consciousness, and how it contributes to building our known selves and the known world. Consciousness is seen as a pure energy or light, which immediately and habitually reaches out to the world and takes the shape of perceptions, which are largely already learned, much as a hand takes the shape of something it holds. Meditation both allows us to witness how this happens, what consciousness can be like before it does so, and the beneficial result. An example is in this Buddhist text:

> When nothing is intended, acted upon or lies latent, there is no support for the establishment of consciousness. Without support, consciousness does not land anywhere. When it does not land and grow, there is no becoming of someone, and . . . that is the end of this entire mass of human suffering. (Bodhi, 2012)

Even in this sublime experience of consciousness when it is not automatically wrapped around preexisting conditioned forms, it is still dependent on basic elements such as senses and therefore it is contingent and cannot be said to be something existing in and of itself, as is expressed in the quote under the title of this article. It is more like the light in a room, which is itself invisible to the eye but when it lands and lights up an object, it is known together with the objects. The subjective awareness—for example the witnessing of consciousness in operation—is still dependent on conditions. An awakening experience, an epiphany, would not be described as within consciousness but would be liberated states of being, which impact and change consciousness.

Here below is some basic guidance on meditation, with a step-by-step guidance on how to release consciousness from its contracted perception of assumed forms, including time and space.

Guidelines for Meditation on Awareness

First Stage: Basic Steps

- Choose a quiet corner or room in the house or a shaded protected spot in nature that will become your favorite and regular place for practice.
- Sit upright and steady, without strain but with an energetic and relaxed dignity, like a king. If on the ground, sit on the edge of a cushion or two to bring your knees closer to the ground and reduce the tendency to roll backward. Your legs can be crossed in any way that is easy. If on a chair let your two feet touch the ground and sit erect.
- Close eyes and gather yourself together, mind and body present in this moment. It feels like a transition or small journey, from distraction and what was, to landing in the midst of our life experience. It has a sense of return and ease, like arriving home. Feel the luxury of simply being.
- Remind yourself that meditation is about being, not doing: there is nothing to be fixed, nothing to be gained or obtained, nothing wrong and no struggle. The only effort needed is to be present as consistently as possible.
- After some time choose the breath as a primary object of mindfulness, and a natural place to consistently rest the attention. Notice how each breath feels, track it as it enters and travels through the body, expanding the stomach, and then its journey out with the stomach contracting again. Let each breath be natural and soft, just as it is, closely embraced with your attention, as if the breath is breathing you.
- Notice how breath and bodily awareness are experienced together. Sometimes move the attention to this sense of embodiment, the sensations of aliveness that move and change in the body, such as the touch of the hands.
- Meditative attention does not need to hunt for a particular "right" experience. It is a direct knowing of what is actually going on, with a radical acceptance.
- When thoughts, commenting, memories, pictures, stories about the past or future or any other mental content or pattern arises, let them appear and pass by like clouds in the sky, in background. If the mental content is insistent and dominant it can be helpful to step back and give it a label such as "judgmental thought." If you identify with the thought and it carries you away with it to the past or future, step out of the train of thinking onto the platform, and return to the breath.
- All kinds of experiences and impressions can arise naturally. They can be at the level of the body, such as stress, tension, pain, ease, or peace. They can be at the level of feelings, such as tiredness, boredom, agitation, sadness, enthusiasm, blankness, irritation, or any other. They are not disturbances or problems that stop meditation, they are the raw material for meditation. Leave the breath and turn your attention to these experience as simple phenomena, that arise, stay for a while, and pass. Try to identify them but not identify with them. Be aware of the actual experience as it is and what if any are the reactions to it. Be aware if it is pleasant or unpleasant. If you are called by feelings or emotions, listen to them, as if you are sitting on the bank of a river listening to the sounds of the water flowing by.
- Be kind to yourself. No need to judge yourself as a success or failure. No need to measure progress. No need to label any psychological or physical experience as problematic. Allow

whatever arises to be just as it is, even if it is difficult, let it be, and let it go. Be aware that everything passes by.

- Gradually get a sense that you are not on automatic pilot, but are able to hold your flowing inner life with kindness, appreciation, and interest. Notice that this way it also empowers you to change it, soften it, or shift it; for example, note the way loving attention to a stress in the body can melt it and reduce its dominance.

Second Stage: Working with Mental Content

- Turn to the stream of thoughts with interest and curiosity. What are they saying, what were the triggers for the current thought, what thoughts follow?
- Stay with the thoughts and explore their transitional and ephemeral nature. Watch the way each thought has a beginning and an end, and see if you can catch the space at the ending of a thought and before the next one.
- A thought may be visual, auditory, or be in the form of words in the mind. Notice the differences. Notice any feelings associated with the thoughts—for example, thoughts about the future, even a simple but repeating "to do" list may have an undercurrent of concern that without such thoughts you will feel lost.
- Begin to explore your place as an observer. Discover that if you can know a thought, you are not the thought but its audience or spectator. Begin to develop a consistent sense of being a fair witness of mental content. In a traditional image, it is like separating an emulsion, such as milk, into its distinct phases of oil and water.
- Repeatedly step back or step out of the identification with thinking, until it becomes easier, a second nature, to take the position of the audience to the movie not the chief actor.
- Reduce any pressure or agenda held by the witness. The witness is just standing back and observing, is simple and free, and uncontaminated by what it is observing.
- Notice how it feels to stay in the witness mind, what happens to the sense of a subject, the sense of well-being, and stillness.

Third Stage: Meeting Consciousness as It Is

- As you keep going with the witnessing meditation, notice that it is dualistic. Experience the position of the witness, a subjective stance that is separate from the object being witnessed.
- Begin to extend the objects to include anything in the sensory field. It can be thoughts, feelings, visual objects, or sounds.
- Particularly explore any sensory fields that point beyond normal limits of habitual boundaries, such as the wind, the space around the body, the sounds that have no special location, and so on. Watch the way the object becomes attenuated, thinner, less solid.
- Continue being creative, playful, and experimental with the apparent boundaries and location of objects. Extend the reach of the awareness out to a sense of endless space—a sense of a big mind, like a balloon of awareness.
- One useful practice is sky gazing. With soft open focus, gaze at the expanse of the sky and let go. Gradually the conscious awareness itself will feel like the sky.

- Along with making the object thinner and less solid, begin to notice how the observer itself is less solid and has less location and subjective stance.
- Using the tool of relaxation, ease, and non-effort, begin to sense that the nature of the witness is just the quality of knowing, which becomes clearer when there is less dualistic sense of solidity in both subject and object.
- Drop into the quality of knowing itself, the essence of witnessing. If questions arise as to who knows the knowing, see them as the natural pressure of the habit of duality. Don't try and work these things out, but just drop out of any loops of that kind.
- This quality of pure knowing or awareness can be there for a moment or for longer. Appreciate it, as if it is a gift from beyond.
- It will come and go, and itself is quite ephemeral; however, it can be supported by various means. One is the sense that the conscious awareness is like the sky, and thoughts or content that arise are like clouds that drift across the sky. The clouds cannot disturb the sky. Similarly, it is like being the screen upon which the scenes of the movie are projected. The screen is unphased or unchanged by the movie.
- Another support is a general sense of presence that has no real object. It is just a sense that we exist, prior to any object that proves it.
- A third possible support is surrender. As soon as we try and catch or grasp pure awareness, we will lose it. So the inner movement is surrender, or dropping back into it, as if we are dropping backward.
- Finally, this is a gradual process, and it can be vague or clearer, long or short. It is helpful not to measure or assess, but to enjoy moments where we are not controlled by object of mind and heart, as moments of freedom and harmony.

There are a range of audio guided meditations, articles, talks and online courses including the various stages described above at www.stephenfulder.com

References

Bodhi, B. (Trans.). (2012). *Anguttara Nikaya (41:7)*. Wisdom Publications.

Bodhi, B. (Trans.). (2000). *Samyutta Nikaya (12:38)*. Wisdom Publications.

Huxley, A. (1946). *The perennial philosophy: An interpretation of the great mystics, East and West*. Chatto & Windus.

Figure 0.3
Michal Rovner. Maagal (Circle), 2020 (detail). Steel vitrine, stone with video projection. Copyright: Michal Rovner.

6 Indirect Phenomenology: Prereflective Self-Awareness and Intersubjective Interaction

Shaun Gallagher

Overview

In this chapter I argue that to understand consciousness in relation to social interaction, we need to consider not just brain mechanisms, but also the extensive and enactive processes of bodily engagement in the social environment. In addition, I suggest that the starting point for such an analysis should not be reflective self-consciousness (see **glossary**), as if that were simply equivalent to consciousness. Rather, I'll argue for the importance of prereflective self-awareness (see **glossary**) as a basic structural feature of consciousness. I provide some of the academic background on how I arrived at these arguments.

Background

When I was a graduate student, I took Julian Jaynes out for a beer after he did a guest lecture at Bryn Mawr College. Jaynes had published his book, *The Origin of Consciousness in the Breakdown of the Bicameral Mind,* the year before and it was great to be able to talk with him as we waited for my own professors to turn up for dinner. Jaynes had controversially argued that ancient humans, before around 1000 BCE, were not reflectively self-conscious because their brains had not fully evolved. Indeed, neurologically, things were such that they suffered (or perhaps benefited) from auditory hallucinations—they would hear voices and interpret them as messages from the gods. Needless to say his book was causing a stir, (Ned Block [1977] called it fascinating, but also preposterous and absurd), and sometime before Francis Crick invented the phrase, Jaynes was clearly defending an "astonishing hypothesis." What impressed me, however, was the way Jaynes went about constructing his argument. To put his theory together Jaynes had to combine neuroscience and psychology with a cultural hermeneutics of ancient sources and texts. It emerged in our conversation that he didn't consult the Greek texts in their original language—he didn't know Greek. But neither did I, and I still appreciated the interdisciplinary nature of his research, despite thinking that Ned Block might be right.

My own nascent ventures into an interdisciplinary mode of philosophy were very much influenced by the work of Maurice Merleau-Ponty, who in his *Phenomenology of*

Perception took an approach that combined classic phenomenology, neurology, and psychology. In contrast to Edmund Husserl, Merleau-Ponty worked on what was to become known as embodied cognition. He was not so much concerned about consciousness per se, but about perception, motor processes, and action, all of which involved a mix of nonconscious and conscious processes. In this case, phenomenology was not enough; one needed to consult the experimental literature, which is what Merleau-Ponty did. Although Jaynes was quite focused on neuroscience, he also considered the importance of ritualistic behavior, and it became clear that he too was emphasizing more than just brain processes. Motoric and action-related processes, as well as cultural factors, played an important role in working out the details of his argument. Let me be clear that at that time I had no idea about what was happening in cognitive science. It was my study of Merleau-Ponty that motivated me to take graduate seminars in neuroscience and physiological psychology, even though my graduate training was primarily in philosophy. Jaynes encouraged me along this line, as did my graduate advisor, José Ferrater-Mora, who fortunately turned up for dinner just as I was running out of things to ask Jaynes.

Later, when the large project of consciousness studies started to emerge in the late 1980s (e.g., Marcel & Bisiach, 1988), and when so many philosophers and scientists started to discuss the "hard problem" of consciousness as David Chalmers (1995) defined it, I picked up on a particular strategy that Chalmers mentioned, one that seemed to summarize my own approach in trying to combine phenomenology and the experimental literature. The strategy, which he called worthwhile but "relatively unambitious," was as he explained, "simply to explain something else."

> Some researchers are explicit that the problem of experience is too difficult for now, and perhaps even outside the domain of science altogether. These researchers instead choose to address one of the more tractable problems such as reportability or the self-concept. Although I have called these problems the "easy" problems, they are among the most interesting unsolved problems in cognitive science, so this work is certainly worthwhile. (Chalmers, 1995, p. 206)

I took this to mean that, in terms taken from my own phenomenological training, we could in some sense bracket the hard problem and concentrate on working out the so-called easy problems, solutions to which could indirectly put us in a better position to address the hard problem. I want to call this a kind of indirect phenomenology. It's the sort of thing that I had been doing in my graduate thesis and in several papers that I published in the 1980s. The idea was to figure out how motoric processes contribute to perception, for example, and to ask what aspects are conscious, what aspects are nonconscious, and how these processes are related.

A bit later I got to know Francisco Varela just around the time he published his paper "Neurophenomenology: A Methodological Remedy for the Hard Problem" (1996). A remedy isn't a solution, and as most commentators have suggested, Varela was not trying to resolve the hard problem, but to dissolve it. Although this was not precisely the strategy I was pursuing, I appreciated the attempt to bring neuroscience together with phenomenology and dynamical systems theory. I think we learn something important from this approach, even if it is not a head-on attempt to solve the hard problem. I subsequently used a neurophenomenological approach to study the experiences of awe and wonder during space flight (Gallagher et al., 2015). This doesn't provide a solution to the hard problem either, but it does tell us something about how brain processes, embodied affective processes, environmental factors, and the diverse cultural backgrounds of our experimental subjects all play a role in making such experiences what they are.

Around the time I met Varela I had already started an interview project with the aim of exploring connections between phenomenology and neuropsychology. Starting in 1998, my interviews with Michael Gazzaniga, Marc Jeannerod, Christopher Frith, Jonathan Cole, Michael Arbib, and Jaak Panksepp were all published in the *Journal of Consciousness Studies*. They, together with an interview with Jacques Paillard, formed the basis of my book, *Brainstorming* (Gallagher, 2008). In addition to some ongoing work with Cole, I had further discussions and interactions with Jeannerod (who had attended Merleau-Ponty's seminars in Paris), and with Frith. A specific encounter I had with Frith motivates the issues discussed in the following sections.

Prereflective Self-Awareness

In 2006 I was invited to be a participating commentator at a debate between Frith and Christof Koch on the topic of "The Origin of Consciousness: Perspectives from Neuroscience," at the Center for Semiotics and Cognitive Studies, at the University of San Marino. In the simplest terms, the disputed question was whether we should think of consciousness as originating exclusively in brain processes (in which case, as Koch [2004, p. 4] maintained, "consciousness is an intensely private matter") or in social processes, as Frith argued.

The etymology of the English term "consciousness" appears to support both views. It derives from the Latin *conscientia*, through the French, *conscience*. In its translation from French to English it splits in two: conscience and consciousness. Conscience, as the moral-social sense of knowing the right thing to do, originates in social (familial or cultural) processes, although it is often portrayed as a private voice. This word clearly retains the original sense of con-science (knowing with [others]). The term "consciousness," however, fits better with the private Cartesian conception of mind. It is what I (in my first-person existence) experience for myself, and what no one else has access to. Robert Fludd, an Oxford alchemist of the seventeenth century, pictured important anatomical aspects of the brain/mind as including sensation, imagination, and intellect—that is, those that constitute the private elements of consciousness. Yet he also included *custos* relating to custom, memory, and motive, extending down the spine for action (see figure 6.1). Custom and action, as public or social aspects seemingly enter into the formula of mind. Thus, Block (1995, p. 227) calls consciousness a hybrid or mongrel concept, and distinguishes between *phenomenal consciousness*, which may be private, and *access consciousness*, which may involve more social factors.

To be more precise, I think there were at least two questions being debated (see Cole et al., 2010; Gallagher, 2010). (1) Why do we have consciousness, or what is it for? I took Frith to be addressing this question. A clue to answering that question is an answer to a further question: When does it happen? But these questions are different from the next question: (2) How does the brain generate consciousness, or what are the mechanisms that can explain the phenomenon of consciousness? This was Koch's question. My own view is that if there is an answer to the second question; that is, if there is a mechanism that can explain

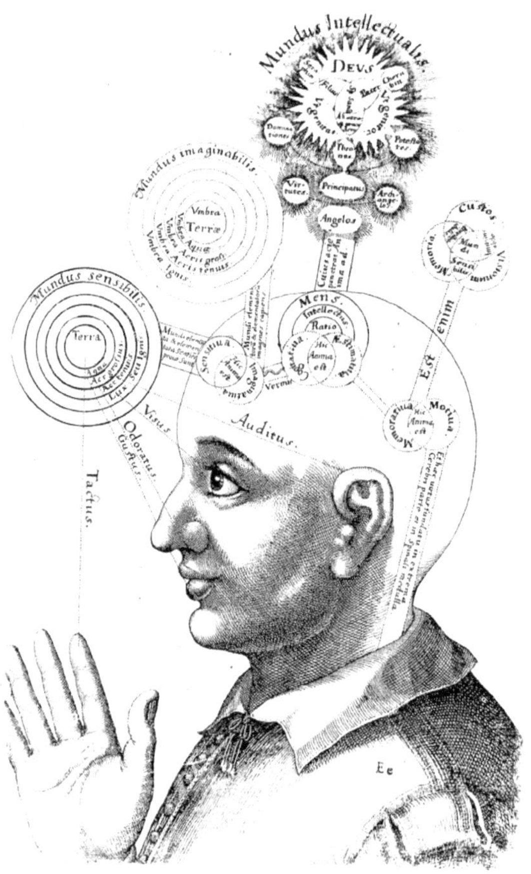

Figure 6.1
From Fludd (1619).

consciousness (and we can take this in a general sense to include the idea that the mechanism might be broad dynamical processes in the brain, as Varela suggests), *and* if we are able to explain it, the first question would still be outstanding, and still interesting. Moreover, whether the answers to all of these questions would add up to an answer to the hard problem is another matter. In any case, I was keen to introduce some phenomenological distinctions into the debate in San Marino.

The first distinction I mentioned was that between the noetic (see **glossary**) act of consciousness (i.e., whether it is an act of perception, or memory, or imagination, or judgment), and the noematic (see **glossary**)/intentional content (i.e., what I experience, and what meaning it has). For phenomenologists, these two aspects of consciousness constitute its intentionality. They are to be distinguished from a third aspect: the phenomenality of my experience (i.e., "what it is like," or the qualitative feel of the experience). For example, I might be perceiving a tiger (rather than remembering a tiger, or imagining a tiger). As I am perceiving the tiger, I'm standing safely in the zoo outside of the tiger's cage, and I take it to be nonthreatening (part of my noematic content), and that may explain why the qualitative feeling I have does not include fear. If the cage suddenly opens, my perception would remain noetically a perception, but otherwise would be very different noematically (no longer a tiger in a cage, but a tiger that could attack me) and phenomenally because fear would permeate my experience. We may be able to say quite a bit about the brain processes that correlate to these changes. Explaining the noematic aspect, however, would require reference to worldly context, and explaining the change in phenomenality would likely involve various extraneural embodied processes. That is, a full explanation of the phenomenal change cannot be simply an increased activation of the amygdala—a change that happens quicker than my ability to reflect on my situation—or even a broader shift in the autonomic nervous system with accompanying bodily-affective changes. Even assuming a reference to the evolutionary background, it would also need to include the experiential foreground—the change in my (ecological) perception of the tiger and the prereflective sense of my own situation. If one accepted these distinctions as clear then, I suggest, they complicate any simple answers to the debated questions.

I've just mentioned the other phenomenological distinction that seems central: the one between prereflective self-awareness and reflective self-awareness. This goes back to the claim made by Jaynes. If we equate consciousness with a relatively higher-order, reflective, self-consciousness, as Jaynes and a significant number of philosophers and cognitive scientists do, then one might think that animals, and young children, walk around (or crawl around, as it may be) in nonconscious states. But not all consciousness is equivalent to reflective self-consciousness.

The idea of prereflective or nonreflective self-awareness can be thought of in several different ways. First, it can be equated with an implicit sense of my own experience as I am living through it, part of the very structure of consciousness. Indeed, if there is something it is like to be conscious of *X*—the taste of a fine wine for example—then there is an implicit reference to the subject who is experiencing this, an implicit experiential sense of what it is like *for me*. This applies equally to nonsensory cognitive processes. There is something it's like for me to be solving a mathematical problem, and this is different from what it's like to be remembering the last time I visited San Marino. The point is that the phenomenality of consciousness depends on prereflective self-awareness, since if there were no "for me" involved, it wouldn't feel like anything to taste the wine, or solve the math problem, or remember San Marino. Part of the experiential structure of phenomenal consciousness involves prereflective self-awareness as its necessary condition. This is not something that depends on reflection. I do not first have to reflectively scrutinize a specific perception or, for example, a feeling of pain, and subsequently identify it as being mine. The mineness is already built in. First-person experience involves an immediate and non-observational self-awareness.

This notion of prereflective self-awareness is a central phenomenological concept to be found in Husserl, Jean-Paul Sartre, Merleau-Ponty, and others. But it is also expressed in analytic philosophy of mind. Thus, for example, Harry Frankfurt writes:

> What would it be like to be conscious of something without being aware of this consciousness? It would mean having an experience with no awareness whatever of its occurrence. This would be, precisely, a case of unconscious experience. It appears, then, that being conscious is identical with being self-conscious. . . . [This claim] does not mean that consciousness is invariably dual in the sense that every instance of it involves both a primary awareness and another instance of consciousness which is somehow distinct and separable from the first and which has the first as its object. Rather, the self-consciousness in question is a sort of immanent reflexivity by virtue of which every

instance of being conscious grasps not only that of which it is an awareness but also the awareness of it. (Frankfurt, 1988, p. 162)

This immanent reflexivity is precisely prereflective self-awareness, and as Frankfurt suggests, if it were missing, you would necessarily be in an unconscious state. If you are in pain, however, you know that it is you who are in pain; you don't have to check to confirm it.[1]

The phenomenology of prereflective self-awareness is not meant to provide a causal account of what makes a mental state conscious. It is not that there are mental states waiting to become conscious, and then, when some extra factor of prereflective self-awareness is added, that mental state suddenly becomes conscious. Rather, the phenomenology indicates a constitutive aspect of consciousness. This does not rule out the possibility of nonconscious perception, as in the case of blindsight (see **glossary**). In the case of blindsight we can simply appeal to the neuroscience as the most reasonable explanation of the blindsight subject's ability to identify visual stimuli, despite lack of visual experience. Nonconscious perception in the visual modality may be considered the processing of sensory information in relevant brain areas, in a way that fails to generate the dynamics required for the emergence of visual experience. Despite the lack of visual experience, the nonconscious neural processes involved may generate a conscious "feeling" that allows the blindsight subject to answer correctly on forced-choice tests about presented visual stimuli. Importantly, these kinds of nonconscious processes go on all the time in our everyday existence. Although some aspects of my movements are experientially available prereflectively (e.g., I am certainly aware that I am running or walking or standing still), many of the detailed aspects of motor control are the result of nonconscious bodily processes. For example, when I reach to grasp something, the visual information that controls the precise shape of my hand is processed nonconsciously (Jeannerod, 1997).

To return to the two kinds of questions at stake in the San Marino debate, these considerations about first-personal prereflective self-awareness and nonconscious processes, considerations that are thus partly phenomenological and partly neuroscientific, can help address the question about how the system works—what some would label the mechanism question. But none of this tells us anything about *why* some aspects of our behavior are conscious, and others nonconscious, or what consciousness is for. Frith's attempt to answer this in social terms does not ignore the role of the brain. Indeed, there is something of a translation involved in Frith's account such that social processes are understood in terms of brain processes. This is the idea of the social brain (Frith & Frith, 2010).

Intersubjective Interaction

Frith's argument starts with an elucidation of the neural correlates of consciousness. He then focuses on a specific instance of consciousness—our consciousness of being in control of our actions, which many authors call the *sense of agency* (see **glossary**). This sets up his answer to the why question: Why do we need to be aware of our actions, especially if we can achieve much of what we want just on the basis of nonconscious processes, and without awareness? His hypothesis is that consciousness, and specifically the sense of agency, play a critical role in social interactions, allowing us to assign responsibility, develop acceptable justifications, and share our experiences in ways that facilitate group decisions. One can take issue with each step in the argument. For example, enactivists (see *enactivism* in the **glossary**) like Alva Noë and Evan Thompson (2004) have questioned the very idea of the neural correlates of consciousness. Others have taken issue with the concept of a sense of agency. For example, José Bermúdez (2010), commenting on Frith's argument, doubts that there is such a thing as a sense of agency. Setting these issues aside, I want to focus on the notion of social interaction and on the best way to conceive of social cognition.

For Frith (writing together with Uta Frith) this is primarily a question about the social brain. The brain has the capacity to process social signals, and this capacity allows us to learn about the world and other people, and thereby create a shared social world. Nonverbal social signals can be both issued and received automatically and nonconsciously, starting in the first year of life. As we develop, we can gain conscious control of such signals, allowing us to modulate the nonconscious processes. Reflective (deliberate, ostensive) self-consciousness "closes the loop" so that "the [interacting] partners each know that they have the intention to communicate" (Frith & Frith, 2010, p. 165). Frith and Frith point to evidence for this conscious control in developmental studies of eighteen-month-old children. This deliberate social signaling involves reflective self-awareness and consciousness of the effect that the signals have on others. Accordingly, they contend, this ability to process social signals develops together with (reflective) consciousness, and is tied to the maturing of a theory of mind (ToM) mechanism and processes in medial prefrontal cortex (MPFC), temporal poles and posterior superior temporal sulcus (STS), the latter area involved in perceiving agency (Frith & Frith, 2004, p. 45). This (self-) conscious mindreading ability is, for them, the central component of social cognition, although they also mention that other processes happen much earlier in development—at three to six months; for example, at six months there is the recognition of self-moving agents in the environment, and the perception of another agent's intention, both connected to processing in STS.[2] Moreover, they mention that even earlier at several weeks of age infants are already interacting with their caregivers, smiling and vocalizing and recognizing kinematic patterns in the movements of others. The implication, however, is that these early processes of interaction are nonconscious. "The ability to react reflexively to movement of gaze as a priming cue for one's own eye movement is likely to be innate as it can already be observed at three months of age. . . . This is different from the voluntary following of the general direction of an adult's gaze, which is not accomplished until around twelve to eighteen months of age" (2004, p. 51).

For Frith, then, the social function of consciousness—that is, reflective self-consciousness—is tied to ToM processes that develop later in infancy (starting at eighteen months) and become manifest at three to four years of age when children can pass the false-belief task. In developmental psychology, however, running parallel to the theoretical discussion of ToM, with its focus on false-belief tests and the maturation of the social brain, others have explored interaction processes in early infancy. Here the work of Colwyn Trevarthen, Daniel Stern (2018), Peter Hobson, Vasu Reddy (2008), and others developed a rich account of primary and secondary intersubjective processes in the first year of life. This research raises two issues for us to consider. First, should we think that the best account of such early social processes can be had by a methodological individualist focus on mechanisms in the social brain rather than a focus on embodied intersubjective interactions? Second, what should we say about consciousness in this connection?

Clearly the brain is an important component in the cognitive system. However, in contrast to the narrow, neurocentric focus of classic cognitivism, the Varela-style enactive approaches emphasize the cognitive system as a whole, inclusive of brain-body-environment, where environment includes the social environment. The notion of primary-intersubjective interaction fits well with this enactive approach. Primary intersubjectivity (starting from birth), involves innate or early-developing embodied, action-oriented abilities in processes of intersubjective interaction (Trevarthen, 1979). These are capacities manifested in action and perceptual experience, as we engage with others. We are able to see what others feel or intend in their bodily postures, movements, gestures, facial expressions, gaze direction, and vocal intonation and we respond with our own bodily movements, gestures, facial expressions, and so on. Such perceptual processes are already at work in early infancy. Infants respond to agents (but not to non-agents) in the environment, and in distinctive ways to human faces and facial gestures. The evidence from various developmental studies is fully consistent with the phenomenological idea of an emerging shared affective intentionality—what Merleau-Ponty calls intercorporeity—and what Gopnik and Meltzoff describe, despite their own theoretical (ToM) tendencies, as innately mapping "the visually perceived motions of others onto our own kinesthetic sensations" (1997, p. 129).

Some of these processes are nonconscious. In the same way that action processes in older children and adults may involve a mix of nonconscious processes and prereflective self-awareness, however, there is no reason to deny the same mix to infants. The type of social interaction described in these embodied-enactive terms depends on a distinction between self and non-self, and a proprioceptive experience of one's own body.[3] The infant is able to differentiate between touching herself, touching something else, or some other agent touching her. Infants in the first year of life already have a sense, from their own self-movement and proprioception (see **glossary**), of what it means to be an experiencing agent—in effect, a prereflective sense of agency. Behavioral evidence for a sense of agency in two- to four-month-old infants includes, for example, the fact that they increase the frequency of specific limb actions that yield an interesting or pleasant effect compared to baseline (e.g., Argueta et al. 2021).[4]

These same prereflective experiences of self-agency play a role, not just in actions but also in primary-intersubjective interactions. As is the case more generally, we have a mix of nonconscious and prereflective processes. In such interaction, agents may unconsciously coordinate or align their movements, gestures, and speech acts (Lindblom, 2015), entering into synchronized resonance with others, with slight temporal modulations in either in-phase or phase-delayed rhythmic covariation (Fuchs & De Jaegher, 2009). In some cases this interaction itself plays an essential role in constituting social cognition (De Jaegher et al., 2010). However, just such interactive processes are not entirely a matter of automatic or nonconscious mechanical procedures; Csibra and Gergely (2009) have shown that the infant, on average, responds to another person's actions only if that person is attending to the infant. Numerous experimental studies of infants demonstrate that *what it is like* (in the sense of a prereflective affective awareness) to participate in genuine interaction with others is for them quite different from *what it is like* to be excluded from the dynamics of interaction (Murray & Trevarthen, 1985; Tronick et al., 1978).

Conclusion

There are two points to emphasize here. First, to understand the role that social interaction plays in consciousness, we need to consider not just internal mechanisms of the social brain, but the extensive and enactive processes of bodily engagement in the social environment. Second, in thinking about consciousness in this context, the starting point is not reflective self-consciousness, as if that were simply equivalent to consciousness. Rather, we need to consider prereflective self-awareness as a basic and important structural feature of consciousness involved in action-oriented and interaction-oriented engagements with the world and with others.

Glossary

Blindsight: The ability of people who are blind due to damage in the visual cortex to recognize or respond to visual stimuli that they do not consciously see.

Enactivism: An approach to embodied cognition that emphasizes action-oriented processes and mind-life continuity.

Noema (noematic): The sense or aspectual meaning of the intentional object of which one is conscious.

Noesis (noetic): The particular character of the intentional, conscious act that allows us to distinguish between an act of perception versus memory versus imagination versus judgment.

Prereflective self-awareness: An implicit, first-order, non-observational, immediate awareness of one's own experiences that does not rely on a reflective act that would make that experience an explicit object.

Proprioception: Sometimes called the sixth sense, proprioception tracks body posture and position. It is the sense one has of where one's limbs are relative to the rest of one's body. It is sometimes defined to include kinaesthesia, the sense of one's bodily movement.

Reflective self-consciousness: Typically, an explicit, second-order, observational consciousness of one's own experience that takes that experience as an intentional object.

Sense of agency: The sense that I am the one who is causing, generating, and/or controlling an action. For example, the sense that I am the one who is causing something to move, or that I am the one who is generating a certain thought.

Notes

1. As William James (in his *Principles of Psychology*) put it, consciousness is always personal. Likewise, philosophers like Ludwig Wittgenstein and Sidney Shoemaker have argued that just this aspect of consciousness is immune to error through misidentification. Wittgenstein's example is that when I experience a pain, I cannot be wrong about who it is that experiences the pain. I do not first consider a pain, and then reflectively decide that this is *my* pain. This personal prereflective aspect of consciousness may indeed be private, as Koch (2004) argues, but it does not rule out the possibility of an empathic experience. In some cases I may see that you are in pain, or indeed, I may see the visible aspects of your pain in your movements or facial expressions. I won't try to address the various complex debates about the notion of empathy here. My point (to be developed in the next section) is only that the prereflective phenomenal quality of consciousness is not completely cut off from social influence. Consider the familiar example of the child who falls and bruises his knee. Whether he experiences the bruise as painful (to some degree), or affectively saturated to the point of generating tears, depends to some extent on the immediate reaction (e.g., facial expression) of his caregiver.

2. Frith and Frith (2004) suggest that intentionality detection is different from goal detection, and that the former requires mindreading (since intention is a mental state) and does not develop before eighteen months. Some scientists take intentionality detection to be operative in the first year of life. If we think of intentionality in terms of motor intentionality (Merleau-Ponty) or intention-in-action (Searle 1983), then it is something that can be perceived rather than inferred. That is, it would not require mindreading.

3. Indeed, the primordial roots of an experiential difference between self-movement and movement caused by another have been traced to nonconscious prenatal processes that involve an intercorporeity even if not an intersubjectivity (Ciaunica et al., 2021).

4. I note that this early sense of agency is not universally accepted (see, e.g., Teichmann et al., 2020) or may develop slightly later starting at three to four months (Zaadnoordijk et al., 2020).

References

Argueta, A., Sloan, A., Jones, N. A., & Kelso, S. (2021). Emergence of agency in infants. *FAU Undergraduate Research Journal*, *10*, 17–17.

Bermúdez, J. L. (2010). Action and awareness of agency: Comments on Chris Frith. *Pragmatics & Cognition*, *18*(3), 576–588.

Block, N. (1995). On a confusion about a function of consciousness. *Behavioral and Brain Sciences*, *18*(2), 227–287.

Block, N. (1977, March 6). Was man really unconscious for centuries? *Boston Globe*, p. A17.

Chalmers, D. J. (1995). Facing up to the problem of consciousness. *Journal of Consciousness Studies*, *2*(3), 200–219.

Ciaunica, A., Constant, A., Preissl, H., & Fotopoulou, K. (2021). The first prior: From co-embodiment to co-homeostasis in early life. *Consciousness and Cognition*, *91*, 103117.

Cole, J. Dascal, M., Gallagher, S., & Frith, C. (2010). Final discussion. *Pragmatics and Cognition*, *18*(3), 553–559.

Csibra, G., & Gergely, G. (2009). Natural pedagogy. *Trends in Cognitive Sciences*, *13*, 148–153.

De Jaegher, H., Di Paolo, E., & Gallagher, S. (2010). Does social interaction constitute social cognition? *Trends in Cognitive Sciences*, *14*(10), 441–447.

Fludd, R. (1619). *Utriusque cosmi maioris scilicet et minoris metaphysica, physica atque technica histori* [The metaphysical, physical, and technical history of the two worlds, namely the greater and the lesser]. Johan-Theodori de Bry.

Frankfurt, H. (1988). Identification and wholeheartedness. In H. Frankfurt (Ed.), *The importance of what we care about: Philosophical essays* (pp. 159–176). Cambridge University Press.

Frith, U., & Frith, C. (2004). Development and neurophysiology of mentalizing. In C. Frith and D. Wolpert (Eds.), *The neuroscience of social interaction: Decoding, imitating, and influencing the actions of others* (pp. 45–76). Oxford University Press.

Frith, U., & Frith, C. (2010). The social brain: Allowing humans to boldly go where no other species has been. *Philosophical Transactions of the Royal Society B: Biological Sciences*, *365*(1537), 165–176.

Fuchs, T., & De Jaegher, H. (2009). Enactive intersubjectivity: Participatory sensemaking and mutual incorporation. *Phenomenology and the Cognitive Sciences*, *8*, 465–486

Gallagher, S. (2008). *Brainstorming: Views and interviews on the mind*. Imprint Academic.

Gallagher, S. (2010). Defining consciousness: The importance of non-reflective self-awareneess. *Pragmatics and Cognition*, *18*(3), 561–569.

Gallagher, S., Janz, B. Reinerman, L. Bockelman, P., & Trempler, J. (2015). *A neurophenomenology of awe and wonder: Towards a non-reductionist cognitive science*. Palgrave-Macmillan.

Gopnik, A., & Meltzoff, A. N. (1997). *Words, thoughts and theories*. MIT Press.

Jeannerod, M. (1997). *The cognitive neuroscience of action*. Blackwell Publishing.

Koch, C. (2004). *The quest for consciousness: A neurobiological approach*. Roberts & Company Publishers.

Lindblom, J. (2015). *Embodied social cognition*. Springer.

Marcel, A., & Bisiach, E. (Eds.). (1988). *Consciousness in contemporary science*. Clarendon Press/Oxford University Press.

Murray, L., & Trevarthen, C. (1985). Emotional regulations of interactions between two-month-olds and their mothers. In T. M. Field, & N. A. Fox (Eds.), *Social perception in infants* (pp. 177–197). Ablex Publishing.

Noë, A., & Thompson, E. (2004). Are there neural correlates of consciousness? *Journal of Consciousness Studies*, *11*(1), 3–28.

Reddy, V. (2008). *How infants know minds*. Harvard University Press.

Searle, J. R. (1983). *Intentionality: An essay in the philosophy of mind*. Cambridge University Press.

Stern, D. N. (2018). *The interpersonal world of the infant.* Routledge.

Teichmann, F., Grosse Wiesmann, C., & Musholt, K. (2020). Do infants have agency?—The importance of control for the study of early agency. ArXiv preprint. https://doi.org/10.31234/osf.io/pm243

Trevarthen, C. (1979). Communication and cooperation in early infancy: A description of primary intersubjectivity. In M. Bullowa (Ed.), *Before speech: The beginning of interpersonal communication.* Cambridge University Press.

Tronick, E., Als, H., Adamson, L., Wise, S., & Brazelton, T. B. (1978). The infant's response to entrapment between contradictory messages in face-to-face interaction. *Journal of the American Academy of Child Psychiatry*, *17*(1), 1–13.

Varela, F. J. (1996). Neurophenomenology: A methodological remedy for the hard problem. *Journal of Consciousness Studies*, *3*(4), 330–349.

Zaadnoordijk, L., Meyer, M., Zaharieva, M., Kemalasari, F., van Pelt, S., & Hunnius, S. (2020). From movement to action: An EEG study into the emerging sense of agency in early infancy. *Developmental Cognitive Neuroscience*, *42*, 100760.

7 The Politics of Consciousness

Yuval Noah Harari

Personal Reflections

In relation to his interest in consciousness, he says, "I believe the question of consciousness is the key to understanding life. The fact that science does not understand consciousness is the greatest lacuna in the scientific worldview. And in the twenty-first century, this lacuna is becoming extremely dangerous. For we are now acquiring the ability to re-engineer life. If we indeed learn how to re-engineer life before we understand consciousness, we might misuse our power in truly terrible ways."

Overview

Scientific studies of consciousness often treat consciousness as a neutral observer of the world. In fact, consciousness is the only thing in the universe that involves suffering, which is the very opposite of neutral observation. While observation tries to capture the present reality as objectively as possible, at the core of any experience of suffering is a rejection of present reality. We might well define consciousness as the capacity to suffer, and to reject reality. This approach also highlights the crucial ethical and political implications of the science of consciousness, as is apparent in debates about abortion, animal rights, and the legal status of artificial intelligence (AI).

The study of consciousness is not just a scientific project. It is also an ethical and political enterprise. Our understanding of consciousness has a deep impact on laws and moral choices. Therefore, scholars of consciousness should be aware of the potential political consequences of what we are researching and publishing. In particular, we should remember that flawed scientific theories can have dangerous political fallout—as happened in the past, for example in the case of erroneous theories regarding sex and gender.

Taking politics into account is not just the responsible thing to do. It can actually advance the scientific study of consciousness. This may sound counterintuitive, because mixing politics with science is usually bad for science. But consciousness is a special case. As explained in the following pages, taking political questions into account might help us make sense of competing scientific theories of consciousness. At present, there are many such theories, and it is not easy to understand how they differ from one another (Melloni et al., 2023; Mudrik et al., 2023). Examining their potential political consequences can clarify their differences.

The Potential to Suffer

As an essential first step, let's start with a definition. What is consciousness? Over the years, numerous complex definitions have been offered, and no scientific consensus has so far emerged (Dehaene, 2014; Dennett, 2018; Harari, 2016; Solms, 2021; Tononi et al., 2016). But from an ethical and political perspective, the definition of consciousness is extremely simple: consciousness is the capacity to suffer. Conscious entities, like you and me, can experience pain and fear (as well as pleasure and joy), which is why we are ethical and political subjects. Stones, tables, and cars do not have consciousness: they cannot suffer, and are therefore not ethical subjects. They are mere objects. Stealing my car is a crime not because it makes the car suffer, but because it makes *me* suffer.

The link between consciousness and suffering manifests itself in numerous political debates. Modern societies assume that political authority ultimately stems from having consciousness and being able to suffer (as well as to be happy). To be included in political debates, you need the ability to experience pain and fear, as well as pleasure and joy. Describing your experiences of pain, fear, and joy is often a necessary prelude for making your voice heard on questions ranging from race and gender to ecology and taxation: "I feel, therefore I am entitled to speak."

This was not always the case. In many premodern societies, political authority had little to do with feelings. Authority did not come from our inner feelings—it emanated from outside us, from either the gods or the laws of nature. Rulers, for example, were allegedly chosen by the gods. Certain behaviors were forbidden because the gods said so, or because the laws of nature supposedly said so. The Ten Commandments, for example, forbade people to murder. Why? Because God said so.

One of the biggest revolutions of the modern era was to shift the source of ethical and political authority away from the alleged laws of the gods and of nature, to the feelings of human beings. Modern-day rulers, at least in democracies, are not chosen by gods—they are elected by humans according to our feelings (Harari, 2016; Taylor, 1992).

The same humanist shift is evident in attitudes to crime. Like ancient societies, modern societies also ban murder. But they no longer justify the ban by appealing to divine commandments. Rather, they forbid murder because it hurts people. This humanist shift from divine decrees to inner feelings is most apparent in the demise of "victimless crimes": crimes that do not cause anyone to suffer. Thus for millennia, Judaism, Christianity, and Islam banned homosexuality and persecuted LGBTQ people. Why? Because God said so, or because scriptures said so, or because, allegedly, homosexuality was a crime against the laws of nature. In traditional Jewish, Christian, and Muslim societies, for one man to fall in love with another man was one of the worst crimes imaginable, sometimes punishable by death.

In recent generations this supposedly divine or natural taboo began to be questioned, in the name of human feelings. The key question raised by thinkers like Jeremy Bentham and

Magnus Hirschfeld has been "Who suffers?" If two men love each other, how does this harm anyone? And if nobody suffers, why should it be considered a crime? Since criminalizing homosexuality caused immense suffering to gay people without benefiting anyone else, it looked less and less sensible. The decriminalization of homosexuality, and the increasing acceptance of LGBTQ people, demonstrate how modern ethics and politics came to consider human feelings to be more important than allegedly divine commandments (Hildebrandt, 2014; Mignot, 2022).

Of course, the humanist shift and its emphasis on feelings did not resolve all ethical and political dilemmas, but it did change the terms of the debate.

For example, consider the following political conundrum, featuring the statue of a famous dead leader, which is placed in a prominent public square. Some people demand the statue's removal, because they see it as an offensive symbol of historical injustices committed against their group. Other people object, claiming that for them it signifies pride in their own history and identity. Assuming that both sides are sincere, the political argument boils down to weighing feelings: If a statue makes some people feel threatened or ignored, while making other people feel proud and valued, how do we weigh these feelings against each other? Whose feelings count for more?

Such conflicts raise the more general question of how we measure feelings, and how we measure suffering. Much of modern ethics and politics revolve around the attempt to measure suffering. I mentioned earlier that raising political questions can help us make sense of the various scientific theories of consciousness. One question to ask about any theory of consciousness is whether it can help us measure suffering, and formulate a scale of suffering. If it cannot, I would doubt whether it truly is a theory of consciousness.

Zero Suffering

One very important part of any scale of suffering is the lowest end of the spectrum. An entity that has zero capacity for suffering falls completely outside the ethical and political domain. For instance, in the aforesaid debate about the statue, we take into account the feelings of people who abhor the statue, and we take into account the feelings of people who admire the statue, but we do not take into account the feelings of the statue itself, because it has none. When it is up on its pedestal, the statue does not feel happy. If you cut off its head and throw it into a ditch, it does not feel pain. Accordingly, the statue itself is not an ethical and political subject. It is merely an object. The presence or absence of consciousness and the capacity to suffer determine the very borders of ethics and politics.

As another example, consider the heated debate about abortion. For many people, this debate revolves around the issue of fetal consciousness. Does a six-week-old fetus have the capacity to feel pain or fear? Is it conscious in any way? How about a ten-week-old fetus? Can it suffer? If not, then many people would conclude that the fetus is not an ethical subject, in which case the question of abortion should focus on the feelings and potential suffering of the mother, and perhaps of other people. If, on the other hand, scientists could demonstrate that a ten-week-old fetus is capable of feeling pain, many people who currently support the right to abortion may start having second thoughts (Berk, 2013; Lagercrantz & Changeux, 2010; Moser et al., 2021; Passos-Ferreira, forthcoming; Warren, 1989).

Accordingly, when we compare the merits of the integrated information theory of consciousness, the global neuronal workspace theory, and the other competing theories, one key yardstick to evaluate them is to consider what they can teach us about fetal consciousness. Can they determine whether a ten-week-old fetus is conscious? Would their advocates be willing to testify about it in court, or in front of the US Congress? And if the theory cannot resolve the debate about fetal consciousness, is it really a theory of consciousness?

Nonhuman Consciousness

Another case in point concerns animal welfare. Can cows, chickens, and fish suffer? If the answer is yes, this has far-reaching implications. It means that these animals are ethical subjects, and that we should think carefully about how we treat these animals in farms, industrial facilities, and university laboratories (Birch, 2020, 2022; Godfrey-Smith, 2016; Gregg, 2013; Panksepp, 2005; Shevlin, 2021; Singer, 2023).

In 2012 a group of prominent neuroscientists signed the Cambridge Declaration on Consciousness, which states that "convergent evidence indicates that non-human animals have the neuroanatomical, neurochemical, and neurophysiological substrates of conscious states along with the capacity to exhibit intentional behaviors. Consequently, the weight of evidence indicates that humans are not unique in possessing the neurological substrates that generate consciousness. Non-human animals, including all mammals and birds, and many other creatures, including octopuses, also possess these neurological substrates" (Low, 2012).

This declaration was one of the reasons why three years later, in 2015, the New Zealand parliament passed the Animal Welfare Amendment Act. The act stipulates that it is obligatory to recognize animals as conscious, and hence attend properly to their welfare in contexts such as animal husbandry (Ministry for Primary Industries, 2015). New Zealand thereby became the first country in the world to legally recognize animals as conscious beings (Browning & Veit, 2022; Robertson and Goldsworthy, 2022). This is a far from trivial act, especially in a country where sheep outnumber humans nearly five to one

Of course, recognizing that animals have consciousness and the capacity to suffer does not mean that we should immediately ban all experiments on animals and all meat and dairy products. The debate continues. But the terms of the debate are changing. Even after accepting that cows, chickens, and fish are conscious beings, some may argue that it is nevertheless ethical to slaughter and eat them, because their capacity to suffer is smaller than the human capacity to suffer, and their feelings should therefore count for less than human feelings. This, however, is an empirical claim. Can the science of consciousness prove or disprove it? To do so, the science of consciousness needs to find a way to measure and scale consciousness, and to measure and scale suffering. The fate of billions of sentient beings may depend on this.

Dangerous Territory

Any attempt to measure and scale consciousness is unlikely to stop with animals. The question will inevitably be raised about humans, too. Perhaps within humankind itself there is a scale of consciousness, and significant differences in people's capacity to feel and to suffer? Do babies and adults have the same level of consciousness? Do people with certain medical conditions have a reduced type of consciousness, or a lesser capacity to suffer? Can various treatments ranging from genetic engineering to brain-computer interfaces produce upgraded humans with a superior type of consciousness (Kurzweil, 2012, 2024)? These are not just interesting scientific questions, but ethical and political issues with potentially very dangerous implications.

Scholars working on theories of consciousness should therefore be very careful about the claims they make, especially regarding the ability to measure and scale consciousness. There are few strong theories in science that do not translate into measurements and scales. If and when we formulate a strong theory of consciousness, it would probably result in a method for measuring and scaling consciousness, which would almost inevitably translate into a *hierarchy* of consciousness. Numerous people already believe in an unproven hierarchy of consciousness between humans and other animals, according to which humans have some kind of superior consciousness and a greater capacity to suffer. This unproven theory justifies the abysmal way humankind treats billions of other animals, and may well have been responsible for more suffering than any other idea in the whole of history. This should alert us to the potential danger of positing a hierarchy of consciousness within humanity itself, which might place some humans above others.

It should also be remembered that a theory of consciousness might be completely wrong, and still have a huge ethical and political impact. History is full of examples of erroneous theories that reshaped the world, like the bogus racist theories that many biologists, anthropologists, historians, and other scholars disseminated in the nineteenth and twentieth centuries. We are now in the initial stage of formulating scientific theories of consciousness. Researchers are merely sowing the seeds of these theories. But even as we sow a seed—we should think of the fruit it could bear. From the very beginning, we should be aware of the far-reaching ethical and political implications of such theories.

No Neutral Ground

Reminding ourselves of the ethical and political implications is one reason why we had better define consciousness as the capacity to suffer. But on a deeper level, this definition also reminds us of the unique nature of consciousness, and of how it differs from all other known phenomena in the universe.

Too often in scientific studies of consciousness, scholars tend to treat consciousness as a neutral phenomenon, focusing on processes like perception, observation, and calculation, which lack strong emotional and ethical content. This "observational" view of consciousness is a by-product of the fact that when scientists try to understand highly complex phenomena, we often start with the least complicated part. Accordingly, when studying consciousness, we seldom begin with the ability to feel love or hate, and instead start with simpler aspects of consciousness, such as the ability to see red (Humphrey, 2006; Jackson, 1982; Ludlow et al., 2004; Niikawa, 2020; Tye, 2002).

Unfortunately, the focus on such simple aspects of consciousness may cause us to adopt a sanitized attitude to consciousness, and conceptualize it as neutral "observational phenomenon." As if consciousness is a kind of camera that just observes the world, and as if "seeing red" is at the core of consciousness, while things like love and hate are peripheral add-ons. This misconception may cause us to misunderstand the consciousness of humans and animals, and also tempt us to ascribe consciousness to nonorganic devices. If to be conscious means to see red and to distinguish red from green, maybe cameras that distinguish between red and green are also conscious? Maybe thermostats that measure temperature and distinguish hot from cold have consciousness (Ciaunica et al., 2021; Safron, 2021; Solms, 2021)?

In fact, what makes consciousness so unique and so important is that it is the least neutral phenomenon in the universe. It is the only thing in the universe—as far as we can tell—that involves suffering. And in many ways, suffering is the opposite of pure observation. While observation tries to capture the present reality as objectively as possible, at the core of any experience of suffering is a rejection of present reality and a preference for something else, that *does not exist* at present. We feel something—but we want to feel something else instead. We feel pain—but we want the pain not to be there.

Suffering is not the pain itself. Pain is a sensation. Suffering is the conscious reaction to that sensation. It is a reaction of rejecting and hating the reality we experience, and preferring something else, which is not there. This rejection of reality, and the craving for something else, can happen not only in reaction to pain. Anger, greed, depression, and other kinds of misery also involve the same kind of reaction. Even a sensation of pleasure might result in suffering, if we react to the pleasing sensation by craving more of it, and becoming disappointed when we do not get more. A child might cry while eating a tasty apple—because they want ice-cream instead. A tycoon with a billion dollars in the bank might be extremely frustrated because they want two billion. The ability to reject reality and prefer something that isn't there, is perhaps the most characteristic feature of consciousness, which is absent from cameras and thermostats, and which is responsible for the misery of humans and other animals.

It is of course an open empirical question whether consciousness *always* involves preferring one thing over another. Can conscious entities ever experience moments of pure observation, when they do not prefer anything and do not reject anything? That is something we need to explore further. But it would be extremely misleading to embark on a study of consciousness without first noting the remarkable ability of consciousness to reject reality, at least *sometimes*. What kind of thing can reject reality?

Computers in Love

Electrons, atoms, molecules, stars, black holes, and galaxies are all neutral entities. They cannot reject reality, and they cannot suffer. They have no consciousness. As for cameras and thermostats, they can measure light and temperature alright, but cameras do not suffer when it is very dark and thermostats do not suffer when it is very hot. When we set an air conditioner to 25°C, and its thermostat detects the room temperature is now 28°C, the thermostat activates the cooling unit in order to lower the temperature. But the thermostat does not generate any feelings of depression or embarrassment from what it measures. It never rejects what it measures. That is why throwing a thermostat into a boiling cauldron is not a crime.

The same goes for computers, at least as of 2024. And this is a crucial point. Thermostats are not revolutionizing the world. Computers are. We are now in the midst of an enormous revolution in AI. Understanding the vital link between consciousness and suffering will help us deal better with a whole host of difficult questions raised by AI (Butlin et al., 2023).

As computers and robots become more intelligent, and actually surpass human intelligence in some fields, numerous practical questions arise about our ethical and political attitude to AI. Should robots have rights? Should AI be authorized to make ethical decisions? Should we, for example, produce autonomous weapon systems that decide on their own initiative who to kill?

Such debates, both within and outside academia, are too often conducted on the basis of misguided assumptions. There is widespread confusion between intelligence with consciousness, which gives the impression that artificial intelligence will inevitably be endowed with consciousness. In fact, however, intelligence and consciousness are very different things. Intelligence is the ability to solve problems and attain goals—find food, evade predators, win a chess game, drive a car safely from point A to point B. Consciousness is the ability to feel things like pain, pleasure, love, and hate. In humans and other animals, intelligence goes hand in hand with consciousness. Humans, chimps, cows, and chickens rely on their feelings in order to solve problems and attain their goals. However, computers solve problems in a very different way from animals, and there is absolutely no reason to think that they are anywhere on the road to developing consciousness (Butlin et al., 2023; Harari, 2016, 2018).

Over the past eight decades there has been an immense advance in computer intelligence, but there has been exactly zero advance in computer consciousness. As far as we know, computers in 2024 are no more conscious than their prototypes were in the 1950s. By 2050 computers are likely to be far more intelligent than humans in a wide range of fields, but they may still have zero consciousness and zero capacity to suffer.

Of course, the history of computer development over the past decades does not rule out the possibility that eventually computers could develop consciousness. However, past developments do indicate that there are several alternative roads leading to intelligence, and only some of these roads involve gaining consciousness. In mammals, the evolution of intelligence involved the evolution of consciousness, too (LeDoux, 2023; Mitchell, 2023). But computers have so far been developing along a very different path, and they may continue along that path far into the future. Just as airplanes fly faster than birds without ever developing feathers, so computers may come to solve many problems much better than humans without ever developing feelings (Harari, 2018).

Unless we remind ourselves of the deep connection between consciousness and suffering, it will be very difficult to deal with the ethical and political implications of the rise of artificial intelligence. If we see consciousness as a neutral phenomenon, we are likely to go on confusing it with intelligence. Then, as computers become more intelligent than humans, our ethical and political systems might end up privileging intelligent computers over conscious humans.

References

Berk L. (2013). *Child development* (9th ed.). Pearson Higher Education.

Birch, J. (2020). Global workspace theory and animal consciousness. *Philos Top, 48*, 21–38.

Birch, J. (2022). The search for invertebrate consciousness. *Noûs, 56*, 133–153.

Browning, H., & Veit, W. (2022) The sentience shift in animal research. *The New Bioethics, 28*(4), 299–314.

Butlin, P., Long, R., Elmoznino, E., Bengio, Y., Birch, J., Constant, A., Deane, G., Fleming, S. M., Frith, C., Ji, X., Kanar, R., Kelin C., Lindsay, G., Michel, M., Mudrik, L., Peteres A.K.P., Schwitzgebel, E., Simon, J., & Kanai, R. (2023). Consciousness in artificial intelligence: insights from the science of consciousness. *arXiv preprint arXiv:2308.08708.*Ciaunica, A., Safron, A., & Delafield-Butt, J. (2021). Back to square one: The bodily roots of conscious experience in early life. *Neuroscience of Consciousness, 2021*(2), article niab037.

Dehaene, S. (2014). *Consciousness and the brain: Deciphering how the brain codes our thoughts.* Viking.

Dennett, D.C. (2018). Facing up to the hard question of consciousness. *Philosophical Transactions of the Royal Society B: Biological Sciences, 373*, article 1755.

Godfrey-Smith, P. (2016). *Other minds: The octopus, the sea, and the deep origins of consciousness.* Farrar, Straus and Giroux.

Gregg, J. (2013). *Are dolphins really smart?: The mammal behind the myth.* Oxford University Press.

Harari, Y. N. (2016). *Homo deus: A brief history of tomorrow*. Harvill Secker.

Harari, Y. N. (2018). *21 lessons for the 21st century*. Spiegel and Grau.

Hildebrandt, A. (2014). Routes to decriminalization: A comparative analysis of the legalization of same-sex sexual acts. *Sexualities, 17*(1–2), 230–253.

Humphrey, N. (2006). *Seeing red: A study in consciousness.* Harvard University Press.

Jackson, F. (1982). Epiphenomenal qualia. *The Philosophical Quarterly, 32*(127), 127–136.

Kurzweil, R. (2012). *How to create a mind: The secret of human thought revealed.* Penguin Books.

Kurzweil, R. (2024). *The singularity is nearer: When we merge with AI.* Viking.

Lagercrantz, H., & Changeux J. P. (2010). Basic consciousness of the newborn. *Seminars in Perinatology, 34*(3), 201–206.

LeDoux, J. E. (2023). *The four realms of existence: A new theory of being human.* Harvard University Press.

Low, P. (2012, July 7). *The Cambridge declaration on consciousness.* https://web.archive.org/web/20131109230457; http://fcmconference.org/img/CambridgeDeclarationOnConsciousness.pdf

Ludlow, P., Nagasawa, Y., & Stoljar, D. (Eds.) (2004). *There's something about Mary: Essays on phenomenal consciousness and Frank Jackson's knowledge argument.* MIT Press.

Melloni, L., Mudrik L., Pitts, L., Bendtz K., Oscar, F. O., Gorska, U., Hirschhorn, R., Khalaf, A., Kozma, C., Lepauvre, A., Liu, L., Mazumder, D., Richte, D., Zhou, H., Blumenfeld, H., Boly, M., Chalmers, D. J., Devore, S., Fallon, F., . . . Tononi, G. (2023). An adversarial collaboration protocol for testing contrasting predictions of global neuronal workspace and integrated information theory. *PLoS One, 18*(2), Article e0268577.Mignot, J. (2022). Decriminalizing homosexuality: A global overview since the 18th century. *Annales De Démographie Historique, 143*, 115–133.

Ministry for Primary Industries. (2015). *Animal welfare amendment act 2015.* New Zealand Government. https://www.mpi.govt.nz/dmsdocument/9247-Animal-Welfare-Amendment-Act-2015-Update-and-early-information-for-animal-ethics-committees-AECs

Mitchell, K. J. (2023). *Free agents: How evolution gave us free will.* Princeton University Press.

Moser J., Schleger F., Weiss M., Sippel K., Semeia L. & Preissl H. (2021). Magnetoencephalographic signatures of conscious processing before birth. *Dev Cogn Neurosci, 49*, Article 100964.

Mudrik, L., Mylopoulos, M., Negro, N., & Schurger, A. (2023). Theories of consciousness and a life worth living. *Current Opinion in Behavioral Sciences, 53*, Article 101299.

Niikawa, T. (2020). A map of consciousness studies: Questions and approaches. *Frontiers in Psychology, 11*, Article 530152.

Panksepp, J. (2005). Affective consciousness: Core emotional feelings in animals and humans. *Consciousness and Cognition, 14*(1), 30–80.

Passos-Ferreira, C. (2023). Are infants conscious? *Philosophical Perspectives, 37* (1), 308–329.

Robertson, I., & Goldsworthy, D. (2022). Recognizing and defining animal sentience in legislation: A framework for importing positive animal welfare through the five domains model. *Monash University Law Review, 48*, Article 244.

Safron, A. (2021). World models and the physical substrates of consciousness: Hidden sources of the stream of experience? *Journal of Consciousness Studies, 28*(11–12), 210–221.

Shevlin, H. (2021). Non-human consciousness and the specificity problem: A modest theoretical proposal. *Mind & Language, 36*, 297–314.

Singer, P. (2023). *Animal liberation now*. HarperCollins. (Original work published 1975)

Solms, M. (2021). *The hidden spring: A journey to the source of consciousness*. W. W. Norton and Company.

Taylor, C. (1992). *Sources of the self: The making of the modern identity*. Harvard University Press.

Tononi, G., Boly, M., Massimini, M., & Koch, C. (2016) Integrated information theory: From consciousness to its physical substrate. *Nature Reviews Neuroscience*, *17*(7), Article 450.

Tye, M. (2002). *Consciousness, color and content*. MIT Press.

Warren M. A. (1989). The moral significance of birth. *Hypatia*, *4*, 46–65.

8 What Is Conscious Vision Good For? Debunking the Devil from the Details

Dominique Lamy and Nitzan Micher

Personal Reflections

Dominique Lamy: *The experimental approach adopted by cognitive psychology researchers (and to some extent also, by most cognitive neuroscience researchers) when investigating consciousness is to target the transition between unconscious and conscious vision. As a consequence, and somewhat ironically, the study of conscious experience is de facto mostly the study of what conscious experience is not. In this sense, what we learn about subjective experience from investigating unconscious processes is akin to what we learn about an object from its shadow: the feel and taste of experiencing a color, a feeling, or a memory are left aside. In addition, these approaches adopt scientific rigor as their main guiding thread and yield research that treads very carefully, in minute steps, often going back and forth, toward solid answers. During the workshop "Perspectives on Consciousness" held in Israel in 2020, I encountered strikingly different approaches, many placing phenomenology at the center of their endeavors, others suggesting bold and sweeping theories without relying on experimental data, and yet others choosing not to take the scientific path in deepening their understanding of subjective experience. At first it felt that all we had in common was a hollow word, "consciousness," and that each of us fills with a totally different meaning. Yet, as the workshop progressed and group as well as one-on-one discussions intensified, the connection between the different approaches and how they can fertilize each another started to emerge.*

For one thing, it became clear to me during the workshop, that in the same way as studying unconscious processing can provide important insights on the functions, if any, of consciousness, a descriptive approach to the subjective conscious experience and its different facets can also shed light on what distinguishes unconscious processing from conscious experience. In addition, the interactions that were made possible by the workshop led us to realize that researchers in different disciplines often use the same words to refer to entirely different concepts, a state of affairs that engenders confusion, misunderstanding, and even distrust. The open and receptive atmosphere that prevailed during the workshop encouraged us to communicate long enough to get to the bottom of misconceptions, and to acknowledge the need to develop a common terminology that should render the work done in each discipline accessible to researchers in different domains.

It is most often only with hindsight that we decipher our motivations. What brought me to the field of consciousness research? As my PhD research focused on the role of attention in perception, it seems natural that I took the small step that led to studying the role of consciousness in perception. Yet, it is an entirely different association that comes to mind when I attempt to answer this question: in my high school days, a discussion flared up during a philosophy class, around the ideas defended by Malebranche, one of Descartes' disciples. Malebranche[1] denied any form of consciousness in animals, and the students who sided with his view challenged those of us who were outraged by it to prove that animals are conscious. Failing to find definitive arguments to demonstrate what felt so obvious was a particularly frustrating experience. I would have been surprised to learn that, many years later, I would spend much of my time trying to figure out how one can determine whether an observer is aware or unaware of the stimulus that hits their retina, and that I would coauthor a paper on the ethical issues involved in using animal models for the study of consciousness (Mazor et al., 2023).

Overview

There is the notorious "problem of consciousness," often called the "mind-brain problem" or the "hard problem." However, to study consciousness empirically, many of us have turned our backs on the issues that are most impregnated with the sense of mystery that underlies the fascination for consciousness. We restricted ourselves to tackling only what is called the "easy problem"—namely, "what processes differ when I am aware of a visual stimulus presented on the computer screen and when I am not?" Yet, even this limited endeavor has yielded no definite answers. A compilation of all the reported neural correlates of visual consciousness leaves out virtually no area of the brain. Likewise, the answers provided by behavioral studies on the extent to which a stimulus can be processed without conscious vision, display considerable differences: they range from excluding any processes beyond basic feature extraction to including complex arithmetic operations. In this chapter, we focus on the methodological issues that may account for many of these discrepancies. In particular, we survey the main intricacies associated with the three ingredients necessary in any research evaluating unconscious vision: a manipulation of conscious perception, a measure of conscious perception, and a measure of visual processing. Although this contribution is strictly methodological, it informs the theoretical conclusions that can be drawn from current research on the function of conscious vision.

The devil is in the details: The details of a matter are its most problematic aspect.
—Oxford Languages

Introduction

Most would agree that our bodily processes run smoothly without us noticing them, and that being unaware of them does not affect us. However, consensus breaks when it comes to sensory information processing. Can a stimulus that we do not perceive consciously, affect our behavior and is its influence qualitatively or quantitatively different when it is consciously perceived? These questions have been debated since the first scientific demonstration of unconscious perception. The fascination they exert hinges on the belief that understanding the limits of unconscious perception could help define the necessary conditions of conscious awareness and answer century-old questions such as "who has consciousness?" and "can we build conscious machines?" Recently, the idea that perception can be unconscious has been crucial in one of the largest endeavors in the field of cognitive neuroscience: the search for the neural correlates of consciousness. Perception without awareness implies that differences between conscious and unconscious processing in the brain and the mapping of these differences may help identify the neural activity that makes us conscious.

However, we must admit that, to date, this research has yielded few definite answers. A compilation of the reported neural correlates of visual consciousness leaves out virtually no area of the brain. Likewise, a survey of the behavioral literature on processes elicited by stimuli outside of conscious vision displays considerable differences.

This chapter focuses on the methodological issues that may explain this situation. We restrict our discussion to issues directly related to behavioral studies of unconscious processing. Thus, we do address the methodological intricacies that arise in studies measuring the neural substrates of conscious and unconscious vision. However, the overlap between the methodological issues in the two fields makes the present discussion relevant to both. As will become clear below, in the field of unconscious perception, the devil is in the details. Thus, although this contribution is strictly methodological, it should inform theoretical conclusions from current research on the function of conscious vision.

Most experiments investigating unconscious processing rely on a deceptively simple rationale called the "simple-dissociation" paradigm: namely, if conscious perception of a stimulus is prevented, and if this stimulus influences behavior in some measurable way, unconscious processing is inferred. The methodological issues we address pertain to three backbone components of these experiments: (1) a *measure of conscious perception*, ensuring that the observer was unaware of the stimulus; (2) a separate *measure of processing*, showing that the invisible stimulus was processed; and (3) a *blinding method* that blocks conscious perception but allows some (unconscious) processing. In the first part of this chapter, we describe the main measures of conscious perception and processing and discuss the main empirical strategies to dissociate them. In the second part, we discuss considerations guiding the choice of the blinding method.

Empirical Strategies in the Study of Unconscious Processing

In scientific research, translating theoretical concepts into operational definitions is essential—for instance, defining addiction to social media as the number of times per hour a participant checks their accounts. The history of unconscious perception research is ridden with debates about what counts as a valid operational definition of conscious experience. The fundamental problem is that we cannot have direct access to another person's subjective experience. As a result, to assess a participant's conscious perception of a stimulus, we can only rely on their report of their subjective experience or infer it from their objectively observed behavior. With *subjective measures* of conscious perception, observers are asked to provide an introspective report of their experience. For instance, they may rate the visibility of a given stimulus on a scale: they are considered to be subjectively unaware of the stimulus whenever they use the lowest rating. With *objective measures*, participants are asked to discriminate a given property of the stimulus, for instance, to determine whether it is red or green, and to guess if unable to decide. Participants are considered to be objectively unaware of the stimulus if their discrimination performance is at chance.

The valid measure of processing without conscious awareness depends on the chosen measure of conscious perception. This has led to three main empirical strategies: the *subliminal-perception paradigm*, the *subliminal-prime paradigm*, and the *liminal-prime paradigm*. We discuss each of these rationales separately (see figure 8.1).

The Subliminal-Perception Paradigm

In a typical subliminal-perception paradigm, a target is presented under conditions designed to prevent its conscious perception. On each trial, participants provide two responses: a subjective report of their conscious experience and a forced-choice discrimination of a property of this target (e.g., "is it a triangle, a square, or a circle?"). Observers are held to be unaware of a stimulus if they report a similar experience when the stimulus is present and when it is absent. The question is whether participants discriminate the stimulus they report not seeing better than by just guessing.

The main criticism of the subjective measures used in the initial study of unconscious perception is called the *criterion* problem: when forced to report our awareness of a stimulus using a binary "yes/no" response, observers may adopt a conservative criterion and categorize weak conscious percepts as "not conscious." This would invalidate any demonstration of unconscious perception. To address this problem, researchers started using more sensitive scales. The most popular is the perceptual awareness scale (PAS Ramsøy & Overgaard, 2004) which includes four levels: 1 = no conscious experience, 2 = brief glimpse, 3 = almost clear image, and 4 = absolutely clear image. With the PAS, observers are held to be unaware of a stimulus only if they use the PAS1 ("no experience") label to rate it (see figure 8.2).

Other subjective scales were developed: they required participants to either rate their confidence in a perceptual decision or place a wager on this decision. These metacognitive measures are useful to evaluate whether observers can distinguish between their own correct and incorrect decisions, but they are not direct measures of perceptual experience: confidence regarding a visual stimulus does not necessarily entail subjective experience. To illustrate, suppose that a visual onset that observers do not experience elicits a reflexive eye movement to its location. Sensory feedback from the eye muscles should provide a reliable indicator of the stimulus location and increase the observer's confidence about their localization responses—however, confident decisions would occur without any subjective perceptual experience.

Using the PAS scale, some studies reported that objective performance falls to chance for the lowest level ("no conscious experience"), suggesting that unconscious perception does not occur. For instance, Overgaard et al. (2008) revisited the most emblematic evidence of unconscious perception: blindsight. In blindsight, patients with primary visual

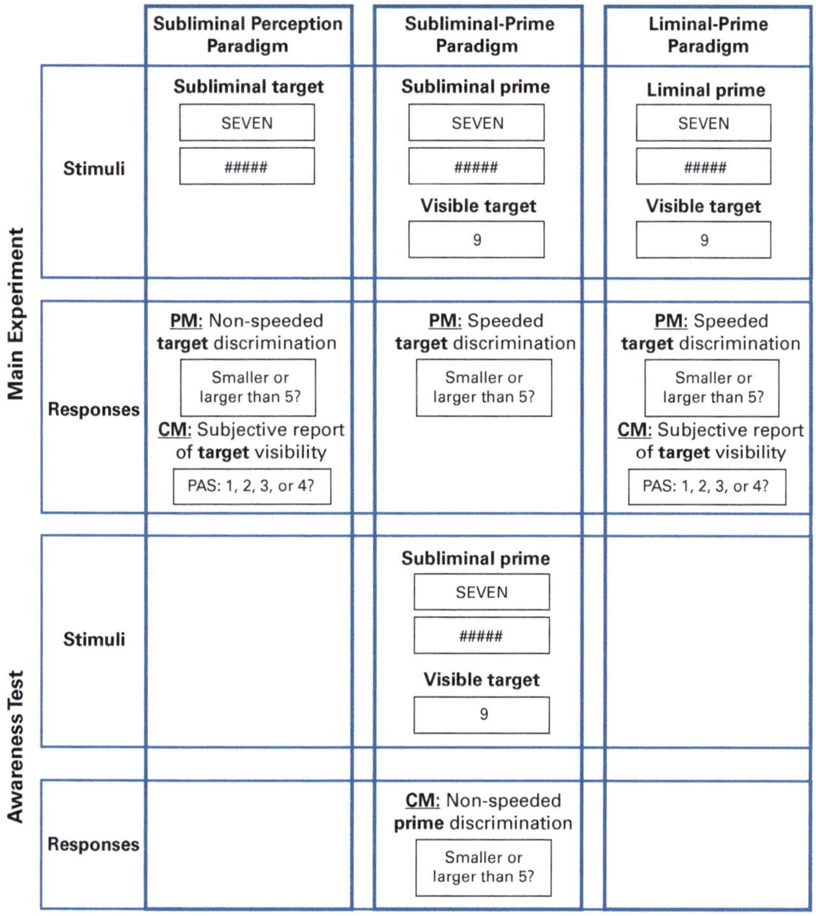

Figure 8.1
Illustration of the procedure in the three paradigms described in this chapter with the same stimuli and general task. Here, conscious perception of the critical stimulus is compromised by having the stimulus immediately followed by a mask (#####). PM = processing measure; CM = measure of conscious awareness; and PAS = perceptual awareness scale.

cortex lesions claim to be blind yet act as if they can see: for instance, they avoid obstacles in their way. In the lab, these patients show above-chance perceptual discrimination for stimuli they claim to have no subjective experience of. Overgaard et al. (2008) showed that preserved discrimination abilities emerged only when the patient reported having "a brief glimpse" but not when she reported "no conscious experience." The authors concluded that blindsight patients have "degraded conscious vision" rather than "unconscious vision."

The Subliminal-Prime Paradigm

Researchers using the subliminal-prime paradigm assume that the criterion problem pervades any subjective measure of conscious perception—even PAS, where observers should use

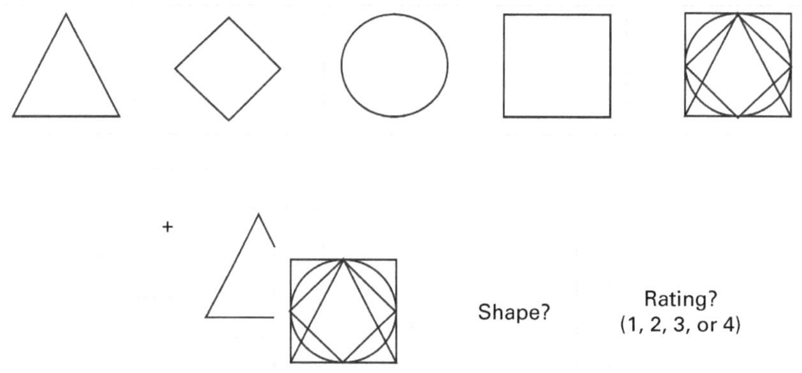

Figure 8.2
Example of the subliminal-perception paradigm. Participants are required to categorize the target (here a triangle), which is immediately followed by a mask. Then, they rate the visibility of the target using the PAS. Unconscious processing is thought to occur in a trial where participants rate the target with the lowest rating (i.e., 1) yet are better than chance at discriminating the target shape (i.e., in this example, if their performance exceeds 25 percent).

intermediate ratings for faint conscious perception and the lowest rating only when they experience no stimulus. Therefore, they seek a criterion-free measure of conscious perception, which they view as the most conservative approach to demonstrate unconscious processing.

A typical subliminal-prime experiment includes two phases: a priming phase and a prime-awareness phase. In both, two stimuli appear in close succession: a prime, the conscious perception of which is prevented by some blinding method (see the next section) followed by a clearly visible target (see figure 8.3). In the priming phase, observers categorize the target (e.g., whether a number is smaller or larger than 5). The prime bears some relationship to the target (e.g., it is also smaller or larger than 5). Prime processing is assessed via the impact of the prime on responses to the visible target—understanding the meaning of the prime should lead to faster responses when the target and prime are both smaller or larger than 5 (congruent condition) than when one is smaller and the other larger (incongruent condition)—an effect known as priming. The prime-awareness phase is similar, but observers have to discriminate some property of the prime instead of responding to the target. If observers perform at chance, one infers that the prime was (also) subliminal in the priming phase and that priming, if found, resulted from unconscious processing. If observers performed above chance, the experiment failed, and no conclusion can be reached. Note that here, above-chance performance denotes conscious perception, whereas in the subliminal-perception paradigm it denotes unconscious processing. Although the subliminal-prime paradigm is currently the most widely used, it has received many criticisms, of which we survey a subset below.

With the subliminal-prime paradigm, ensuring that the prime is never consciously perceived is critical but it is challenging to set prime parameters that obliterate conscious perception without degrading stimulus to a point where it generates no (unconscious) processing at all. As a result, some participants may show above-chance discrimination performance during the awareness test. To ensure that priming cannot be accounted for by these "aware" participants, their data are often excluded (e.g., Sklar et al., 2012). Recently, however, Shanks (2017) sharply criticized this practice and demonstrated that that it creates statistical artifacts.

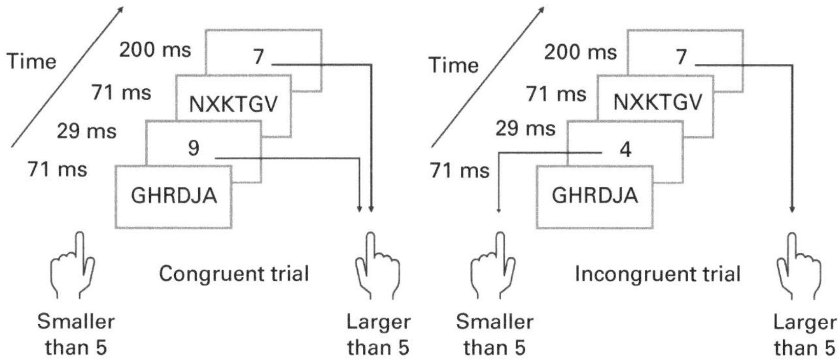

Figure 8.3
Example of the subliminal-prime paradigm (e.g., Dehaene et al., 1998; Kunde et al., 2003). A prime (here, 9 or 4) appears in a "sandwich" between two random masks, followed by a target (here, 7). During the first (priming) phase, participants categorize the *target* number as larger or smaller than 5, as fast as possible. During the second (awareness test) phase, participants categorize the *prime* as larger or smaller than 5. In this paradigm, unconscious processing is thought to occur if participants are at chance on the awareness test but there is a *priming* effect—that is, participants respond faster when the prime and target are congruent (i.e., elicit the same response, right panel) than when they are incongruent (i.e., elicit opposite responses, left panel).

Many other criticisms result from the *task comparability* problem, originally formulated by Reingold & Merikle (1988). In a nutshell, these authors suggested that for any dissociation between a direct measure (i.e., awareness) and an indirect measure (e.g., priming) to be valid, the two measures should be obtained under comparable experimental conditions. The subliminal-prime paradigm falls short of this methodological criterion on several accounts, the list of which is constantly updated.

First, the number of trials is typically much lower in the awareness test than in the priming phase, making it is statistically easier to demonstrate priming than above-chance performance. Second, the priming-phase task (categorizing a clearly visible target) is much easier than the awareness test (categorizing a barely visible prime). As a result, participants may be less attentive or motivated in the awareness test than in the priming phase (Pratte & Rouder, 2009). Consistent with this claim, prime categorization performance is poorer when the low-visibility primes are presented alone (as in typical awareness tests) than when the task is made more engaging by intermixing high-visibility primes. Thus, the two-phase subliminal-prime paradigm overestimates unconscious processing because it underestimates conscious perception of the prime.

However, fulfilling the task comparability criterion may also generate problems. For instance, to ensure stimulus comparability, the same prime-target sequences are presented in the two phases. Thus, during the awareness test, participants must ignore a clearly visible target and categorize a subliminal prime. The target should elicit the response that was repeatedly associated with it during the priming phase, thereby contaminating the responses to the prime. As the target is randomly either congruent or incongruent with the prime, such contamination would explain why chance performance on the prime categorization task is observed—even if participants have some awareness of the prime. Consistent with this conjecture, prime-discrimination performance is higher when a neutral stimulus (that cannot contaminate responses to the prime) replaces the target (Peremen & Lamy, 2014b, exp.1; Vermeiren & Cleeremans, 2012). This issue is even worse in studies that measure both priming (by collecting speeded target-discrimination responses) and prime awareness (by collecting non-speeded prime-discrimination responses) on each trial (e.g., Koivisto & Rientamo, 2016). Again, responses to visible targets are likely to largely contaminate responses to weakly perceived primes.

The Liminal-Prime Paradigm

The subliminal-perception and subliminal-prime paradigms aim to uncover the cognitive processes that can operate without conscious perception. However, by focusing exclusively on unconscious processing, these paradigms provide no information about the extent to which conscious perception modulates these cognitive processes. For instance, an 8-millisecond priming effect with a subliminal prime (e.g., Van Den Bussche et al., 2009) is taken to demonstrate that semantic processing of a stimulus outside conscious awareness is possible. However, how sensible would this conclusion be if the priming effect was 100 milliseconds when the prime is consciously perceived? One would have to conclude that conscious perception plays a decisive role in semantic processing. However, comparing priming when the stimulus is strongly degraded and when it is intact would be misleading: any difference could reflect the role of the stimulus's strength rather than the role of conscious perception per se. How can the impact of conscious and unconscious processing be compared without conflating conscious perception with the physical strength of the stimulus?

The liminal-prime paradigm was developed to compare conscious and unconscious processing under identical stimulus conditions (e.g., Avneon & Lamy, 2018; Van Den Bussche et al., 2013). It relies on the PAS (Ramsøy & Overgaard, 2004) as the measure of conscious perception and on an indirect measure of processing, typically priming. The parameters of the prime stimulus (e.g., its brightness and duration) are set to make its perception liminal, with its visibility varying from total invisibility to clear perception across trials. On each trial, participants first make a rapid response to the target and then rate the prime visibility. Thus, conscious perception of the prime and the impact of this prime on responses to the target (i.e., priming) are both assessed on each trial and can be compared under the same task and stimulus conditions. Priming at the lowest and highest PAS levels are taken to index unconscious and conscious processing, respectively. This paradigm was used to investigate the role of conscious perception in various cognitive functions, from low-level perceptual processing (Kimchi et al., 2018; Lamy et al., 2015) to semantic processing (Avneon & Lamy, 2019; Micher & Lamy, 2023). It proved sensitive enough to show that some processes are independent of conscious perception, others are enhanced by conscious perception, and yet others are entirely contingent on it. For instance, the sudden appearance of an object in our visual field grabs our attention, irrespective of whether we clearly perceive this object or have no conscious experience of it. By contrast, keeping track of the changes occurring in an object is not possible if this object is not consciously perceived (Lamy et al., 2015).

Despite several advantages (see Lamy et al. [2019] for review), liminal-prime paradigm suffers from several potential problems. For instance, if observers provide random visibility ratings, priming should be similar across these ratings, a finding that would seriously underestimate the role of conscious perception. To address this problem, catch trials, in which *no prime* is presented, are interspersed among prime-present trials. Finding that a participant uses high-visibility ratings (signaling that they clearly saw the prime) on prime-absent trials indicates that this participant's introspective reports are not reliable.

Another limitation is that the claim that conscious and unconscious effects are measured under identical conditions is misleading: something must vary in order to for visibility ratings to vary. According to higher-order theories of consciousness (e.g., Lau & Rosenthal, 2011), subjective perception may vary in the absence of any change in first-order representations. However, according to theories such as the global neuronal workspace (e.g., Dehaene et al., 2006), trial-to-trial differences in noise or fluctuating attentional states explain why a liminal stimulus's access to conscious awareness varies. Likewise, priming effects may reflect consequences of conscious processing: to report the stimulus after it is consciously perceived, one must encode it in a temporary memory store called working memory. Therefore, any difference in a stimulus's impact on behavior when this stimulus is consciously perceived versus missed may reflect the role of these factors (e.g., noise, attention, encoding in working memory) rather than the role of conscious perception (see, e.g., De Graaf et al. [2012] for similar considerations regarding the difficulty of disentangling the neural correlates of consciousness from the neural correlates of its prerequisites and consequences). Resolving this issue remains a challenge.

Direct-Indirect Match, Partial Awareness, and Levels of Processing

Our conscious percepts are multilevel and multidimensional (e.g., Kouider & Dupoux, 2004). Thus, for instance, a degraded stimulus may allow observers to consciously perceive its color, but not its identity. If degradation is even stronger, observers may be unaware of the stimulus's mere presence. Researchers disagree about the property that observers should be unaware of for unconscious processing to be established.

With subjective measures (i.e., in the subliminal-perception and liminal-prime paradigms), the answer is clear: invisibility is defined as participants' inability to determine whether the critical stimulus was present or absent. In this sense, these paradigms are conservative. With objective measures (i.e., in the subliminal-prime paradigm), the answer is less consensual. The problem arises because these measures probe participants' ability to discriminate only one property of the prime (e.g., its identity), leaving open the possibility that they may consciously perceive other properties (e.g., its color or shape). Asking participants to decide whether the prime is present or absent would guarantee that no property of the stimulus is consciously perceived. However, very few researchers use this stringent test, because it would require degrading the critical stimulus very strongly and might result in throwing the baby out with the bathwater: it would cut into unconscious processing itself.

Several authors argued that only conscious perception of the property that governs priming is relevant for showing that unconscious priming occurred (e.g., Breitmeyer et al., 2015; see Reingold & Merikle [1988] for the first recommendation that the direct and indirect indices should measure the same discrimination). For instance, if a masked prime (either pleasant or unpleasant) influences how fast observers decide that a visible target is pleasant or unpleasant, the appropriate awareness test would be whether observers can discriminate whether the prime is pleasant or unpleasant better than chance. Notably, however, such high-level unconscious processing seems to be contingent on conscious perception of lower-level properties. For instance, Gelbard-Sagiv et al. (2016) showed that face-identity processing can occur when participants are unable to discriminate the face identity above chance but does not occur when participants have no conscious access to any feature of the suppressed face (location or color). They concluded that high-level processing is tightly bound with at least some level of conscious awareness.

The foregoing discussion documents different viewpoints with regard to the information participants should be unaware of in order to conclude that unconscious processing of this information occurred. However, the information that observers actually use (or "criterion content") is not always the information experimenters assume they should be using. As a result, priming that appears to reflect high-level processing may in fact reflect lower-level processing. Here are two striking illustrations of how such misinterpretations may arise.

Kouider and Dupoux (2004) ingeniously used pseudo-words that resembled real color words (e.g., "gener" for "green") as primes in an ink-color discrimination task ("is the color red or green?"). They showed that word and pseudo-word primes influenced participants' responses to the target equally strongly, even when participants were unable report whether the prime was a word or a non-word. However, when the prime was further degraded so that participants could not report any of its letters, the prime no longer influenced performance. The authors concluded that awareness of the letters accounts for what appeared to be unconscious processing of a word's meaning: participants consciously perceived individual word fragments or letters and illusorily reconstructed the whole word, based on the expectations promoted by the task. Because the number of possible targets was small, the expectation associated with individual letters was strong.

Likewise, Damian (2001) challenged demonstrations of unconscious high-level (semantic) processing. He argued that when prime stimuli also appear as targets, "acquired mappings between targets and response keys are also applied to subliminally presented primes" (Damian, 2001, p. 158). In that case, perceptual processing of the prime suffices to elicit the associated response. For instance, if observers repeatedly respond to a visible target digit (1, 4, 6, or 8) by using their right hand when this digit is larger than 5 and their left hand when it is smaller than 5, stimulus-response associations quickly develop: shallow perceptual processing of the stimulus "1" suffices to automatically activate the left hand. Priming, then, reflects learned, low-level, stimulus-response associations. Supporting this argument, Damian (2001) showed that response priming does not occur for primes that never appear as targets (but see, e.g., Naccache & Dehaene [2001]).

Blinding Methods

Researchers have designed an impressive number of noninvasive methods to prevent stimuli from accessing visual awareness. In a recent review, Breitmeyer (2015) counted twenty-four different blinding paradigms, and the list keeps growing. The oldest and most common of these methods is backward masking: the critical stimulus is presented briefly and followed by a different stimulus (the mask) at the same location (e.g., Dehaene et al. 1998). Newer methods have also gained popularity, for instance, continuous flash suppression (CFS) wherein highly dynamic visual noise (sometimes called "Mondrian" in reference to its resemblance with the painter's colorful geometric shapes) is presented to one eye and can reliably suppress conscious awareness of an image presented to the other eye (e.g., Peremen & Lamy, 2014a; Sklar et al., 2012; Stein et al., 2020). More recently, a tachistoscope capable of displaying images for *micro*seconds was recently used to prevent conscious perception of unmasked stimuli (Beauny et al., 2020)

In principle, if one can show that the critical stimulus is processed despite the fact that it escaped conscious perception, the method used to produce invisibility should not matter. Accordingly, up until ten years ago, these methods were used somewhat interchangeably, and in the first systematic review of the main blinding methods, Kim and Blake (2005) were mostly guided by practical considerations in their evaluation of the strengths and weaknesses of each method.

However, researchers came to realize that the choice of the blinding method strongly affects how much unconscious processing can be measured and may therefore have far-reaching theoretical implications. Dehaene et al. (2006) proposed the first theoretically relevant distinction between different blinding methods. A review of the neural correlates of visual awareness revealed that the contrast between consciously perceived and invisible stimuli yielded seemingly incompatible patterns of neural activations across studies. To resolve these apparent inconsistencies, the authors suggested that two types of unconscious vision can be distinguished: subliminal vision and preconscious vision. According to their taxonomy, a stimulus is subliminal if its invisibility is achieved by limiting its sensory (bottom up) strength (e.g., with added visual noise or low contrast), and it is preconscious if sensory stimulation is strong, but invisibility is achieved by withdrawing attention from it. Crucially, they argued that unconscious processing is more reliable, longer lived, and occurs at a higher range of levels for preconscious than for subliminal stimuli. Thus, whether one uses a blinding method that induces perceptual or attentional blindness should have a substantial impact on the conclusions drawn about how deeply a stimulus can be processed without conscious perception.

In recent years, several researchers have attempted to provide a more detailed classification of blinding methods, aimed at assessing how late in the processing hierarchy each method blocks access to conscious vision. The guiding principle underlying such classification is that to maximize the chances of identifying processes that can be performed without consciousness, one should use the blinding procedure that impairs processing the latest. To illustrate this point ad absurdum, blindfolding observers to prevent conscious perception would be a bad strategy because it would also thoroughly prevent unconscious processing (Peremen & Lamy, 2014a). Breitmeyer (2015) sketched a functional hierarchy of the levels of unconscious visual processing that are indexed by the different blinding methods. He mainly relied on two types of empirical evidence. On the one hand, he adopted a meta-analytic strategy to determine whether a given level of unconscious processing (e.g., processing of the stimulus's meaning) is more often reported with some blinding methods than with others. However, this strategy generated noisy outcomes. For instance, with CFS, some studies reported unconscious semantic processing of multiple-word expressions (i.e., high-level unconscious processing [Sklar et al., 2012]), whereas other studies found that perceptual organization based on the gestalt principle of similarity (i.e., a much lower-level process) does not survive suppression (Kimchi et al., 2018). A likely reason for these disparities is that studies using the same blinding method may differ widely in the measure of conscious perception they use (as explained in the previous sections): studies relying on more liberal measures of invisibility are more likely to report high-level unconscious processing than studies relying on more stringent measures. On the other hand, Breitmeyer (2015) also reviewed attempts to compare different blinding methods within a single study (e.g., Breitmeyer et al., 2008; Peremen & Lamy, 2014a), thereby circumventing the noise inherent to aggregating very disparate studies using the same blinding method. The outcome of Breitmeyer's review is a hierarchy of blinding methods that encourages researchers to maximize their chances to uncover unconscious processing by using the methods that lie at the top, and relinquishing the methods that lie at the bottom (because they allow only minimal unconscious processing).

The picture may be more complicated, however. Each blinding method is associated with a number of critical parameters. For instance, how efficiently backward masking prevents conscious perception largely depends on the relative brightness and duration of the critical stimulus and of the mask. Processing stimuli rendered invisible by backward masking is likely to also vary with these parameters. Thus, the idea that suppressive effects of one blinding method reliably precede or follow that of another method in the processing hierarchy (Breitmeyer 2015) rests on the assumption that parameter variations only affect the *probability* that the method prevents conscious access, but not the *functional level* at which it interferes with processing. There is no systematic empirical investigation to date showing that this is indeed the case.

Moreover, the fact that parameters matter and that they differ for different methods, mandates that some "fair" way be found to compare different methods: some common measure is needed to ensure that the methods are equally efficient at preventing conscious perception. Peremen and Lamy (2014a) suggested use of the liminal-prime paradigm and selection of parameters for which the critical stimulus elicits the same proportion of null-visibility ratings on the PAS scale (Ramsoy & Overgaard, 2004) with the two compared methods. Using this rationale, they found that CFS interrupts processing earlier than backward masking.

Concluding Reflections

The scientific journey on which many researchers have embarked in the last thirty years aims at eroding the deep sense of mystery that surrounds consciousness. What is it that makes us conscious? How is consciousness implemented in the brain and who has it? These questions have meshed with older ones, no less fascinating or mysterious, that revolve around the unconscious. Is there a zombie inside us—that is, a collection of mechanisms that are associated with physical states that drive our actions but generate no mental states (Chalmers, 1996). What are its powers? What additional abilities does consciousness grant us over this zombie?

In this review, we showed that the first step often taken to answer these questions is deceptively simple: it consists of asking what information we are able to process from visual objects that are consciously invisible to us. Yet, each and every methodological aspect of the experimental studies designed to answer this simple question has been a matter of fierce debate. Often times, these endless discussions about methodology make it appear as though the essential questions we set up to ask are buried under heaps of details. However, behind what may appear as mere quibble over the most valid operational definitions of conscious awareness and unconscious processing, lies a fundamental question: Can we exclude subjective experience from the study of conscious awareness? Many researchers have opted to do so, based on the premise that subjective methods do not stand up to scientific standards of rigor, unlike methods based on third-person, objective measures. In this review, we showed that the opposite may be true. Objective measures of conscious perception, at least in their present form, are not exempt of methodological pitfalls. Although there is currently no agreed method to investigate unconscious processing and how it differs from conscious processing, paradigms that rely on sensitive measures of subjective experience seem to offer the best prospect for progress.

Note

1. "In dogs, cats, and other animals, there is neither intelligence nor a spiritual soul in the usual sense. They eat without pleasure; they cry without pain; they believe without knowing it; they desire nothing; they know nothing; and if they act in what seems to be an intelligent and purposive manner, it is only because God has made them fit to survive, and has constructed their bodies in such a way that they can organically avoid—without knowing that they do so—everything that might destroy them and that they seem to fear" (Malebranche, 1689, translated from Huxley, 1896, pp. 218–219).

References

Avneon, M., & Lamy, D. (2018). Reexamining unconscious response priming: A liminal-prime paradigm. *Consciousness and Cognition, 59*, 87–103. https://doi.org/10.1016/j.concog.2017.12.006

Avneon, M., & Lamy, D. (2019). Do semantic priming and retrieval of stimulus-response associations depend on conscious perception? *Consciousness and Cognition, 69*, 36–51. https://doi.org/10.1016/j.concog.2019.01.010

Beauny, A., de Heering, A., Muñoz Moldes, S., Martin, J. R., de Beir, A., & Cleeremans, A. (2020). Unconscious categorization of submillisecond complex images. *PLoS ONE, 15*(8), e0236467. https://doi.org/10.1371/journal.pone.0236467

Breitmeyer, B. G., Kiefer, M., & Niedeggen, M. (2015). Exploring the visual (un)conscious. *Consciousness and Cognition, 35*, 178–184. https://doi.org/10.1016/j.concog.2015.04.021

Chalmers, D. J. (1996). *The Conscious Mind: In Search of a Theory of Conscious Experience*. Oxford University Press.

Damian, M. F. (2001). Congruity effects evoked by subliminally presented primes: Automaticity rather than semantic processing. *Journal of Experimental Psychology: Human Perception and Performance, 27*(1), 154–165. https://doi.org/10.1037/0096-1523.27.1.154

De Graaf, T. A., Hsieh, P. J., & Sack, A. T. (2012). The "correlates" in neural correlates of consciousness. *Neuroscience and Biobehavioral Reviews, 36*(1), 191–197. https://doi.org/10.1016/j.neubiorev.2011.05.012

Dehaene, S., Changeux, J. P., Naccache, L., Sackur, J., & Sergent, C. (2006). Conscious, preconscious, and subliminal processing: a testable taxonomy. *Trends in Cognitive Sciences, 10*(5), 204–211. https://doi.org/10.1016/j.tics.2006.03.007

Dehaene, S., Naccache, L., Le Clec'H, G., Koechlin, E., Mueller, M., Dehaene-Lambertz, G., Van De Moortele, P. F., & Le Bihan, D. (1998). Imaging unconscious semantic priming. *Nature, 395*(6702), 597–600. https://doi.org/10.1038/26967

Gelbard-Sagiv, H., Faivre, N., Mudrik, L., & Koch, C. (2016). Low-level awareness accompanies "unconscious" high-level processing during continuous flash suppression. *Journal of Vision, 16*(1), 1–16. https://doi.org/10.1167/16.1.3

Huxley, T. H. (1896). *Method and results: Collected essays (Vol. 1)*. Appleton.

Kim, C. Y., & Blake, R. (2005). Psychophysical magic: Rendering the visible "invisible." *Trends in Cognitive Sciences, 9*(8), 381–388. https://doi.org/10.1016/j.tics.2005.06.012

Kimchi, R., Devyatko, D., & Sabary, S. (2018). Can perceptual grouping unfold in the absence of awareness? Comparing grouping during continuous flash suppression and sandwich masking. *Consciousness and Cognition, 60*, 37–51. https://doi.org/10.1016/j.concog.2018.02.009

Koivisto, M., & Rientamo, E. (2016). Unconscious vision spots the animal but not the dog: Masked priming of natural scenes. *Consciousness and Cognition, 41*, 10–23. https://doi.org/10.1016/j.concog.2016.01.008

Kouider, S., & Dupoux, E. (2004). Partial Awareness Creates the "Illusion" of Subliminal Semantic Priming. *Psychological Science, 15*(2), 75–81. https://doi.org/10.1111/j.0963-7214.2004.01502001.x

Kunde, W., Kiesel, A., & Hoffmann, J. (2003). Conscious control over the content of unconscious cognition. *Cognition*, *88*(2), 223–242. https://doi.org/10.1016/S0010-0277(03)00023-4

Lamy, D., Alon, L., Carmel, T., & Shalev, N. (2015). The Role of Conscious Perception in Attentional Capture and Object-File Updating. *Psychological Science*, *26*(1), 48–57. https://doi.org/10.1177/0956797614556777

Lamy, D., Ophir, E. A., & Avneon, M. (2019). Studying the Benefits and Costs of Conscious Perception with the Liminal-Prime Paradigm. In *Transitions between Consciousness and Unconsciousness* (pp. 118–147). Routledge. https://doi.org/10.4324/9780429469688-5

Lau, H., & Rosenthal, D. (2011). Empirical support for higher-order theories of conscious awareness. *Trends in Cognitive Sciences*, *15*(8), 365–373. https://doi.org/10.1016/j.tics.2011.05.009

Mazor, M., Brown, S., Ciaunica, A., Demertzi, A., Fahrenfort, J., Faivre, N., Francken, J. C., Lamy, D., Lenggenhager, B., Moutoussis, M., Nizzi, M. C., Salomon, R., Soto, D., Stein, T., & Lubianiker, N. (2023). The Scientific Study of Consciousness Cannot and Should Not Be Morally Neutral. *Perspectives on Psychological Science*, *18*(3), 535–543. https://doi.org/10.1177/17456916221110222

Micher, N., & Lamy, D. (2023). The role of conscious perception in semantic processing: Testing the action trigger hypothesis. Consciousness and Cognition, 107(November 2022), 103438. https://doi.org/10.1016/j.concog.2022.103438

Naccache, L., & Dehaene, S. (2001). Unconscious semantic priming extends to novel unseen stimuli. *Cognition*, *80*(3), 215–229. https://doi.org/10.1016/S0010-0277(00)00139-6

Overgaard, M., Fehl, K., Mouridsen, K., Bergholt, B., & Cleeremans, A. (2008). Seeing without seeing? Degraded conscious vision in a blindsight patient. *PLoS ONE*, *3*(8), e3028. https://doi.org/10.1371/journal.pone.0003028

Peremen, Z., & Lamy, D. (2014a). Comparing unconscious processing during continuous flash suppression and meta-contrast masking just under the limen of consciousness. *Frontiers in Psychology*, *5*(AUG), 969. https://doi.org/10.3389/fpsyg.2014.00969

Peremen, Z., & Lamy, D. (2014b). Do conscious perception and unconscious processing rely on independent mechanisms? A meta-contrast study. *Consciousness and Cognition*, *24*(1), 22–32. https://doi.org/10.1016/j.concog.2013.12.006

Pratte, M. S., & Rouder, J. N. (2009). A task-difficulty artifact in subliminal priming. *Attention, Perception, and Psychophysics*, *71*(6), 1276–1283. https://doi.org/10.3758/APP.71.6.1276

Ramsøy, T. Z., & Overgaard, M. (2004). Introspection and subliminal perception. *Phenomenology and the Cognitive Sciences*, *3*(1), 1–23. https://doi.org/10.1023/b:phen.0000041900.30172.e8

Reingold, E. M., & Merikle, P. M. (1988). Using direct and indirect measures to study perception without awareness deserves thanks for developing the software and for keeping the hardware in working order. *Perception & Psychophysics*, *44*(6), 563–575. https://link.springer.com/content/pdf/10.3758%2FBF03207490.pdf

Shanks, D. R. (2017). Regressive research: The pitfalls of post hoc data selection in the study of unconscious mental processes. *Psychonomic Bulletin and Review*, *24*(3), 752–775. https://doi.org/10.3758/s13423-016-1170-y

Sklar, A. Y., Levy, N., Goldstein, A., Mandel, R., Maril, A., & Hassin, R. R. (2012). Reading and doing arithmetic nonconsciously. *Proceedings of the National Academy of Sciences of the United States of America*, *109*(48), 19614–19619. https://doi.org/10.1073/pnas.1211645109

Stein, T., Utz, V., & van Opstal, F. (2020). Unconscious semantic priming from pictures under backward masking and continuous flash suppression. *Consciousness and Cognition*, *78*, 102864. https://doi.org/10.1016/j.concog.2019.102864

Van Den Bussche, E., Notebaert, K., & Reynvoet, B. (2009). Masked primes can be genuinely semantically processed a picture prime study. *Experimental Psychology*, *56*(5), 295–300. https://doi.org/10.1027/1618-3169.56.5.295

Van Den Bussche, E., Vermeiren, A., Desender, K., Gevers, W., Hughes, G., Verguts, T., & Reynvoet, B. (2013). Disentangling conscious and unconscious processing: A subjective trial-based assessment approach. *Frontiers in Human Neuroscience*, *NOV*. https://doi.org/10.3389/fnhum.2013.00769

Vermeiren, A., & Cleeremans, A. (2012). The validity of d′ measures. *PLoS ONE*, *7*(2), e31595. https://doi.org/10.1371/journal.pone.0031595

9 Toward the Mathematics of Consciousness

Oded Maimon and Moshe Klein

Personal Reflections

We are working on the theory presented here because of revelations from our direct subjective experience and our capabilities in abstraction. For us, consciousness research confronts the biggest enigma in science today. Many scientific issues are important, but comprehending consciousness will be a significant achievement in our understanding of humanity, the environment, and the universe.

Currently, we lack the language to express consciousness and the mathematical foundations that would enable research to progress. We hope the ideas presented here are an important contribution to the science of consciousness. We believe that our theory will find many applications.

We believe that pure consciousness was never born; it always spontaneously exists. Consciousness then evolves to develop the experiences in our lives.

Overview

Consciousness exists but is elusive; it is the unknown key to understanding humanity and the universe. To research consciousness scientifically, we need a solid, rigorous mathematical language, which is built on proper foundations. This article aims to begin the work on this significant task by presenting a new paradigm that introduces a new type of number and develops a new form of logic, termed "soft logic." The roots of such an approach can be found in the works of various scientists and philosophers, including Kant. The novel mathematical model and rigorous analysis suggested here provide a bridge between the inner experience, the outer world, and the interaction between them.

Introduction

Consciousness is everywhere and nowhere. It exists but it is elusive. It is known privately and collectively. It is currently one of the great enigmas of science. It emerges and enacts, but was here before it emerged or acted. It is private and directed inward but also external. Understanding the paradox that may appear here is one of the inherent demands of the new science of consciousness. Can it be addressed mathematically? How do we capture all these aspects scientifically and verbally? How can we represent processes in consciousness, which also have no time and space dimensions, yet are active and evolving? What is the scientific foundation and purpose for studying consciousness? What is our a priori knowledge, and how do we derive our knowledge of reality from it? Do we need a new logic and new representation of basic scientific concepts, even a new numbering system, and new operations (including basic ones, like addition and multiplication)? What can be scientifically known without losing the great mystery of the universe?

One needs fearless curiosity to research the foundation of consciousness, and to suggest a new science of mathematics for this purpose. To be unafraid of paradoxical logic that might appear, or of a mystery that exists in science and is part of the endeavor. Formal algebraic representations should be involved, together with geometrical visual representations. The geometry of consciousness is not only visual but encompasses all senses, which are associated with spatial and temporal experiences that are unified in consciousness.

In our work, we humbly take on this significant task, developing from its foundations a new rigorous theory that we term "soft logic." We are developing a new concept of modeling and dealing with the uncertainty of time and space. We then develop a frame of reference with two different paradigms and suggest a way to combine them. This constitutes new, fundamental mathematics, with many potential usages. Applications of this theory are relevant to the combination of cognitive and physical worlds and related theories—for example, in psychology and economics, where internal behavioral and external factors share in the decision-making process.

The motivation and reason for this research is our hope that this theory will be a step toward improving our understanding of humankind and the environment.

The Philosophical Background of Soft Logic

This section briefly describes the philosophical background that inspired us and helps explain the mathematical theory of "soft logic," a language through which we aim to bridge two worlds: the real world (which is beyond the understanding of humans according to some philosophers) and the inner world of phenomena, appearances, representations, and processes as perceived by our reasoning and senses. We do this by defining a new number called a "soft number" (SN), which is composed of multiples of zero and a part that is real (the real numbers, $\mathbb{R}$, as we know them). This yields two types of representations in an integrated model. Two of the important philosophical foundations for this are Immanuel Kant's epistemology and Salomon Maimon's doctrine of differentials—the approach to infinitely small sizes.

Kant (1724–1804) argued that human consciousness has access only to "phenomena," which are the world of sensible objects in time and space (Kant 1838). According to him, inaccessible to human consciousness are the objects of these phenomena, the "things in themselves" or "noumena." Thus, the object is derived from the phenomena (this is Kant's revolution). Despite their inaccessibility, consciousness must assume their existence (Kant did not use this term directly). Here an irreconcilable gap arises in Kant's theory: What is the essence of the relationship between the things in themselves and the phenomena?

Kant's approach to epistemology was revolutionary because he posited the structure of human thinking as a central factor in human perception of the world. Our knowledge of external objects is not founded on conclusions based on those objects; rather, Kant argued, there are synthetic a priori intuitions—namely space and time—that alone enable us to be conscious of objects and the medium that enables them. Kant also contends that there are categories of reasoning that human consciousness uses to make judgments about the impressions it receives through space and time. Kant maintains that mathematics is synthetic a priori knowledge derived from the same source as the intuitions of space and time; it is a priori because it does not rely on experience, but it does require synthetic construction through the act of intuition. Kant holds that the source of our potential in mathematics is the same as that for formalizing objects in time and space.

The greatest critique of Kant's epistemology in his time was that of Salomon Maimon (Maimon, 2001, 2010). Maimon in his "Essay on Transcendental Philosophy" analyzes, criticizes, and develops a new theory of Kant's weakest spot—namely "the things in themselves." He argued that a noumenon itself is an infinitesimal world of infinitely small-size relationships that are mathematical derivatives of preexperiential processes of human consciousness. Every object, before it is perceived by human awareness, exists in an infinitesimal size, having no quantitative size, only a qualitative one. The sensory images that we observe as zero sizes (but distinct) do not yet form awareness, which is formed by an additional act of the power of imagination:

> Considered in itself as a quality, every sensible representation must be abstracted from all quantity, whether extensive or intensive. For example, the representation of the color red must be thought without any finite extension, although not as mathematical but rather as a physical point or as the differential of an extension. (Maimon, 2010, p.19)

Salomon Maimon argued that there are two types of infinitely small sizes. The first arises when a series of sizes or numbers become smaller and tends toward zero without ever reaching zero. This is called a "potential infinitesimal." The second infinitely small size occurs when a quantity of nothing changes, and this is called an "actual infinitesimal."

In conclusion, it appears that one of the most important issues in epistemology is understanding the essence of the relationship between the world of things as they are and the phenomenological world. The nature of consciousness (which holds both aspects, i.e., the differential and the phenomena) is that we are compelled to assume that there is some kind of relationship between these two worlds. According to Kant, mathematics is the result of a synthesis between internal processes taking place in one's consciousness, along with observations of the world. In other words, mathematics essentially describes the possible affinity between the two worlds. According to Maimon, the phenomena of objects are the result of a mechanism of infinitesimal differentials that exist in human consciousness.

Soft logic is a new language of mathematics that facilitates and rigorously describes the two worlds, and bridges between the two worlds. The world of objects is mathematically represented as an axis of real numbers, while the world of inner consciousness is described by the zero-axis, which is composed of multiples of the number zero. The relationship between them describes a new type of number called "soft numbers," which are part real and part multiples of zero (expressing the ultimate, infinitely small sizes). One of the main theorems that result from soft logic analysis is the relationship between the infinitesimal and the function itself. For these ideas to be clear, we shall define them rigorously.

The French mathematician, physicist, and philosopher Blaise Pascal proved the existence of the physical vacuum (Adamson, 1995). We point out a rich mathematical world that exists in the abstract vacuum, which is infinitesimal and infinite at the same time (Ehrlich, 2007). Continuing Salomon Maimon's philosophical approach, we develop this world as an expansion of the number zero from a single point to a continuous line of zeros. This chapter presents an axiomatic theoretical development of the world without time and space (where entanglement processes and non-locality exist), which is tightly coupled with the regular phenomenological world (in a noncommutative way). The axioms themselves are deeply inspired by the philosophical approaches introduced above. We also show the relevance of this work to the current dynamic reality in which we live (e.g., the swift addition of cyberspace to our core lives). One of the manifestations of this reality is the integration of the real world and the digital world (Belk, 2013), each having its own rules. An accurate description of these interwoven worlds requires the development of a mathematical language that can bridge the two worlds, which ostensibly run in parallel but are related and affect each other (though governed by different physical laws).

The familiar Möbius strip (Starostin & Heijden, 2007) is one geometric model that results directly from our theory and is suitable for describing the mutual influence of the two different worlds. The real world is on one side of the strip, while the world of information is on the other. Locally, the strip has two sides, but globally it has only one (unity of differences). Similarly, soft logic naturally encompasses opposites and paradoxes.

The mathematician Felix Klein developed a geometric-algebraic model called "blow-up" (Griffiths & Harris 1978). By blowing up the point of origin of some coordinate system, a Möbius strip is formed.

In soft logic, we blow up the zero point by distinguishing between the various multiplications of zero and creating a zero-axis. Using five axioms, we define soft numbers and develop their algebra, which we prove almost creates a mathematical field (omitting the mysterious zero-axis, where there is no inverse). A field is a set of elements that are closed under two operations, addition and multiplication. These operations satisfy certain axioms. In a field, every element has an inverse under multiplication, despite the zero element. In this chapter, we briefly describe the soft coordinate system, the algebra of soft numbers, the relationship with differential and integral calculus, the expansion of soft functions, and their presentation on soft curves. To conclude, we present several practical avenues for future research using soft logic. The full theory is presented in references (Fivel et al., 2022; Klein & Maimon, 2016a, 2016b, 2019, 2020, 2021).

The next section lays the foundation of soft logic. The rest of the chapter aims to answer, in a concise way, the questions and issues raised in the first section.

Axioms and Definitions

We start by stating the axioms upon which we develop the soft logic theory. We define the 0-axis, the 1-axis, the processes (basic addition and multiplication operations), and the relationship between the two worlds (bridge numbers, which are noncommutative, meaning that the influence of the axis on each other is not symmetric). There is also a specific point (the absolute zero) that will later emerge as a "rabbit hole," a mystery that does not obey any rule (including those that apply to the rest of the space). We will show that it serves as a torsion point, which allows the existence of some paradoxical properties to exist. We conclude this section with the presentation of a new number that captures both realities, as we perceive them, and the infinitesimal property of reality (the essential relation of reality).

According to traditional mathematics, the expression 0/0 is undefined, although any real number could represent this expression, since $a * 0 = 0$ for all real numbers a. This observation opens a new scope and area for investigation.

According to Pradip K. Datta, the special vision of the two zeros quotient was among Ramanujan's unique capabilities (Datta, 2013). The Indian mathematician Ramanujan discovered almost 4,000 original mathematical formulas. He did not prove the formulas that he discovered. He simply sketched them, as if creating a drawing. When he was asked how he knew that they were correct, he answered that the Indian goddess Namakel whispered them into his ear. Ramanujan is undoubtedly a mathematician who inspired an intuitive way of thinking with direct experience. Can we learn from his way?

The exercise of dividing zero by zero (0/0) demonstrates that there is a space of numbers in which everything is possible simultaneously (because there is no time and space, but there are distinct events and operators). Soft logic describes a mathematical language that bridges the real world and the symbolic world. The constant and ongoing distinction between the real and symbolic worlds creates "two eyes" that make it possible to navigate from a starting point and arrive at the desired solution or new formula.

We assume the existence of a continuum of distinct multiples $a\bar{0}$, where a is any real number and $\bar{0}$ symbolizes an object that may be called a "soft zero," while its multiples are also called "soft zeros." We denote by $\bar{\mathbf{1}}$ the real number 1 when all other real numbers are conceived as its multiples.

We also introduce the term absolute zero as a bold value: $\mathbf{0} = 0\bar{0}$.

Let a, b be any real numbers and $a\bar{0}, b\bar{0}$ be two corresponding soft zeros as defined above.

Axiom 1 (Distinction): If $a \neq b$ then $a\bar{0} \neq b\bar{0}$.

In soft logic, we extend the zero from a point to a line (axis). This creates a distinction between different multiples of $\bar{0}$ (per the logic of identity, since a and b are not the same). This axiom relies on the basic identity concept in philosophy. For example, $a = 3$ and $b = 4$, $3\bar{0} \neq 4\bar{0}$, are two points on the zero axis. Both of them are 0, but they are different points.

From this axiom we derive the first definition of soft logic.

Definition 1 (Order): If $a < b$ then $a\bar{0} < b\bar{0}$.

Axiom 2 (Addition): $a\bar{0} + b\bar{0} = (a + b)\bar{0}$.

A continuum of zeros is a concrete model of the infinitesimal. In soft logic, we develop a model, which locates infinite numbers of zeros on a straight line. We define the addition of multiples of $\bar{0}$ as the addition of their corresponding real factors. For example, if $a = 3$ and $b = 4$ then the addition on the zero axis $3\bar{0} + 4\bar{0}$ creates another point, which is $7\bar{0}$.

The $\bar{\mathbf{1}}$ axis behaves regularly: $a\bar{1} + b\bar{1} = (a + b)\bar{1}$.

Axiom 3 (Nullity): $a\bar{0} * b\bar{0} = 0$.

Numbers on the zero axis "collapse" under multiplication operation. The addition has significance and meaning, but multiplication does not make any distinction whatsoever (these are the two extremes of wave interference for superposition operations: addition and nullity). For example, the above two points $(3\bar{0}, 4\bar{0})$ result in an absolute zero point if multiplied. This is like two infinitesimals that are squared and thus become smaller, approaching zero.

Thus multiplication represents a destruction process, while the addition is an enhancement process.

Conversely, $a\bar{1} * b\bar{1} = ab\bar{1}$, $a\bar{0} * b\bar{1} = ab\bar{0}$, and $b\bar{1} * a\bar{0} = ba\bar{0}$.

These equations describe the basic operations of a hybrid process.

Let us proceed and define a value that combines a value from the zero line ($\bar{0}$-axis) with a value from the real line ($\bar{1}$-axis), and denote it as $a\bar{0} \perp b\bar{1}$. Note that although the symbol between the two values reminds one of perpendicularity, it nevertheless has quite a different meaning—that of combining or "bridge making," and we call it the "bridge sign," which differs from the regular addition symbol (+).

Axiom 4 (Bridging):

There exists a bridge between a zero axis and a real axis and vice versa, denoted by a pair of a bridge numbers and its mirror image on either side of the bridge sign: $((\boldsymbol{a\bar{0} \perp b\bar{1}}, \boldsymbol{b\bar{1} \perp a\bar{0}}))$.

This is an existence axiom, stating that there is some special connection between two lines in the coordinate system. It implies that even two seemingly contradictory concepts (e.g., one concept lacks time and space, while the other has them) may have an affiliation and connection (e.g., our night dreams and daily reality). The next axiom is a property of this connection.

Axiom 5 (Non-commutativity): $a\bar{0} \perp b\bar{1} \neq b\bar{1} \perp a\bar{0}$.

Note that the multiplication of a zero-axis number with a regular real number, as defined in axiom 3, is commutative, but for the subsequent parts of the theory the order of these numbers is important. This means that the influence of the 0-axis on the 1-axis is not necessarily the same as vice versa. This property can be paralleled to the fact that the influence of a dream on reality is not necessarily the same as the influence of reality on the dream.

Following the philosophical inspiration, this concludes the axiomatic part, and now we proceed to further construct the theory. The rest of the theory is based on these five axioms (and no more), and from this starting point, we can develop the rest (some are shown in this chapter).

Definition 2 (Soft numbers):

A soft number is a construction of the following form:

$$a\bar{0} \dot{+} b\bar{1} = \{(a\bar{0}) \perp (b\bar{1}); (b\bar{1}) \perp (a\bar{0})\},$$

where *a, b* are any real numbers.

Any soft number denominated as $a\bar{0} \dot{+} b\bar{1}$ has a component from the real line or $\bar{1}$ line, and a component from the $\bar{0}$ line (later on, we will omit the line above the numeral 1, or even the 1 itself, but the line over the numeral 0 is essential). Notice the new operation "$\dot{+}$" that we define here. The dot above the addition sign implies that, unlike a regular sum, a soft number does not have one single value, but rather *two* distinct values, as presented in its definition.

In our theory, the bold **SN** denotes the set of all soft numbers:

$$\mathbf{SN} = \{a\bar{0} \dotplus b : a, b \in \mathbb{R}\}.$$

This defines all possible soft numbers (any a and b that are real). It is important to remember that any soft number in this set has two values, as each number consists of a bridge number $a\bar{0} \perp b$ and its mirror image of the bridge sign $b \perp a\bar{0}$, which have different meanings by the noncommutativity axiom.

The Coordinate System of Soft Logic

In this section, we proceed to define a new soft coordinate system, as presented in figure 9.1. This creation of a geometrical space allows the visualization of various processes. This system also enables us to present soft numbers in an infinite strip, and create the Möbius function, which shows the geometry of the space. The Möbius representation allows local antagonistic points to become the same globally, thus mathematically capturing one of the mysteries of consciousness, where differences may be transformed into unitedness (concepts of duality and non-duality can coexist, depending on the observer's context).

In figure 9.1, the middle vertical line is the 0-axis that includes all the multiplications of 0: $x\bar{0}$. The right-side vertical line is the 1-axis that includes all the multiplications of 1: $y\bar{1}$. The left vertical line is the 1-axis that includes all the multiplications of 1: $y\bar{1}$ but in the opposite direction. This coordinate system allows us to present all of the real numbers and all of the soft zeros. We can now use this coordinate system for representing various soft numbers. Given a soft number $C = x\bar{0} \dotplus y$, where x and y are any real numbers, we represent

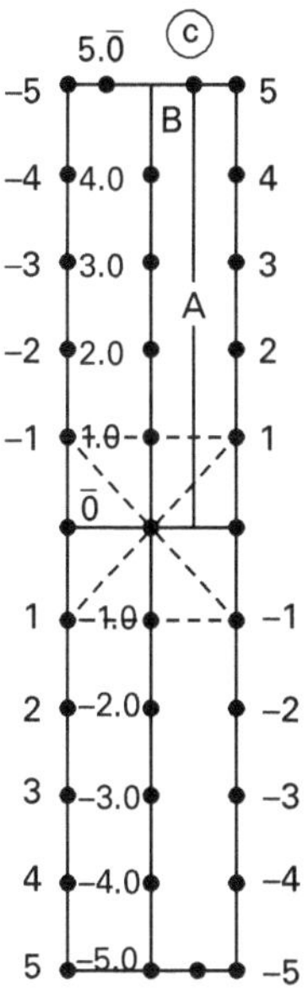

Figure 9.1
Soft coordinate system. It has the 0-axis in the center with the various multiples of the number 0. Parallel to it on the right is the 1-axis with the multiples of the number 1 and parallel to it on the left again is the 1-axis but in opposite direction.

both elements of a soft number as two points, on the other side of a page. The point $x\bar{0} \perp y$ appears on one side of the page and the point $y \perp x\bar{0}$ is the same point but on the other side of the page:

$$C = \{c, c'\} = \{x\bar{0} \perp y, y \perp x\bar{0}\}$$

where $c = x\bar{0} \perp y$ is located in front of the page, and $c' = y \perp x\bar{0}$ is located *behind* the page.

As the infinite strip, presented (partially) in figure 9.1, is intended for the presentation of soft numbers, we call it a soft numbers strip (SNS).

Definition 3: Let C be any point on the SNS:

- The *height* of point C is the vertical distance from C to the horizontal segment with the absolute zero at its center. The A denotes the height with a sign.
- The *width* of point C is the horizontal distance from C to the 0-line with a sign and is denoted by B.

The definitions above provide every point C on the SNS with two parameters, A and B, where

$$-\infty < A < \infty, -1 \leq B \leq 1.$$

The condition $A > 0$ is satisfied in the positive part of the SNS; $A < 0$ in its negative part, or correspondingly, above and below the horizontal segment containing the absolute zero, while on this segment $A = 0$.

For the second parameter, B, there is $B = 0$ on the 0-axis, $B = 1$ on the right line bounding the SNS, $B = -1$ on the left line bounding the SNS, and otherwise $-1 < B < 1$.

Given $A \in \mathbb{R}, B \in [-1,1]$ we define $C(A, B)$ as follows. For $0 \leq B \leq 1$, then $c = (1 - B)A\bar{0} \perp BA\bar{1}$ is located in front of the page; and $c' = BA\bar{1} \perp (1 - B)A\bar{0}$ is located *behind* the page

$$C = x\bar{0} \dotplus y = \{c, c'\}$$

For $-1 \leq B \leq 0$, $c = (1 + B)A\bar{0} \perp BA\bar{1}$ is located in front of the page; and $c' = BA\bar{1} \perp (1 + B)A\bar{0}$ is located *behind* the page

$$x\bar{0} \dotplus y = \{c, c'\}$$

Note that if we use |B|, the two equations above are the same.

So, in brief, and in terms of the absolute value of B, we show that $c = (1 - |B|)A\bar{0} \perp BA\bar{1}$ is located in front of the page; and $c' = BA\bar{1} \perp (1 - |B|)A\bar{0}$ is located *behind* the page

$$x\bar{0} \dotplus y = \{c, c'\}$$

Hence, by a coefficient comparison of the real part and the soft part in

$$x = (1 - |B|)A$$

$$y = BA$$

or equivalently, after solving the above (A, B) using some arithmetic of absolute value and sign functions:

$$A = (|x| + |y|)sgn(x)$$

$$B = \frac{y \cdot sgn(x)}{|x| + |y|}$$

These equations define a map between the Cartesian plane XY to a point on the Möbius strip. We call this map the "soft Möbius map," which is depicted in figure 9.2. The significance of this map is that it transforms the points on the plane into soft numbers that are represented as points on the Möbius strip. Each colored region is mapped to a corresponding region on the Möbius strip.

Analysis of Soft Logic

In this section, we mention only some of the theoretical results that make soft logic a viable mathematical field and enable a detailed investigation of the philosophical ideas discussed in the introduction. Specifically, we show the results that tie the philosophical ideas of infinitesimals to the theory of soft logic. Namely, we obtain the result that any function (which represents phenomena) is related to its derivative in the zero-axis in our paradigm (thus supporting the theory of Salomon Maimon).

First, the algebra of soft logic is demonstrated. We show that it is possible to define a group (**SN**,+), and a ring (**SN**,+,×) with a set of soft numbers, and that this ring is almost a field (we say "almost" because the inverse is undefined only when $B = 0$). This "almost" property is related to the mystery of soft logic, which allows a paradox to be involved in a rigorous theory. We briefly present four results.

1. *Basic arithmetic of soft numbers.*

The addition operator of two soft numbers is defined with the following rule:

$$(a\bar{0} \dotplus b) + (c\bar{0} \dotplus d) = (a + c)\bar{0} \dotplus (b + d)$$

With this definition, the set **SN** is a group under addition (**SN**,+), when the + sign represents the addition operation for this group.

The multiplication of two soft numbers is defined by the following rule:

$$(a\bar{0} \dotplus b) \times (c\bar{0} \dotplus d) = (ad + bc)\bar{0} \dotplus (bd).$$

This operation is commutative and satisfies the laws of associativity and distribution. With these two operations, + and ×, soft numbers create the ring (**SN**,+,×), in which the + and × signs represent the addition and multiplication operators of the ring.

The inverse of a soft number exists when $b \neq 0$, and is defined as:

$$(a\bar{0} \dotplus b)^{-1} = \left(-\frac{a}{b^2}\bar{0} \dotplus \frac{1}{b}\right).$$

Therefore, (**SN**,+,×) is almost a field, an algebraic structure with two operations, addition and multiplication, which satisfy certain rules ("almost" because the inverse is undefined only when $b = 0$).

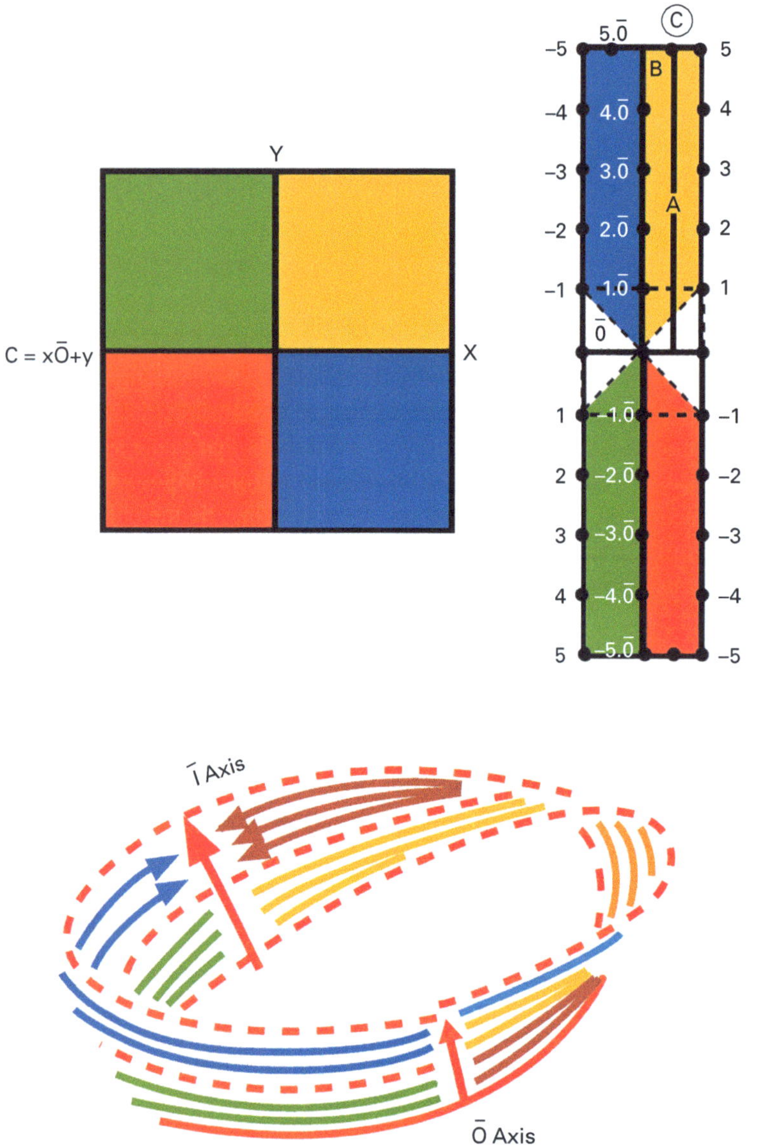

Figure 9.2
Soft Möbius map. The soft coordinate system has a twist that creates the Möbius strip and therefore enables correspondence between soft numbers and points on the surface of the Möbius strip.

2. *The power of soft numbers* (which is different from regular powers, because of the third axiom).

We arrive at the following definition for a function of a soft variable that arises from the real differentiable function *f*(*x*), for any soft number $a\bar{0} \dotplus x$:

$$\bar{f}(a\bar{0} \dotplus x) = af'(x)\bar{0} \dotplus f(x).$$

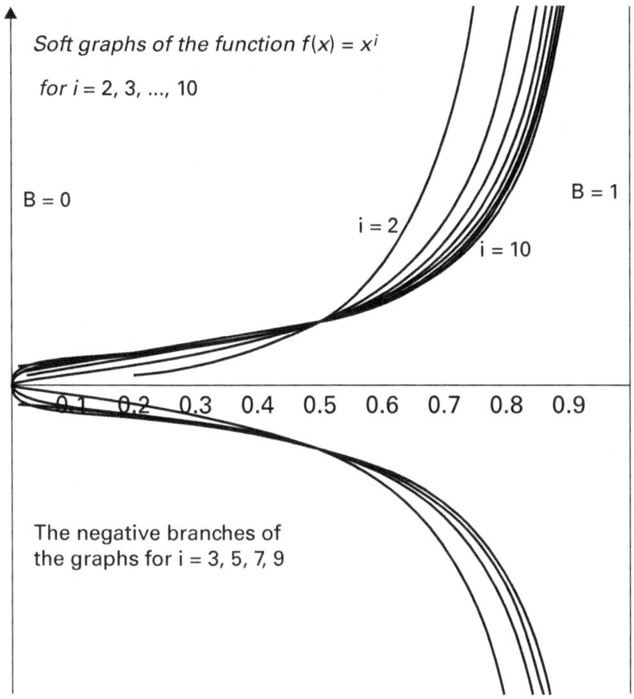

Figure 9.3
The soft curve $\hat{f}(x) = x\bar{0} \dot{+} x^i$.

This is one of the most important results regarding the philosophical background mentioned in the beginning as it ties the derivative of the function with the function itself when considering functions of soft numbers.

3. *Soft curves.*

We present an extension of a real function $f(x)$ to a soft function of a real variable, $\hat{f}: R \to SN$, performed by the formula

$$\hat{f}(x) = x\bar{0} \dot{+} f(x) .$$

4. *Soft probability and entropy.*

In the classical probability theory, in continuous random variables, there is no distinction between the probability of an event involving strict inequality and an event with nonstrict inequality. Moreover, a probability density function (PDF) involving equality in a certain value, collapses to zero, without distinguishing among the values of the random variable.

We develop soft probability by incorporating soft numbers into probability theory. We develop a probability involving equality as a "soft zero" multiple of a PDF. This further supports the theory of the actual existence of infinitesimals.

Suppose X is a continuous random variable (CRV), then for all real numbers

$$F_X(x) \overset{(a)}{=} \Pr(X \leq x) \overset{(b)}{=} \Pr(X < x)$$

where (a) is a cumulative distribution function (CDF); (b) indicates no distinction between "≤" and "<"; $\Pr(X = x) = 0$ is the equality collapses to zero on CRVs; $f_X(x) = \frac{d}{dx}\{F_X(x)\}$ is a PDF; and $\Pr(x < X \leq x + \Delta x) \approx f_X(x)\,\Delta x$ is the probability approximation for X adjacent x up to a small Δx.

Our motivation is to distinguish between $\Pr(X \leq x)$ and $\Pr(X < x)$ in a probability of CRV, and to obtain some information from the probability $\Pr(X = x)$ that will not collapse to zero.

Observation:

$$\Pr(X = x) = f_X(x) \cdot 0;$$

$$\Pr(X \leq x) = \Pr(X = x) + \Pr(X < x) = f_X(x) \cdot 0 + \Pr(X < x).$$

Approach: The implementation of soft numbers on probability theory for CRVs so that each "zero probability" object with the form $\Pr(X = x)$ is an element in a zero-axis based on its PDF $f_X(x)$ that achieves the distinguishing between $\Pr(X \leq x)$ and $\Pr(X < x)$ (we call it *Ps*—for soft probability notation).

We also extend this notion of soft probabilities to the classical definitions of complements, unions, intersections, and conditional probabilities, and to the expectation, variance, and entropy of a CRV. This extension provides information regarding a CRV being within a discrete set of numbers, such that its probability does not collapse completely to zero. The entropy thus defined also differs from classical entropy. This is leading to a new version of the known decision tree models (Rokach & Maimon, 2008).

Concluding Reflections and Further Research Topics

The mathematical model developed here aims to provide a rigorous description for modeling the phenomena as well as the noumena (in its infinitesimal interpretation). The Möbius strip is a simple geometrical model that demonstrates one special aspect of our model because it has two sides locally and one side globally.

In the article, we described the fundamentals of soft logic and the soft coordinates system, which has a built-in twist that creates a Möbius strip. We believe that a mathematical model of consciousness should use soft numbers.

As future plan we are looking at modeling rituals, where we assume that an additional zero-axis appears (for the duration of the ritual). We are currently studying this model to see if it is suited for describing ceremonies and rituals, where synchronization occurs with the emergence of collective consciousness. To date, we have developed the model of consciousness for one person, and plan to continue with the interaction of multiple people, which means developing the mathematics of relationships and synergy in a group of N people participating in a collective ritual (as a special kind of cooperation game in the SNS multispace). We observe collective situations ranging from superorganisms in nature to ecstatic rituals (e.g., Sufi rituals), where collective traits, such as group synchronization, emerge. Our goal here is to develop a mathematical model of processes in which consciousness alters during collective rituals with components of self-organization when the connection of consciousness is linked to the synchronic physical connection occurring in the ritual.

As part of our model, we assume a concept in which the whole is greater than the sum of its parts with the existence of another hidden dimension that expresses their mutual recognition, and the possibility of collective recognition that naturally emerges. We intend to construct a model of what occurs in the physical space during rituals and in participants' consciousness.

Another application that we are currently studying is the use of soft logic to represent the privacy paradox. The privacy paradox is a known conflict in the behavior of people in the cyberworld, which is the gap between people's declared intentions to maintain their own privacy in the digital world and how they actually behave. The magnitude of the paradox also depends on their technological literacy.

In this chapter, we treated explicitly processes beyond time, space, and senses. We also integrated these processes with known phenomena. This is a major leap forward in formal studies of consciousness. We hope that this work will advance the science of consciousness by establishing a method of rigorous mathematical modeling and creating a common expressive language for further exploration of this subject.

Acknowledgments

This research was supported by the Koret Foundation grant for Smart Cities and Digital Living 2030 granted to Stanford University and Tel Aviv University. We would like to express our gratitude for this support.

References

Adamson, D. (1995). Blaise Pascal: Mathematician, physicist, and thinker about God. Springer.

Belk, R. W. (2013). Extended self in a digital world. *Journal of Consumer Research*, 477–500.

Datta, P. K. (2013). National mathematics year: A tribute to Srinivasan Ramanujan. *Science and Culture*, *79*(3–4), 158–162.

Ehrlich, P. (2007). Review of the continuous and the infinitesimal in mathematics and philosophy. In J. L. Bell (Ed.), *The bulletin of symbolic logic* (Vol. 13, pp. 361–363).

Fivel, O., Klein, M., & Maimon, O. (2022). Decision trees with soft numbers. *International Journal of Circuits, Systems and Signal Processing*, *15*.

Griffiths, P., & Harris, J. (1978). *Principles of algebraic geometry*. John Wiley & Sons.

Kant, I. (1838). *Critique of pure reason* (F. Haywood Trans.). William Pickering.

Klein, M., & Maimon, O. (2016a). The mathematics of soft logic. In *IOP second international conference on mechanical engineering and automation science, Japan*.

Klein, M., & Maimon, O. (2016b). Soft logic and soft numbers. *Pragmatics and Cognition*, *23*(3), 473–484.

Klein, M., & Maimon, O. (2019). Axioms of soft logic. *p-Adic Numbers, Ultrametric Analysis and Applications*, *11*(3), 205–215.

Klein, M., & Maimon, O. (2020). The dynamics in the soft number coordinate system. *Journal of Advances in Mathematics*, *18*, 1–17.

Klein, M., & Maimon, O. (2021). Fundamentals of soft logic. *New Mathematics and Natural Computation*, *17*(3), 703–737.

Maimon, S. (2010). *Essay on transcendental philosophy* (N. Midgley, H. Somers-Hall, A. Welchman, & M. Reglitz, Trans.). Continuum International Publishing Group.

Maimon, S. (2001). Essay towards a new logic or theory of thought, together letters of Philaletes to Aenesidemus. In G. di Giovanni, & H. S. Harris (Eds.), *Between Kant and Hegel: Texts in the development of post-kantian idealism* (pp. 158–203). Hackett.

Rokach, L., & Maimon, O. (2008). Data mining with decision trees: Theory and applications. In *Series in machine perception and artificial intelligence* (Vol. 69). World Scientific Publishing.

Starostin, E., & Heijden, G. (2007). The shape of a Möbius strip. *Nature Mater*, *6*, 563–567. https://doi.org/10.1038/nmat1929

10 Waves And Flames in the Creative and Conscious Human Brain

Rafael Malach

Personal Reflections

I am including here, consistent with the theme of this book, a personal account of my journey researching the mysteries of neuronal manifestations of conscious and subconscious processes in the human mind.

I would like to explain why I think that a personal account of this research might be helpful. The main reason is that I firmly believe that—despite major disagreements, widely dispersed perspectives, and different fields of expertise—all of us who find the mystery of brain and mind exhilarating ultimately struggle with similar puzzles and conceptual conundrums. One can therefore hope that a personal account will not really be so personal after all but rather resonate with similar puzzles faced by others. Such a personal account could also help provide a more accessible account of the rationale and motivation behind the questions and experiments I have conducted.

Before moving to the research, some words are needed to describe my overall perspective on these issues and how they fit within my field of expertise. By training, I am an experimental neuroscientist who, for many years, has studied mainly neuronal processes of the visual system. Unlike theoreticians and philosophers (whose work I truly admire and find very valuable), as an experimentalist I am interested, first and foremost, in the outcome of planned experiments. There is a Yiddish saying that nicely captures this perspective: "Mensch tracht und Gott lacht" or "Man plans, and God laughs." We can come up with the most elegant and compelling plan (theory) of the workings of the human brain, only to get it totally trashed by the results of experiments. In fact, in my (perhaps unfortunate) experience this has been the rule rather than the exception. Even worse, it is often the case that a theory I was infatuated with was not demolished, but the brain simply revealed something totally unexpected, a result that was not planned and that necessitated a complete shift into a new direction. While acknowledging the crucial importance of theoretical work, for an experimenter what is most valuable is not how well reasoned a philosophical or theoretical model is, but rather the quality of the experimental findings and how clear and error free they are.

In the case of the brain mechanisms of human consciousness, on which I will focus here, we are unfortunately at a painfully early stage because the methods available for obtaining signals from the human brain are far behind the revolutionary progress in animal models, where novel methods allow both control and recordings of large neuronal populations at an unprecedented degree of resolution. To my mind this, rather than theoretical or philosophical limitations, is what hinders progress in our search for a neuroscientific theory of human consciousness.

Overview

We wake up in the morning, open our eyes, and immediately an amazingly rich and diverse movie starts running in our head, a movie that includes sights, sounds, smells—the panoply of our universe of conscious experiences. This universe is created so effortlessly and naturally that we take it for granted and rarely spend time marveling at it. No less intriguing and mysterious is the subconscious turmoil that goes on below our awareness threshold, which to a large extent drives our behavior and emotions. A subject also explored extensively in the theories underpinning psychoanalysis. Intriguingly, subconscious processes often underlie our moments of insight and creativity, as exemplified by those striking cases in which a solution unexpectedly pops into our heads.

However, there is an additional layer of mystery associated with both conscious and subconscious processes. This mystery stems from the realization that they are completely linked to the physical processes that go on in our brains where masses of electrical pulses are exchanged among millions of tiny brain cells. This brain-mind link is so baffling that it almost appears magical—yet it is one of the most rigorously established scientific truths revealed through numerous neuroscience experiments. While we cannot solve the magic of the mind-brain link, we neuroscientists are in the midst of an exceedingly challenging and thrilling journey to uncover the principles and rules that underlie the association between physical brain states and the myriad of conscious experiences, as well as, and not less importantly, the unconscious processes that go on in our brain.

In this chapter, I will attempt to share with you a personal account of my struggle with these issues as a neuroscientist and my attempts to explore them experimentally. The experimental work led me to suggest certain principles that seem to be key to understanding conscious experience and its subconscious counterparts in the human brain. First and foremost, among these principles is the importance of cooperative group activity—the joining together of neurons into a cohesive group. Such cooperative action is achieved through recurrent, mutual exchange of signals among the members of each neuronal group. When this cooperative action is high and fast—akin to the ignition of a neuronal "flame"—we become conscious of the content for which the ignited neurons are specialized.

Complementary to these rapidly rising bursts of ignition are ultraslow waves of activity. These waves are irregular, spontaneous, and of low profile. Intriguingly, these waves constitute a subconscious exploration process that often leads to spontaneous behavior and creative insights. Thus, our research leads us to the humbling insight that much of our spontaneous and creative behaviors can be ultimately traced to our subconsciousness. Our conscious experience is the flame of a unitary, definitive event that bursts out of the myriad of subconscious slow waves of creative explorations.

Everything you can imagine is real.
—Pablo Picasso

Our Universe of Conscious Experience

Waking up in the morning, you find yourself in a vast universe of possible experiences—from rich and colorful visual images, to sounds and music, feelings, memories, thoughts. These experiences stream in your mind with no interruption, in what William James coined as the stream of consciousness (James, 1890), until you fall asleep. And then the stream resurfaces, but in a more bizarre and fascinating format, in what we call dreaming. The various contents of our conscious experiences are essentially limitless, from seeing a speck of light to the most abstract and complex thought. However, despite this unbounded diversity, conscious experiences share certain common aspects that unify them and justify naming them as members of a single universe.

By far the more obvious of these is the fact, often taken for granted, that this experiential universe is the only universe we have, and it is exclusively ours! Metaphorically, we are the only ones that can purchase a ticket to our private consciousness show. Of course, we can hear vivid and detailed reports about the experiences of others, and even be empathetic to them, but we can never directly experience those other universes. Another striking property of our conscious experience is its absolutely certain nature; nobody can try and deny that our conscious experiences exist for us.

Finally, no effort is required to start rolling our private movie: it is the most natural and in fact unstoppable stream that can ever exist. Just be awake, and we are inevitably plunged into the movie theater. Perhaps this immediacy and ease by which conscious experiences surface make their emergence appear to us so mundane and self-evident. We rarely stop to ponder where these experiences come from, or how they are generated within us. They simply are there—center stage of all our life.

However, once we start exploring the nature and origin of conscious experience, and especially their truly incomprehensible link to the staggeringly complex neuronal networks in our brain, only then does it dawn on us that every moment of wakeful, conscious experience is the most amazing, almost miraculous, occurrence that we can conceive of. It is linked, in a manner that we are still not able to comprehend, and perhaps never will, to the exchange of millions of tiny electrical pulses that occur in our brain every time we have a conscious experience.

But no less amazing is our subconscious universe—those processes that are so influential in our life, as deeply demonstrated in, for example, Freudian psychoanalytical theories—and yet we are, by definition, unaware of their existence. Here I will focus on one manifestation of our subconscious processes: our spontaneous behaviors and creative insights. A paradigmatic case of such subconscious activity is an idea that suddenly, without warning, pops in your head and often offers insightful solutions to nagging and difficult problems. In truly profound cases, these spontaneous insights lead to world-changing creations of genius in philosophy, art, and science. Similar to conscious experience, these subconscious creative processes are associated with unique physical states in our brains. But what these neuronal processes are, and in what way they differ from those linked to our conscious experience, remains a challenging and ongoing topic of research.

Below, and following the personal theme, I will describe a few central questions that as a neuroscientist I have struggled with in my own research, grappling with the mystery of conscious and subconscious processes and their relationship to neuronal activity in the human brain.

Consciousness Matters to the Brain!

About twenty years ago I decided to study brain mechanisms of conscious experience. At that time I was conducting research using noninvasive brain imaging technology (called functional magnetic resonance or fMRI) on visual processes, specifically those that are found in brain areas associated with object vision (Grill-Spector & Malach, 2004). These areas are located in the cerebral cortex—a rather uniform-looking sheet with many undulations that engulf the entire brain just below our skull. The cerebral cortex is thought to constitute the central brain organ that underlies the diverse contents of our conscious experience.

Together with my student (now professor) Kalanit Grill-Spector, we conducted an experiment aimed, almost playfully, at finding the neuronal signature of the then popular (and later discredited) subliminal "Coca-Cola" effect—that is, the behavioral impact of pictures (or ads) presented too fast to be consciously perceived (Grill-Spector et al., 2000). Using such briefly presented invisible pictures, we were somewhat disappointed to discover that, in contrast with visible images that generated massive activation of nerve cells in the relevant high cortical visual areas, no such brain responses could be detected in these areas when participants failed to see them (see figure 10.1, top).

In retrospect, it is embarrassing to realize how fixated I was, prior to this experiment, on the belief that visual brain centers merely "represented" in many clever ways the physical world we look at, regardless of whether we are consciously aware of this visual world or not. It is not that I was rejecting the possibility that consciousness matters to brain activity, I simply didn't think about it! This blindsight of mine nicely illustrates the wisdom inherent in the statement that in science, asking the right questions is not less important than getting answers.

This experiment was followed by another, conducted with the student (now professor) Uri Hasson (Hasson et al., 2001) using the famous Rubin vase-face illusion (see figure 10.1, bottom), and together these experiments sent me a clear message: consciousness matters to the brain. Or, to put it differently, the transition from nonconscious to conscious awareness is not merely a philosophical issue or an ill-defined psychological phenomenon but actually constitutes a massive physical event—namely, the eruption of brain activity involving millions of nerve cells in specific cortical locations that are specialized for the unique contents of each conscious experience. Subsequent research revealed that this correspondence between bursts of activity in specific cortical loci and conscious content was extremely wide-ranging. It could be found in perception of faces and objects (Malach et al., 1995), recollections (Norman et al., 2017, 2019; Gelbard-Sagiv et al., 2008), and extending even to abstract thoughts (Berkovich-Ohana et al., 2020). Having the initial observation as a starting point and ignoring dire warnings from close friends that consciousness was a hopeless and doomed-to-fail experimental endeavor, I decided to make the topic of consciousness and the brain a central focus of my research.

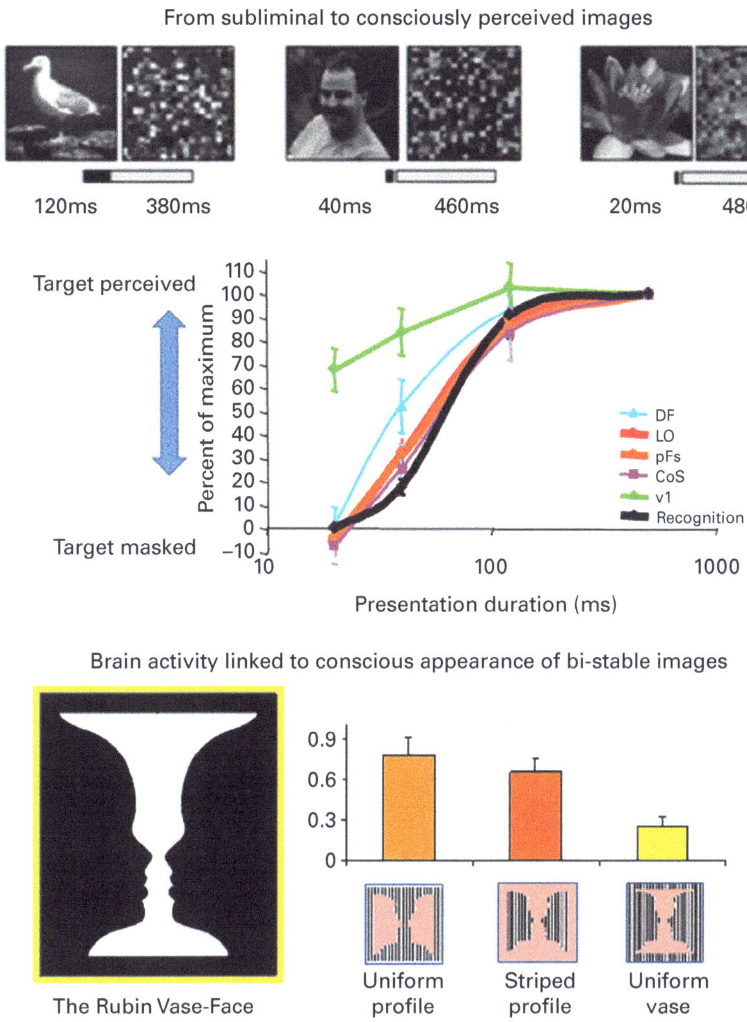

Figure 10.1
Top: Neuronal manifestations of conscious experience in the human brain. In the top panel there is a tight link between perceptual awareness (black line) and cortical activity in high-order visual areas that are sensitive to the target images (blue and red lines) during a "backward masking" experiment in which target images were presented briefly (fraction of a second) followed by a meaningless picture (mask). By gradually extending the duration of the targets one can transition from seeing only the mask to also seeing the target prior to the mask (indicated by the black line). Note the lack of activity when the targets are invisible in the high brain areas. By contrast, this activity could be detected in low brain areas; for example, in V1 (green) the earliest processing stage of the visual cortex (Grill-Spector et al., 2000) probably reflecting some response to the mask in these areas.
Bottom: Brain activity is modulated by the content of conscious experience. When viewing a bi-stable image we perceive it as alternating between seeing two face profiles to seeing a central vase (left). When participants were induced to experience the profiles, higher brain activity was found in cortical areas selective to faces. The activity was reduced in these areas when participants were consciously experiencing the vase instead (Hasson et al., 2001)- despite the fact that optically- there was no change in the image presented.

Losing the Self

But if you want to study manifestations of consciousness in brain activity, you must begin by forming some notion of what consciousness actually is: What are the common characteristics that underlie all conscious experiences? Otherwise, you may end up studying a specific subset of conscious experiences, which is great but will not advance our quest aimed at finding *general* brain principles that underlie all conscious phenomena of any type.

Unfortunately, trying to think about this deceptively simple question—What constitutes consciousness?—immediately plunged me into a deep conceptual deadlock. The more I tried to envision what consciousness is, the more my mind started spinning in an endless regression. Nothing new here of course. The problem of what consciousness is has a long history of debate and controversy in philosophy and psychology circles. My particular brand (as many before me) was that I got stuck on the problem of the role of the self in conscious experience. Perhaps the problem was exacerbated for me because in Hebrew the word for consciousness is "Mudaut" or "Hakara" and both imply knowing and recognizing. The logical conclusion from knowing and recognizing is that there must be someone there—call it the witness—that observes and recognizes the conscious experience. But the moment you start with this "trivial" insight, you are locked into a slippery regression, because the next obvious question is when you think about this "witness" who then is observing this original witness? And this chain can go on and on in an infinite regression. The unique problem for me, as an experimenter, was how one can conceive of any meaningful experiment when your target of research is entangled in what was coined as a "strange loop" by Douglas Hofstadter in his book *Gödel, Escher, Bach.*

Neuronal Signatures of "Losing the Self"

Fortunately, at some unexplained insightful moment, it dawned on me (long after it was discovered, for example, in Buddhist thought) that conscious experience is possible, and in fact becomes more vivid, when our self-awareness is absent from the scene (Malach, 2007; Malach & Josipovic, 2006). I will momentarily digress here and note that the term "self" is itself a poorly defined and multifaceted (and debated) term. Here I use the term "self" with a reference to its rather intuitive manifestation that may be termed the "observing self" (see **glossary**)—that is, the notion we often have that we are a separate entity that witnesses or observes the outside world.

However, contemplating these experiences, I realized that there are clear instances when the observing self is absent. A striking instance of such an "experience without an experiencer" is the state of absorption. Consider the experience of watching a truly riveting action movie. Can you genuinely claim that while being totally absorbed in, say, a car chase scene, you were at the same time experiencing yourself or introspecting about yourself as a witness in any manner? In other words, did you consciously experience a "knower" that was observing the movie during these totally absorbing scenes? The answer is emphatically "no," as hard as it may be to accept intuitively. We, if defined as self-aware observers, are absent from the movie especially when we are fully immersed in it! If you think about it carefully, all experiences in which you become self-conscious—for example, becoming aware that you just saw a wonderful movie, reporting about it to your friends, evaluating how scary or entertained you were—come *after* the riveting experience is already over.

So, I was now convinced that consciousness is not dependent on an observing witness, and this made the problem far more tractable experimentally. Furthermore, this realization could be easily translated into a straightforward experimental question. Indeed, this led to my first intentionally consciousness-related functional brain imaging experiment that was conducted together with my student Ilan Goldberg (now a neurologist) (Goldberg et al., 2006).

The idea behind the experiment was exceedingly simple: if indeed sensory absorption leads to "losing the self" then we should be able to observe the brain-correlate of this self-loss through brain imaging technology. Thus, we constructed an experiment that created a perceptually absorbing state inside the fMRI scanner and contrasted this state with an introspective state specifically aimed at enhancing self-awareness. To our delight the results turned out even more striking than we expected: not only did the absorption state fail to

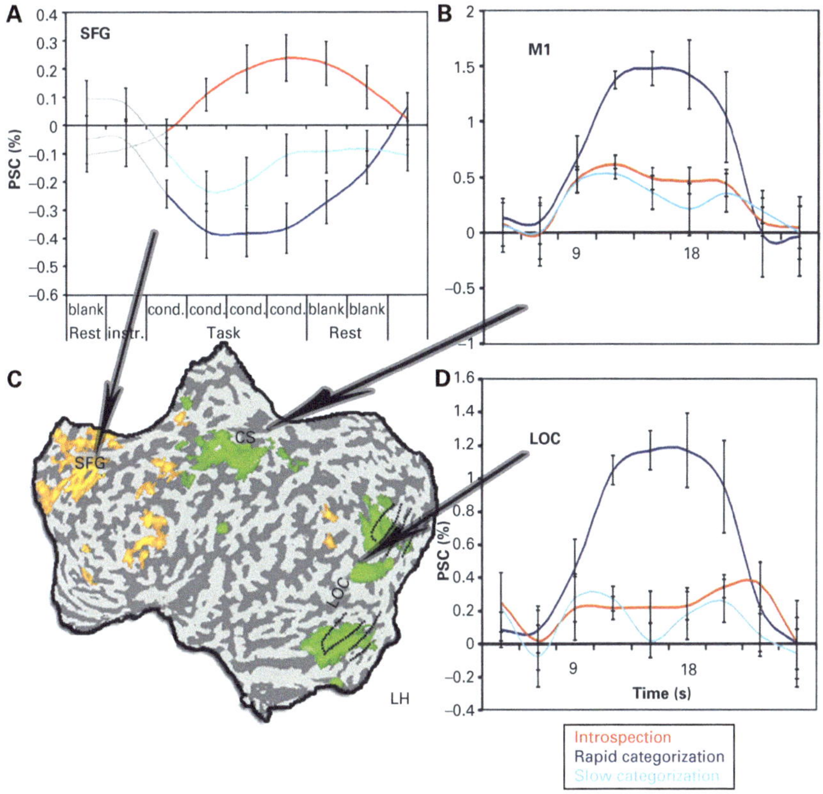

Figure 10.2
Neuronal signatures of "losing the self." Neuronal activity sampled using brain imaging technology (fMRI) in three central parts of the human cortex during an experiment that contrasted a highly engaging, immersive visuo-motor task with an introspection task aimed to enhance self-awareness. Note that during introspection, a cortical area in the front of the brain was activated (shown in orange on an unfolded map of the cortex), but this area was suppressed below the resting baseline when participants were immersed in their visuo-motor task. By contrast, visual and motor areas (shown in green on the map) were, as expected, highly activated during the visuo-motor experience. These results reveal an antagonistic relationship between self-awareness and visuo-motor immersion when participants are fully absorbed by their perceptual and motor *here and now* conscious experiences (Goldberg et al., 2006).

activate frontal cortical regions that specialize in observing-self function, but it actually suppressed the neuronal activity in these areas significantly below the resting-state baseline (see figure 10.2). This provided clear experimental support to the idea that conscious experience does not depend, and often actually suppresses, self-awareness in the human brain.

Neuronal Strange-Loops

Still, I felt that there is something profound and compelling about the idea that consciousness must imply some "self-knowing" or "observing" element. As scientists we are tempted to think that what we observe about the brain as experimentalists is what counts. But this is, of course, a delusion. What counts, for the brain, is what its own elements are "observing"—what information the neurons residing inside an individual's brain are getting from each other. The most intuitive implementation of such inner knowing or observing is to search for a brain analog of the observing-self idea. Could it be that there is a special part of the brain—say the frontal lobes—that serves as the observer and "reads" the activity generated in, for example, the visual areas located at the back of the brain? But who, then, observes the frontal lobes? We are back in the familiar consciousness spinning mode. Furthermore, the experiments we conducted appeared to argue against this notion (see also, Noy et al., 2015).

But there seems to be one construct that actually embodies in its own architecture and dynamics such infinite self-observing regression—namely, the recurrent loop. Metaphorically, such a loop can be envisioned by two mirrors reflecting each other or a self-devouring snake. In such a "strange loop," questions like "who mirrors whom?" and "who is biting whom?" become meaningless. What is the essential consequence of a true recurrency is precisely the loss of information about cause and effect of the individual elements. We cannot tell anymore which neurons were activated by which. A truly recurrent loop cannot be subdivided into separate elements of an observer and observed or into discrete time steps. This is in stark contrast to, for example, artificial, digitally-based, recurrent loops in which each logical step is well-defined and synchronized via a central clock. It is interesting to note, and probably not a coincidence, that this holistic physical state emerging out of recurrent loops seems to share parallel properties with the holistic, indivisible, nature of conscious experience.

Remarkably, in the cerebral cortex such recurrent loops are exceedingly widespread and implemented massively and ubiquitously by neuronal groups (called cell ensembles by Donald Hebb [1949]). In each and every part of the human cortex, millions of neurons mutually excite each other through dense networks of local synaptic connections (see top panels figure 10.3; Amir et al., 1993). These intrinsic networks are, intriguingly, one of the defining anatomical characteristics of the cerebral cortex when compared to other parts of the brain. In essence, one can say that in such recurrent loops the neurons read or observe their own activity. Could it be that this self-reading is the physical implementation of a knower or an observer? If true, the very same circuits that are being read are the ones that do the reading.

All these notions are interesting and thought provoking, but for an experimenter the crucial question remained: What does the brain, and experimental findings have to say about these speculations? To get some answers, I and my student Dr. Lior Fisch collaborated with Professor Itzhak Fried in recording brain signals from patients undergoing clinical diagnosis for epilepsy. As part of the clinical procedure, the neurosurgeon inserts tiny recording electrodes directly into the brain of these patients in order to identify epileptic foci. While strictly conducted for clinical aims, the recordings provide a rare opportunity to perform cognitive experiments in between the clinical diagnostic sessions. This allowed the recordings of neuronal activity directly from the cortex of conscious participants—a method that is incomparable in sensitivity and space-time resolution to any other brain recording method available in humans.

Fisch and I also conducted an image detection experiment similar to the one I ran with Grill-Spector, comparing brain activity when the patients saw a briefly presented masked target with when they failed to see it (see figure 10.3, bottom panel; Fisch et al., 2009). What we saw can be best captured by the metaphor of igniting a fire. Consciously perceiving the target was associated with a high intensity burst of neuronal activity that, critically, continued for hundreds of milliseconds after the brief visual image (lasting only a few tens of milliseconds) was terminated and masked. This indeed resembled igniting a neuronal fire. As is the case in all natural ignitions, the fire keeps going long after the match that started it has already burned out. Interestingly, somewhat parallel metaphors that have been suggested by others for cortical dynamics include percolations and avalanches (Shriki et al., 2013). Thinking about the mechanism that may lead to such neuronal ignitions readily led to positive-feedback recurrency, which is often the mechanism underlying ignitions in nature, such as igniting a campfire for example. In the case of transiting from invisible to visible images, this positive-feedback recurrency appears to be implemented by an anatomically localized group of neurons that excite each other, just like the twigs heating each other in a campfire (Fisch et al., 2009; Moutard et al., 2015). Such mutual activation can metaphorically be envisioned as an entity that observes and feeds on itself.

In conclusion, conscious experience of a specific content—for example a visual image—appears in our mind when a localized group of neurons, specialized for this content, recurrently activate each other and ignite a self-amplified neuronal flame. I hasten to state that additional and more direct experiments are needed to further verify the hypothetical role of such localized recurrent activity in conscious experience but at least the hypothesis makes testable predictions that so far appear compatible with experimental results and that may be further investigated in future work.

Haphazard Subconscious Processes Underlying Freedom and Creativity

I started my independent research career as a neuroanatomist studying the structure of neurons' communication lines. I vividly remember stumbling on a puzzling geometrical principle that emerged from the data and for which I had no explanation. After searching in vain for an answer for a few days, I woke up one morning and the answer was right in front of me—crystal clear. The solution was so immediately obvious and so easy that I joked when writing the paper reporting this insight (Malach, 1994) that I do not deserve to be the author—somebody inside my head was making a lot of effort and succeeded in solving the problem, subconsciously and without my knowing. I am sure many of you readers have experienced such a spontaneous, unexpected appearance of a new insight, a thought, or a solution to a problem. Such spontaneous events can range from merely an unexpected recollection of a name you were trying to come up with, to an epiphany that may lead to a great theory or

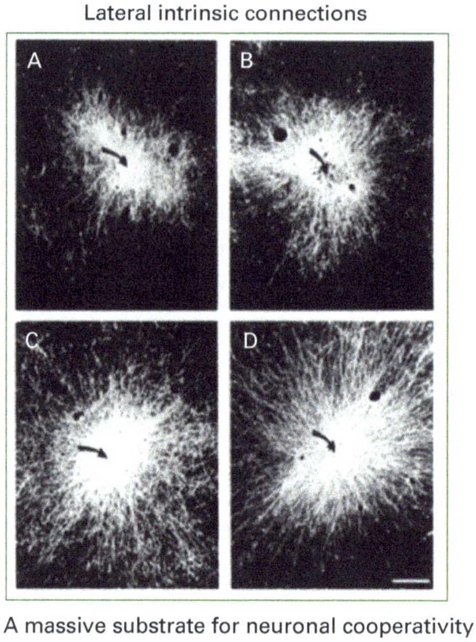

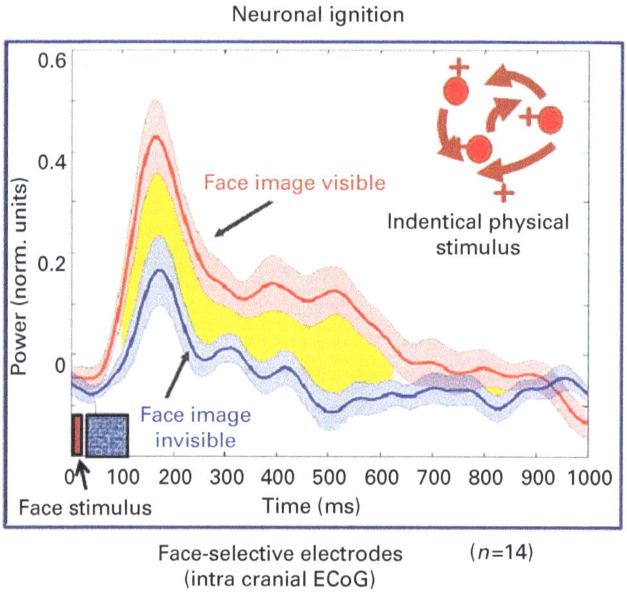

Figure 10.3
Neuronal cooperative ignitions underlying conscious experience.
Top panel: Massive "halos" of local, intrinsic connections, in different visual cortical areas. Local interconnections between neighboring cortical neurons are one of the defining characteristics of the cerebral cortex, which is thought to be the central organ underlying conscious experience. The pictures depict the intrinsic connections emerging from a small group of neurons (marked by arrows) in different visual areas. Note the massive, local, radiating network of connections, which are found in every cortical location and area. These dense networks offer ample opportunity for recurrent interactions between groups of neighboring neurons (Amir et al., 1993).
Bottom panel: Neuronal ignitions linked to conscious experience in the human brain. Neuronal activity recorded in face-selective cortical centers in patients undergoing intracortical recording (see **glossary**) while they viewed briefly presented face images that were sometimes seen and sometimes failed to be detected (red and blue lines, respectively). The experience of consciously seeing the target face is linked to higher and more sustained neuronal activity, specifically in face-selective neurons that outlast the original target image presentation by hundreds of milliseconds. These effects are the typical signatures of igniting a "neuronal flame." It is hypothesized that this ignition is mediated via recurrent excitatory interactions within a local group of neurons (red small arrows) (Fisch et al., 2009).

work of art: a "eureka" experience often described by the geniuses who created them. So, what is going on neuronally? What kind of subconscious brain processes ferment inside our heads, below the surface of our consciousness, bringing about such wonders of spontaneity and creativity?

My personal story in this line of research begins with an unexpected discovery made together with (now professor) Yuval Nir (Nir et al., 2006). By chance, we stumbled on one of the most robust and widespread aspects of cortical activity that has been discovered: the fact that cortical neurons spontaneously generate rich and incessant patterns of ultraslow and highly irregular activity waves that often take seconds to develop. We have discovered these slow cortical activation waves in the visual system but, as shown by numerous other studies, they appear in each and every cortical area when these areas enter an idle mode—that is, are not engaged in a well-defined task (Moutard et al., 2015; Nir et al., 2008). By following intracortical recordings in auditory cortex of patients while they rested in a quiet environment, we were able to show that these internally generated ultraslow fluctuations are manifested in the activity modulations of individual neurons as well as mass activity of large neuronal groups (see top panels in figure 10.4).

What we realized right from the start was that these slow activity waves go on below the consciousness surface; this has been subsequently confirmed by other researchers. Simply put, we are not aware of the turmoil of these slowly undulating activations that go on in our brains all the time (Ramot et al., 2016, 2011). On the other hand, a large body of research has revealed that the patterns of activity produced by these spontaneous waves have meaning: they appear to reflect personality traits, expertise, habitual behaviors, and in unfortunate cases, also brain disorders and pathologies (Harmelech & Malach, 2013; Hahamy et al., 2015; Wilf et al., 2017). In short, one can say that just as our genome reflects our genetic makeup, the patterns of spontaneously emerging activity waves in some sense reflect or manifest our personal traits: who we are. However, a major puzzle remained: What could be the function of these incessant, subconscious and constantly changing patterns of activity? What are they good for?

There is an intriguing, and to my mind, not coincidental, correspondence that may suggest an answer. If one considers the phenomena of spontaneous insights and creativity I described above and the nature of the slow spontaneous waves in cortical activity, then there is a striking parallel. Both creativity and the spontaneous waves appear to be driven from internal sources—that is, they are self-generated rather being determined by an outside command or instruction, and both run their course below our conscious awareness. Both creativity and the spontaneously emerging patterns depend on our stored expertise, talents, and personal knowledge. Finally, and perhaps most tellingly, creativity can be manifested in essentially all aspects of human behavior and, correspondingly, the spontaneous activation patterns emerge in each and every part of the cerebral cortex.

Could it be that these slow spontaneous waves in cortical activity are the very dynamo that drives our creative and free behaviors? Only very recently, together with my student Rotem Broday-Dvir, we found a way to test this intriguing possibility (Broday-Dvir & Malach, 2021). Our hypothesis is illustrated in figure 10.4. The idea is rather simple: when we are faced with a free or creative decision, the spontaneous activity fluctuations explore, below the awareness threshold (dashed horizontal line), the landscape of possible alternatives

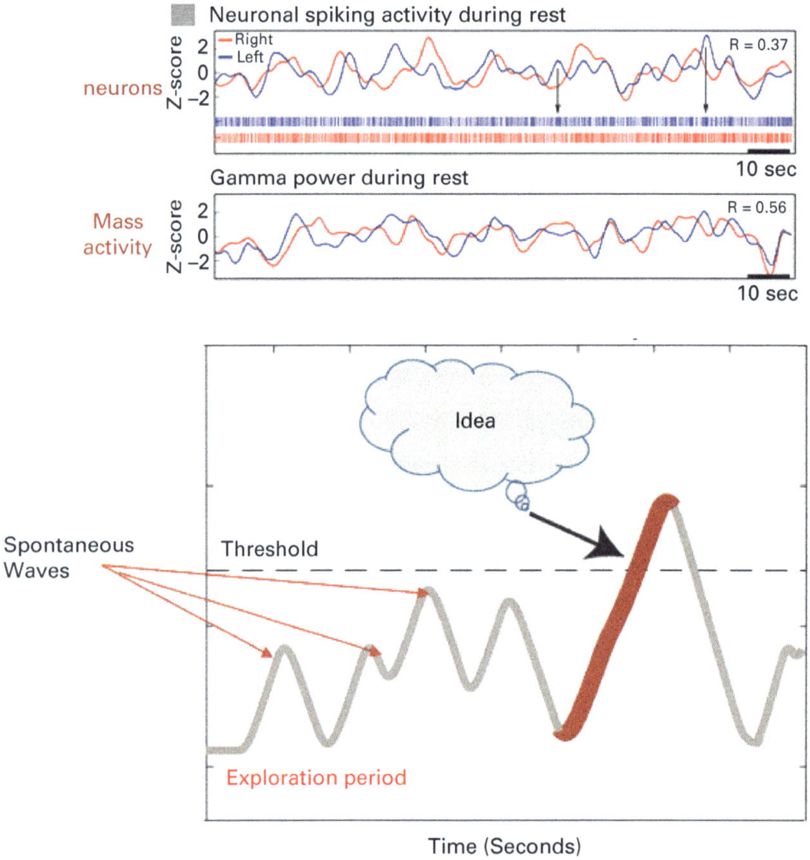

Figure 10.4
Ultraslow spontaneous waves in the human cortex underlie creative insights.
Top panel: Slow spontaneous waves of neuronal activity in the cortex. Examples from two neurons recorded in the auditory cortex of a patient during a silent period. Note the five- to ten-second durations of the irregular waves of neuronal activity that are correlated across the two hemispheres, in the complete absence of any auditory stimulation. Such ultraslow spontaneous waves are among the most ubiquitous and persistent phenomena associated with the idle, no-task state in the human cortex (Nir et al., 2008).
Bottom panel: Spontaneous waves as a universal cortical mechanism driving free and creative ideas. The schema illustrates the proposal that the spontaneous waves randomly and subconsciously explore the landscape of potential solutions—and when they hit on an optimal answer, they cross the consciousness threshold leading to the ignition of a new, creative idea. (Broday-Dvir & Malach, 2021; Moutard et al., 2015).

by activating the relevant but latent neuronal groups (indicated in red arrows in the bottom panel of figure 10.4). When the optimal alternative is hit upon in this stochastic exploration process, the spontaneous wave crosses the awareness threshold (red curve in the figure) and at that eureka moment we become aware of our free decision or creative idea. It is interesting to note that this upswing of the spontaneous wave leading to a free decision shows close similarity to what has been termed the readiness potential in the domain of free movements. It thus confirms and extends the much-debated notion, proposed by Libet (2005) of a subconscious process leading to free decisions. However, the new perspective is that this anticipatory wave is driven by a unique combination of stochastic and deterministic components—that is, a biased randomness that can be metaphorically compared to a loaded dice. Importantly, we have recently been able to obtain experimental support to this notion (Broday-Dvir & Malach, 2021).

Adopting a bird's eye view of these findings, I notice two important themes that, not surprisingly, can be found in numerous domains in nature and biology (for further reading, see Malach [2012]). The first is the critical importance of cooperative group activity: the exchange of information and holistic binding of neuronal assemblies that appears to be a fundamental aspect underlying the emergence of all our conscious experiences. Crucially, there is no "privileged" cortical site for consciousness: every cooperative assembly in the cortex can underlie a unique content of experience (Malach 2021). Specifically, conscious experience does not depend on an observing self or a high-order reflective process. The cortex is a democracy and wherever an assembly of neurons ignites in a recurrent, cooperative burst, a conscious experience will be linked to these local dynamics. However, critically, the content of the experience, be it a visual image, a sound, or a thought about ourselves, will be uniquely determined by the cortical locus of the ignited activity.

The second theme is the powerful and constructive role of spontaneous and subconscious processes in creativity, which echoes beautifully the lessons learned from biological evolution. Just as in creative idea generation, it is an astonishing fact that random noise (mutations), when combined with orderly, informative structures (the genetic code), gives rise to astounding evolutionary processes that among many amazing feats led to the emergence of human beings—an organism that can now think and communicate these ideas through the cooperative action of their nerve cells. This is both a stunning and humbling realization.

Glossary

Cerebral cortex: A two- to three-millimeter thick sheet of neurons typically arranged in six layers. The cortex covers the entire brain with many undulations and is considered a central organ underlying conscious experiences. The cortex is subdivided into lobes, such as the frontal lobe in the front of the brain and occipital lobe in the back.

Cortical areas: The cortex is made up of numerous subdivisions called cortical areas. Each of these areas specializes in a specific task or category. For example, visual areas specialize in the visual domain, motor areas in action, and so on. Sets of areas are connected into larger cortical networks and systems underlying general functions, such as visual identification or action generation.

Functional magnetic resonance imaging (fMRI): A noninvasive brain imaging technique based on magnetic scanners. The method provides whole-brain maps of neuronal activity during a wide range of cognitive tasks as well as different brain states. The method traces the neuronal activity through its indirect effect on slow changes in blood flow and oxygenation.

Hierarchy of visual areas: The visual system is built out of a number of visual areas that are arranged in a hierarchy of growing abstraction. From early visual areas whose neurons specialize in elements of the visual image such as oriented lines to higher-order visual areas specializing in entire templates, such as faces, body parts, or visual motion.

Intracranial recordings: An invasive clinical procedure in which, strictly for diagnostic purposes, typically in neurosurgical treatment of intractable epilepsy, the neurosurgeon implants tiny metal probes directly into the cortex of the patients. This method enables direct recordings of neuronal activity signals in conscious human patients.

Neurons: The basic building blocks of the brain. These tiny cells are essentially communication centers collecting inputs from and sending outputs to other neurons. The neurons signal each other by emitting short electrical pulses called spikes. When a neuron is activated, it increases the rate of its spikes. A neuron is said to be specialized for a specific content—for example, an image of a face—when it is preferentially activated by this content.

Observing self: The awareness of yourself as a separate entity that is witnessing or reflecting upon the external environment.

References

Amir, Y., Harel, M., & Malach, R. (1993). Cortical hierarchy reflected in the organization of intrinsic connections in macaque monkey visual-cortex. *Journal of Comparative Neurology, 334*(1), 19–46.

Berkovich-Ohana, A., Noy, N., Harel, M., Furman-Haran, E., Arieli, A., & Malach, R. (2020). Inter-participant consistency of language-processing networks during abstract thoughts. *Neuroimage, 211*, 116626. https://doi.org/10.1016/j.neuroimage.2020.116626

Broday-Dvir, R., & Malach, R. (2021). Resting-state fluctuations underlie free and creative verbal behaviors in the human brain. *Cerebral Cortex, 31*(1), 213–232. https://doi.org/10.1093/cercor/bhaa221

Fisch, L., Privman, E., Ramot, M., Harel, M., Nir, Y., Kipervasser, S., Andelman, F., Neufeld, M., Kramee, U., Fried, I., Malach, R. (2009). Neural "ignition": Enhanced activation linked to perceptual awareness in human ventral stream visual cortex. *Neuron, 64*(4), 562–574. https://doi.org/10.1016/j.neuron.2009.11.001

Gelbard-Sagiv, H., Mukamel, R., Harel, M., Malach, R., & Fried, I. (2008). Internally generated reactivation of single neurons in human hippocampus during free recall. *Science, 322*(5898), 96–101.

Goldberg, I. I., Harel, M., & Malach, R. (2006). When the brain loses its self: Prefrontal inactivation during sensorimotor processing. *Neuron, 50*(2), 329–339. https://doi.org/10.1016/j.neuron.2006.03.015

Grill-Spector, K., & Malach, R. (2004). The human visual cortex. *Annual Review of Neuroscience, 27*, 649–677. https://doi.org/10.1146/annurev.neuro.27.070203.144220

Grill-Spector, K., Kushnir, T., Hendler, T., & Malach, R. (2000). The dynamics of object-selective activation correlate with recognition performance in humans. *Nature Neuroscience, 3*(8), 837–843.

Hahamy, A., Behrmann, M., & Malach, R. (2015). The idiosyncratic brain: Distortion of spontaneous connectivity patterns in autism spectrum disorder. *Nature Neuroscience, 18*(2), 302–309. https://doi.org/10.1038/nn.3919

Hahamy, A., Wilf, M., Rosin, B., Behrmann, M., & Malach, R. (2021). How do the blind "see"? The role of spontaneous brain activity in self-generated perception. *Brain, 144*(1), 340–353. https://doi.org/10.1093/brain/awaa384

Harmelech, T., & Malach, R. (2013). Neurocognitive biases and the patterns of spontaneous correlations in the human cortex. *Trends in Cognitive Sciences, 17*(12), 606–615. https://doi.org/10.1016/j.tics.2013.09.014

Hasson, U., Hendler, T., Ben Bashat, D., & Malach, R. (2001). Vase or face? A neural correlate of shape-selective grouping processes in the human brain. *Journal of Cognitive Neuroscience, 13*(6), 744–753. https://doi.org/10.1162/08989290152541412

Hebb D, (1949) The organization of behavior. EmphNew York . *Wiley*

James, W. (1890). *The principles of psychology* (Vol. 1). Dover Publications.

Libet, B. (2005). Mind time. The Temporal Factor in Consciousness, *Harvard University Press*

Malach, R. (1994). Cortical columns as devices for maximizing neuronal diversify. *Trends in Neurosciences, 17*(3), 101–104.

Malach, R. (2007). The measurement problem in consciousness research. *Behavioral and Brain Sciences, 30*(5–6). https://doi.org/10.1017/S0140525x0700297x

Malach, R. (2012). Neuronal reflections and subjective awareness. In *Being in time: Dynamical models of phenomenal experience* (Vol. 88, pp. 21–36). https://doi.org/10.1075/aicr.88.02mal

Malach R. (2021). Local neuronal relational structures underlying the contents of human conscious experience. *Neuroscience of Consciousness.* https://doi.org/10.1093/nc/niab028

Malach, R., & Josipovic, Z. (2006). Perception without a perceiver—In conversation with Zoran Josipovic. *Journal of Consciousness Studies, 13*(9), 57–66.

Malach, R., Reppas, J. B., Benson, R. R., Kwong, K. K., Jiang, H., Kennedy, W. A., Ledden, P.J., Brady, T.J., Rosen, B.R., & Tootell, R. B. (1995). Object-related activity revealed by functional magnetic resonance imaging in human occipital cortex. *Proceedings of the National Academy of Sciences of the United States of America, 92*(18), 8135–8139. https://doi.org/10.1073/pnas.92.18.8135

Moutard, C., Dehaene, S., & Malach, R. (2015). Spontaneous fluctuations and non-linear ignitions: Two dynamic faces of cortical recurrent loops. *Neuron, 88*(1), 194–206. https://doi.org/10.1016/j.neuron.2015.09.018

Nir, Y., Hasson, U., Levy, I., Yeshurun, Y., & Malach, R. (2006). Widespread functional connectivity and fMR1 fluctuations in human visual cortex in the absence of visual stimulation. *Neuroimage, 30*(4), 1313–1324.

Y. Nir, R. Mukamel, I. Dinstein, E. Privman, M. Harel, L. Fisch, H. Gelbard-Sagiv, S. Kipervasser, F. Andelman, M. Y Neufeld, U. Kramer, A. Arieli, I. Fried and R. Malach (2008) Interhemispheric correlations of slow spontaneous neuronal fluctuations revealed in human sensory cortex. *Nature Neuroscience*, *9* (11) 1100–1108

Norman, Y., Yeagle, E. M., Harel, M., Mehta, A. D., & Malach, R. (2017). Neuronal baseline shifts underlying boundary setting during free recall. *Nature Communication*, *8*(1), 1301. https://doi.org/10.1038/s41467-017-01184-1

Norman, Y., Yeagle, E. M., Khuvis, S., Harel, M., Mehta, A. D., & Malach, R. (2019). Hippocampal sharp-wave ripples linked to visual episodic recollection in humans. *Science*, *365*(6454). https://doi:10.1126/science.aax1030

Noy, N., Bickel, S., Zion-Golumbic, E., Harel, M., Golan, T., Davidesco, I., Schevon, C.A., McKhann, G.M., Goodman, R.P., Schoeder, C.E., Metha, A.D., & Malach, R. (2015). Ignition's glow: Ultra-fast spread of global cortical activity accompanying local "ignitions" in visual cortex during conscious visual perception. *Consciousness and Cognition*, *35*, 206–224. https://doi.org/10.1016/j.concog.2015.03.006

Ramot, M., Grossman, S., Friedman, D., & Malach, R. (2016). Covert neurofeedback without awareness shapes cortical network spontaneous connectivity. *Proceedings of the National Academy of Sciences of the United States of America*, *113*(17), E2413–2420. https://doi.org/10.1073/pnas.1516857113

Ramot, M., Wilf, M., Goldberg, H., Weiss, T., Deouell, L. Y., & Malach, R. (2011). Coupling between spontaneous (resting state) fMRI fluctuations and human oculo-motor activity. *Neuroimage*, *58*(1), 213–225.

Shriki, O., Alstott, J., Carver, F., Holroyd, T., Henson, R.N., Smith, M.L., Coppola, R., Bullmore, E. and Plenz, D. (2013). Neuronal avalanches in the resting MEG of the human brain. *Journal of Neuroscience*, *33*(16), 7079–7090.

Wilf, M., Strappini, F., Golan, T., Hahamy, A., Harel, M., & Malach, R. (2017). Spontaneously emerging patterns in human visual cortex reflect responses to naturalistic sensory stimuli. *Cerebral Cortex*, *27*(1), 750–763. https://doi.org/10.1093/cercor/bhv275

Figure 0.4
Michal Rovner. Blabla Notebook, 2005. Still frame from a projection on notebook. Copyright: Michal Rovner.

11 Do Quantum Theory and Human Consciousness Share Common Ground?

Zohar Maliniak

A change of meaning is a change of being.
—David Bohm

Personal Reflections

The conference I attended in Neve Shalom, which addressed the study of consciousness, was especially important to me. It provided me with an opportunity to engage in an enlightening and inspiring meeting with a wide range of individuals and scholars for whom the mystery of human cognition and consciousness is a burning question. This meeting of philosophers, neurobiologists, scholars of culture, physicists, educators, and longtime practitioners of spiritual and meditative practices forced me—and I think the other participants too—to rethink consciousness, cognition, realism, first-person, and third-person experiences. It also provoked us to think about the possibility of connecting the different "languages" and perspectives more meaningfully, and to try and create a theoretical picture of human consciousness that may contribute to cooperative and innovative research.

Overview

Does quantum theory contains anything that touches on the hidden possibilities of human existence? Is there anything in it that points to a hidden but perhaps accessible understanding of human consciousness in general, and my experience of daily life in particular? These questions have preoccupied me for almost forty years, since I was a young man in my twenties. At first it was part of a personal search for "truth," and later in the context of my studies in physics and philosophy. My turning point came when I encountered the words of the physicist David Bohm and the philosopher Jiddu Krishnamurti, both in the conversations that they held together and in words that Bohm wrote about the meaning of quantum theory and the possible links between modern science, creativity and order (see **glossary**)–both visible and hidden. Suddenly, things came together for me—the intense journey I took in those years of self-inquiry together with a group of other people, the study of physics that often seemed completely unconnected to my life, feelings, or thoughts. Even the philosophical thinking that was often remote from me and the fundamental problems that occupied me suddenly connected into a living, breathing, resonating, and inspiring essence. Since then, I find within myself, my thoughts, my experiences, a close connection between quantum physics and my own personal experiences. In this article, I will briefly point to some parallels between the way I view the quantum world (through David Bohm's interpretation from 1952a) and the studies I have conducted on human consciousness over many years. I will conclude by raising the question: Is all this coincidental and personal, or does it represent a possible new direction for the physical, phenomenological, and neurobiological study of the mind and consciousness?

The Quantum World According to David Bohm

The question of the nature of realism in physics has provoked wide-ranging discussions about whether it is possible to describe speak or think about the "world" even when we do not observe or measure it (Einstein in one of his criticisms of Bohr's interpretation remarked with bitter cynicism, "Do you really believe the moon is not there when you are not looking at it?"). When discussing quantum theory, many scientists and philosophers use the Zen koan that asks: If a tree falls in a forest and no one is around to hear it, does it make a sound? And in the context of quantum theory, they could ask: If you did not measure an electron, how can you determine it exists or describe it? Furthermore, even if you measured it a moment ago, all you can say is that it existed as an electron a moment ago. What is it now?

David Bohm was a Jewish American physicist (1917–1992) whose work encompassed quantum theory and its interpretations. He completed his doctorate under Robert Oppenheimer and had a close relationship with Albert Einstein. He left the United States in 1950 under suspicion of being a communist and was called upon to give testimony before the McCarthy committee. One of the strange and best known phenomena of quantum theory is named after him and his student Yakir Aharonov, the Aharonov-Bohm effect. The interpretation that Bohm proposed for quantum theory was almost universally rejected by leading physicists in the 1950s, but in the 1960s, it served as a major source of inspiration for J. S. Bell and his theorem of inequality. In the last twenty years, more and more physicists and philosophers view his interpretation as important and valuable.

David Bohm thought that quantum mechanics phenomena are not contrary to our daily intuitions and can be described and imagined in our minds and in our language. It is about the ontology of the world outside. We just have to accept a radical change in our perception of that world.

The picture that Bohm suggested was very simple but also had a new and dramatic aspect to it. According to Bohm, in the quantum world we have both point-like particles responsible for the glitter on the florescent screen in the double-slit experiment described in figure 11.1—and also guiding waves (see **glossary**). The wave is revealed indirectly and it is an "implicate order" (see **glossary**) of the observed phenomenon. Based on a study that Bohm conducted of the world of quantum phenomena, together with the mathematics that describes and serves the field, he realized that the wave that guides the particles has a new and surprising quality: it is a wave whose intensity does not decrease with distance. What is important in the idea of the guiding wave is its global form rather than its energy, which is negligible. The wave immediately "entangles" distant particles—for example, in the Einstein-Podolsky-Rosen (EPR) paradox (Einstein, 1935)—which makes the quantum world and the phenomena it describes non-local and connected (see **glossary**).

To elucidate further, Bohm suggested the following metaphor: Imagine a ship navigating itself by means of radio waves. The ship sails on the power of its engines (the ship is analogous to the quantum particle according to Bohm's interpretation), but its precise navigation is accomplished thanks to radio waves (the radio waves are analogous to the quantum field that drives the particle). In the image that Bohm suggested, the guiding wave has an extremely low amount of energy, with the vast majority of the energy being derived from the particle's energy. What the guiding wave contains is information, and indeed Bohm did eventually called the quantum wave an "information field." For the sake of clarification, I have included figure 11.1, which shows the quantum field and the possible trajectories of particles in the double-slit quantum experiment.

In time, Bohm formulated a number of metaphors that in his eyes characterized the world of quantum phenomena. These metaphors laid out very clearly the dramatic difference between what we obtain in the quantum world compared to the picture of the world that arises and takes shape based on classical Newtonian mechanics.

I will summarize these metaphors as follows (they can each be expanded and we can point to phenomena and experiments in which they are actualized; however, I will not do so in the context of this chapter):

1. The whole precedes its parts and the parts cannot be understood except through their affinity to the whole.
2. Each part of the whole contains all the information of the whole.
3. The universe is entangled and non-local (quantum entanglement, see **glossary**). One might think of the universe or parts of it as being connected "telepathically," and any change that occurs in one part of the universe immediate affects another part of the universe that may be very distant from the first part (this situation contradicts one of the main principles of the special theory of relativity, which determines that no information can travel faster than the speed of light. But such entanglement does occur and has been measured in many EPR experiments).
4. Noncontinuous transitions and movement occur at the same time ("quantum jump"). This is a phenomenon that occurs, for example, in the passage of electrons between the different levels of energy in the "shells" of an atom.
5. The observer cannot be separated from the observed.
6. Context is a crucial partner in the creation of a phenomenon, and properties (e.g., of particles) cannot be discussed detached from the context in which they were measured and observed.

The picture of the quantum world as it emerges from these metaphors, and here I will use a concept about which Bohem wrote a great deal, the concept of quantum order, is fundamentally different from the concepts of order that emerge from Newtonian mechanics and the phenomenological world of classical physics. In the classical model, the world is perceived as a huge machine made up of basic parts, and in order to understand this machine,

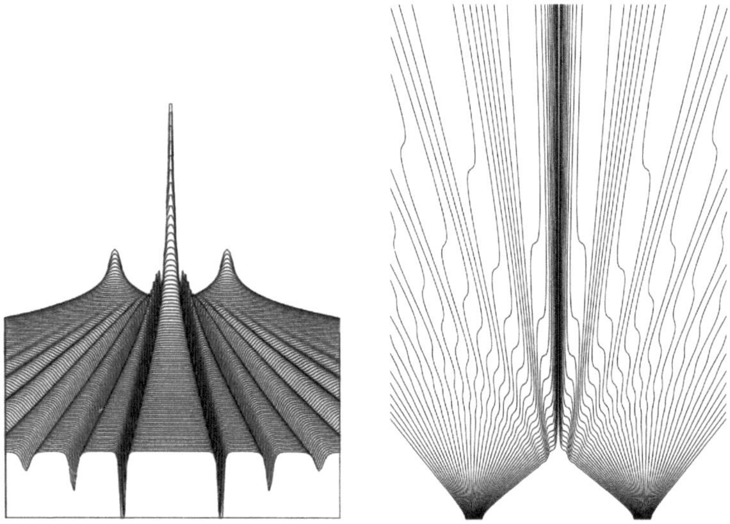

Figure 11.1
Simulation of quantum potential and possible trajectories of electrons in a double-slit experiment according to Bohm and Hiley's interpretation.
(A) Simulation of the quantum potential in the double-slit experiment according to the interpretation of David Bohm's guiding wave. A non-local quantum field that has "troughs and peaks" spreads through the slits. The electron will move according to the structure of the quantum field in troughs and peaks and will not be able to pass from one area to another because of the "destructive" interference of the field.
(B) A streams of electrones pass through two narrow adjacent slits. The possible trajectories of the electron particles emerge (their density is determined by the intensity of the field of the quantum potential in each location). The electron will move on only one of the trajectories, and this trajectory will be determined by the particle's starting location in the "electron gun," which is not precisely known. In the areas of high density of the trajectories, when we conduct numerous measurements, we will find more glitter on the florescent screen located behind the splits. In areas with low density of trajectories fewer hits will be found on the florescent screen. In Bohm's interpretation, this explains the probabilistic nature of the results and their distribution (Bohm & Hiley, 1993).

it has to be disassembled into its basic parts to understand the interactions between the parts, and then, based on this understanding, the whole is "put back together" again. The metaphors that deeply underlie this worldview are as follows:

1. The whole is the sum of its parts, and an in-depth understanding of the whole can be obtained by taking its basic parts apart, understanding the basic parts and the interaction between them.

2. The observer can be separated from the observed and the observer's effect on the observed world is negligible in many cases.

3. The context in which the experiment is conducted has nothing to do with the results of the experiment, which are objective and take a "snapshot" of the real, physical world as it is.

4. Transitions from one state to another are continuous and causal.

5. The universe is local and information between different points is transferred at a finite speed. In other words, the transfer of effect from one point in space and time to another point distant from it occurs at a speed that does not exceed the speed of light and the transfer is continuous.

What I found so fascinating when I learned about the picture that Bohm suggested for the phenomenological quantum world was the question that had ignited Bohm's imagination already in the early 1960s and perhaps even earlier, as Pylkkanen writes (Pylkkanen, 2010, 2016; Pylkkanen et al., 2014): Does this picture of the world have a similar parallel in human consciousness? In other words, does the human mind and consciousness have a "hidden order" whose structure resembles the principles of quantum theory as I have formulated above?

I have studied this question in different variations for ten years together with a group of more than one hundred people. This investigation was not conducted in accordance with accepted academic methods, but through personal and group experiences accompanied by a detailed written report and long conversations that combined findings, feelings, thoughts, and philosophy about the possibility of human existence based on very different principles than those drawn from classical physics.

In the second part of this chapter, I will share with the reader some of my observations and insights that were common to me and the other members of the group, regarding the phenomenological investigations we conducted.

What Implications, According to Bohm, Might Quantum Theory Have for Cognitive Theory?

First, I found that many of my intellectual and psychological insights come to me discontinuously, in the form of an image that takes shape all at once, often "out of the blue," and without a clear causal connection. This seems to resonate and greatly resemble the "quantum jump" so typical of the world of microscopic quantum particles according to David Bohm's picture of the world.

Another noteworthy point is that often, when I try to observe myself, either in action or while sitting silently, it is done by splitting myself into a part in me that observes the phenomenon/behavior that I seek to observe and the phenomenon I am observing. The problem with this kind of self-observation is that it fundamentally changes the phenomenon I am observing. This is not "objective" observation whereby I investigate myself. It is diametrically opposed to the metaphor so prevalent in the world of Newtonian mechanics that the observer can be separated from the observed without any significant effect on the object of observation. Often, when I investigate a phenomenon outside myself—for example, when I try to listen and be attentive to the students I am teaching, if my attention comes from a place of division and separation between me and them—it profoundly and significantly affects the object of my observation and their behavior. I could often discern just how accurate Krishnamurti was when he said, "The observer is the observed." What I see on the outside—aggression, alienation, apathy—is actually a mirror image of my own attitude toward the student or the class.

There is a further, more subtle matter I often noted: when I say or feel something unambiguously, such as I love you, I hate you, I'm angry at you, the very articulation of the sentiment changes the situation. Before the words are articulated, the situation is mixed in a kind of "superposition"—that is, where two states exist simultaneously: I love you but also I don't love you a bit, I hate you but I also don't hate you a bit. When a dramatic and unambiguous statement is made, it changes the situation and "casts" it into a single state. There is a considerable resemblance as I see it between the situation in quantum theory just before the system measures a quantum system that is in a state of superposition until the moment when the measurement "casts" the system into a single value (what is known in the terminology of quantum mechanics as "the collapse of the quantum wavefunction" when a "robust" measurement takes place). But together with experiences of the kind described above, I have had many experiences of attention and perception, in which my attention could "hold" two opposing thoughts or feelings at the same time. In quantum terminology, it actually makes a lot of sense to see my consciousness as being in a situation of superposition of states that appear to contradict each other.

For example, when I enter into a state of deep meditation or attentiveness, alone or with other people, I often find myself in a state of consciousness in which my ego dissolves. I feel within myself a boundless space that I take with me to an unknown space. Rather than frightening me, it fills me with a magical openness. In this state, I feel that my consciousness is like a flowing river and that I am part of that flow. This sensation feels like freedom, and it seems that it was not I that created this state of consciousness, that I did not "do" my ordinary inner activity, and then the "other" state floated to the surface. In this state, the metaphor resonates within me. Everything flows and the feeling in those states, especially when other people are present, is that my brain is not "private," but rather that there is a "field" of intelligence, that it has meaning and is creative and new and that field is "guiding" me and the group of people.

I think that there is a great resemblance between the psychophysical picture I have just described and the picture that Bohm suggested and considered regarding what happens in quantum mechanics—that an actual particle is guided by a holistic quantum field whose activity is non-local. Moreover, I distinctly remember an experience I shared with a large group of people who were intently immersed in deep concentration and attentiveness. It was their concentration and attentiveness that dramatically changed the situation—that changed my sense of alienation into love and fullness and meaning. I clearly remember another situation thirty-five years ago, when I was on a retreat for a few days. During the retreat, I felt depressed, despondent, and tense. I was overcome with negative feelings and gloomy thoughts. I had a small notebook with me in which I monitored my feelings and thoughts. I recorded them carefully and thoroughly because I wanted to interact closely and intimately with those negative feelings. On the last day of the retreat, the whole group sat in a large shaded tent, about one hundred women and men, and I dared to read aloud the report in which I described the "stream of feelings and thoughts" I had experienced during those difficult days. The tent fell deeply silent, and at the end of my reading, I saw many red eyes filled with tears. But what surprised me most of all was that I myself underwent an internal transformation. My unpleasant feelings of alienation transmuted into a sense of great joy and love for all those around me. None of those present did or said anything; it was merely their deep attentiveness and emotional availability to genuinely see my inner suffering that suddenly and profoundly transformed my feelings and thoughts. In my eyes, this resonated with the metaphors—every part of the whole is the whole—and when the observer does not separate himself from what he is observing, it creates a state of healing, compassion, and love that may completely transform the state of consciousness and heal inner suffering.

I had another unique experience many years before the studies I shared in the 1980s with the same group. It happened when I was in the army while participating in a very difficult military exercise that caused me great anxiety because it included an element that would involve a long encounter with claustrophobic fears I had experienced as a young child. When the moment of truth arrived, I remember that I felt overwhelmed by terrible anxiety at what was going to happen, when suddenly, out of nowhere, all those feelings dissipated and were replaced by a sense of peace and happiness. It was a surprising state of consciousness the like of which I had never encountered before. This state lasted for a few days! I had done nothing to bring it on myself. Perhaps the only thing I did was not to resist the huge wave of anxiety that washed over me.

After that exercise, I did not share the feelings I experienced with my friends. The feelings appeared strange and inexplicable, but they and other similar situations led me to look for the possibility of living this way continuously and coherently. This experience was a kind of epiphany that appeared suddenly, without any preparation, and clearly demonstrated to me that it is undeniable that I could have an existence beyond my regular state of being, one in which there is joy, contentment, and acceptance of good and evil, difficulty and ease, an existence devoid of ordinary human suffering. The experience came unintentionally and very abruptly, and when it was over, I tried with all my emotional and intellectual might to bring it back. I was very disappointed and disheartened when I couldn't. Gradually, I realized that what was involved was a state toward which there is no clear path, that I couldn't simply follow someone else who was there and mimic their actions. Over the years, the question arose in my mind: What are the conditions that encourage it and can prepare the ground to return to that enlightened state?

I would like to share another observation I have had numerous times over the years in cases when I and a friend (it often happens to me with my wife) tried to analyze occurrences and conflicts we were involved in. I have become convinced over the years that any attempt to analyze and understand the conflict by means of simple causality—that the reason for the disagreement was this or that—in most cases creates a never-ending process that snowballs and only intensifies the dispute, and that the occurrence is in fact complete and unique. I realized that only by observing the occurrence in real time, without separating myself from the conflict or situation, I would be able to see myself as responsible for the occurrence (not always solely responsible). Only then does a new perspective emerge in some cases and a "solution," in which there is not someone who is right and someone who is wrong, emerges to completely transform the situation. This kind of situation reminds me of when we say, "You can't have one without the other," and it resonates for me with the previously mentioned metaphor—The whole precedes its parts and the parts cannot be wholly understood except through their affinity to the whole (of the whole situation that includes within it me, my wife, and the subject of the conflict).

In these situations of deep attentiveness or meditation, I have found that there is a different "order" for me and for the consciousness of other people. Most of the time I live within myself in a perception that behaves more or less according to a mechanistic worldview, one in which I am at the center, and the reality that I experience is based on separation and a divide between me and it. But there are moments when my consciousness "jumps" to another stream and structure of order that surprisingly shows a great resemblance to the quantum order and quantum metaphors of David Bohm that I presented in brief earlier on in this chapter. It is, in short, the great dependence on context, the great power of unwitting attentiveness, the emergence of an impersonal and non-local field of intelligence, the possibility of being in a state of nonseparation in which there is a superposition of states of consciousness and emotion.

Final Insights

This is the question I would like to raise at the end of this article: Is this similarity between David Bohm's metaphors regarding quantum theory and my states of consciousness as I have described here and which I have experienced, researched, and undergone over years, both alone and with fellow journeyers, accidental and random, or does human consciousness (and perhaps the human brain) also have a quantum order? As of today, the possibility that this kind of order might exist is ignored by most contemporary researchers of the neurology of the brain.

I think that the similarity I have presented here points to a quantum order of the human mind and consciousness and certainly also of a (limited) consciousness of animals and plants. I believe that if research and attention are directed in the directions of the kind I have briefly noted here, new phenomena and surprising results may be discovered that have ethical implications and the strength of correction and compassion.

Based on my experience, there are two basic reactions to this proposal. One will claim that consciousness may have certain quantum features but the brain is a causal and deterministic machine. The other will claim that after all, the brain is made up of building blocks that are neurons, and neurons behave in accordance with classical physics, and therefore, it makes sense to describe and study them according to classical physics and biology, which are controlled by a mechanistic Newtonian worldview.

In my view, these responses do not provide satisfactory answers. I maintain that the body and consciousness, or as they are known in the literature—mind-body—are two perspectives of a single, hidden structure, and if consciousness points to quantum effects, indications of this will also be found in the structure and activity of the brain. Penrose and Hammerhof (1996) tried to suggest a detailed possibility of this kind, which at the moment does not appear productive to most brain scientists. In my humble opinion, questions such as these can be examined seriously only if neurobiologists and philosophers that engage in brain research and the study of cognition ask new questions and adopt a more quantum and less mechanistic approach—shifting away from what currently (covertly) dominates our modes of thought and investigation of the world and consciousness.

Further Reflections

The question then arises: What do science, and most certainly the science of physics and the study of the brain, have to do with consciousness and the subjective experience of the individual (for the purpose of this discussion, my experience and that of my fellow journeyers in the study we conducted)? Isn't contemporary science itself a daring attempt aimed at establishing objective knowledge that is not dependent on the observer or on feeling, opinion, or intuition?

For me, physics is an area of activity and thinking that deals with the material world and as such, it is important to understand the truth of the order of the material universe, and to do so based on the following question: What does all this mean to me as a human being?

I do not consider myself one who views physics as a science solely of results and predictions. The meaning and interpretation of those results, the emerging worldview and order embodied in phenomena are no less and perhaps even more important than the ability to make predictions, conduct sophisticated mathematical manipulations, make a technological "leap forward," and add a few more digits after the decimal point.

As soon as I perceive science in this way, it immediately has implications for the fact that science is not merely a function of experimental results of one kind or another or of mathematical "proofs." Instead, it contains an openness to interpretation that a priori contains a dimension of freedom on the part of the subject, of giving a place of honor to feelings and intuitions and the power of the imagination that is not forced upon us by the experimental findings (this is exactly how Albert Einstein viewed scientific activity). Of course, a serious and in-depth interpretation must take findings and experimental results seriously, and thoroughly integrate them into the interpretations logically and coherently.

I look at the universe as an open and creative being that cannot be fully comprehended by means of theories and mathematics, and any description that science manages to mold and carve out is destined one day to be replaced by a broader, more inclusive, and more detailed description. Each new replacement of a theory is dramatic and represents a paradigm change. Science is not moving toward a complete, reductive, or asymptotic truth. Rather, every good theory is one that fits its field and provides a clear, simple, and coherent description, but also inspires other fields of science and life.

Comprehensive theories, like Newtonian mechanics, Darwin's theory of evolution, the theories of special and general relativity, and genetics apply far beyond the world of the phenomena that they describe, and deeply permeate and impact the perceptions of society and people of themselves. They go on to resonate in human areas that may be very remote from the original fields in which they were created and practiced, such as in the fields of sociology, psychology, economics, pedagogy, education, and more.

For me, physics is about realism, and the "difficult" question is what kind of realism (if any) is possible in a world of quantum phenomena. When I write realism, I refer to what there is in the world, whether or not we observe and measure it. In other words, like when the Newtonian order of classical physics tells us something of great significance about the world beyond us and that which Darwin's theory of evolution observed. This raises the question of whether the quantum world can be described realistically. It is clear based on the world of phenomena and results that if this is a world of realism, it is a realism of a new and different nature.

In my view, the "new" realism that the world of quantum phenomena is pointing to combines and blends the subjective, sensory, and internal with the objective and external. There is a great deal to explore in order to understand how this realism occurs in non-quantum areas too—for example in human society, biology (do trees and plant communicate and collaborate?), genetics (perhaps in epigenetics), and brain research.

Acknowledgments

Many thanks to Or Raphael for his great help, without which this article would not have seen the light of day.

Glossary

Order: A central concept in David Bohm's metaphysics. Bohm took an approach to the concept of order that differed from the accepted approach. In his view, everything and every system has an order with certain features. He did not wish to give an explicit and detailed definition of the concept, but gave examples from different and sundry areas—the order in plant growth, the order in music, the order in human society, the order embodied in Aristotle's philosophy, Newtonian order, quantum order, and so on. As Bohm himself stated in the context of order: in specific order we look for different similarities and similar differences.

Within what Bohm called order, he identified and described two modes. One was the explicate order and the second was the implicate order. The explicate order is what is currently overt and accessible to us—for example, a round table that we can see and touch. The implicate order is present and active, but enfold and hidden (somewhat like our subconscious). In the example of the table, the molecules and atoms that make up the table are its implicate order. An implicate order today may be unfold and become an explicate order tomorrow. In his view, the basic structure of explicate order and implicate order is preserved at all levels of the descriptions of nature. The implicate order of quantum theory may be quantum field theory or elementary particles.

Einstein-Podolsky-Rosen (EPR) paradox: In 1935, Einstein, Podolsky and Rosen wrote a paper in which they suggested a way to "overcome" Heisenberg's uncertainty principle, which states that we cannot know both pairs of properties of a particle with perfect accuracy at the same time (or in quantum language, to know with perfect certainty the values of commutative operators)—position and speed, energy and time, perpendicular directions of spins (Einstein, 1935). This paper challenged the school of thought led by Niels Bohr, known as the Copenhagen interpretation. Over the years, a incisive and bitter dispute arose as to who better described the world of quantum phenomena. In 1964, J. S. Bell made a significant breakthrough when he formulated Bell's inequality (Bell, 1987), which for the first time provided a theoretical possibility to test whether the proposal made in the 1935 article was true or the accepted predictions of quantum mechanics. In 1980, Alain Aspect demonstrated for the first time experimentally that EPR's claim was contradicted by the results of sophisticated experiments, which have since been replicated dozens and hundreds of times with no change in conclusion (Aspect, Grangier, & Roger, 1982).

Guiding wave: In Bohm's 1951 interpretation of quantum mechanics, he maintained that in the quantum world, there is both a wave and a particle. For example, the electron is a particle that has an actual mass and position at any given moment (even when we are not observing it), and it moves in a trajectory determined by a guiding wave. The guiding wave is a new wave in physics and it has unique properties. Bohm maintained that its existence can be deduced from the Schrödinger equation. One of the features of the guiding wave is that its intensity does not decrease with distance. A further feature is its holism and its ability to instantly entangle particles that are far apart.

Quantum entanglement: A phenomenon whereby pairs of quantum particles that may be a great distance apart from one another engage in a kind of "telepathy," so that any change in the observation of one of the particles immediately affects the observation values of the "twin" particle that may be more than a thousand kilometers away, so as to "contradict" Einstein's theory of special relativity. The first to formulate a mathematical inequality that would make it possible to investigate this entanglement was John Stewart Bell in 1964. In 1980, Alain Aspect provided the first confirmation that this phenomenon did indeed occur in the quantum world (Aspect, Grangier, & Roger, 1982). This phenomenon led some physicists to maintain that quantum theory is non-local.

Self-inquiry group in Jerusalem: From the early 1980s, a group numbering at its height about 150 people was active in Jerusalem. The group, called the Group of Self-Inquiry or Teaching of Oneself, was led by the late Yosef Safra, a unique, enigmatic, charismatic, and controversial figure. It engaged in an intensive investigation of the human consciousness of its participants from the perspective that the consciousness of those individuals is governed by a universal set of laws fundamentally different from the scientifically accepted approaches. The intention for the investigation was to lead to a radical transformation in the individual and in the individual's perception of the self and the world. Among the principal sources of influence guiding the group were George Gurdjieff, Jiddu Krishnamurti, and David Bohm. In 1989, a large part of the group together with Safra established a settlement in Israel's Arava region called Neot Smadar, which exists to this day. Neot Smadar aims to be a living, breathing school where people can explore themselves authentically and in cooperation with people like themselves. I was part of that group from the early 1980s and I was also a member of the founding group of Neot Smadar, and I lived and worked there for four years.

Wave function ψ: In 1926, Schrödinger proposed the famous Schrödinger equation that describes how wave functions evolve over time. It is a linear partial differential equation that links the change in time (first derivative of time) to the change in position of a quantum particle(s) (second derivative of place).

References

Aspect, A., Grangier, P., & Roger, G. (1982). Experimental realization of Einstein Podolsky Rosen Bohm Gedanken experiment: A new violation of Bell's inequalities. *Physical Review Letters*, *49*(2), 91–94.

Bell, J. S. (1987). *Speakable and unspeakable in quantum theory*. Cambridge University Press.

Bohm, D. (1952a). A suggested interpretation of the quantum theory in terms of "hidden" variables I. *Physical Review*, *85*, 166–179.

Bohm, D. (1952b). A suggested interpretation of the quantum theory in terms of "hidden" variables II. *Physical Review*, *85*, 180–193.

Bohm, D., & Hiley, B. (1993). *The undivided universe*. Routledge.

Einstein, A. (1935). Can quantum-mechanical description of physical reality be considered complete? *Physical Review*, *47*, 777–780.

Pylkkanen, P. (2010). The philosophical relevance of Bohm's interpretation of quantum theory. *A lecture he gave in Vallico di Sotto, Italy*

Pylkkanen, P., Hiley, B. J., & Pattiniemi, I. (2014). Bohm's approach and individuality. https://arxiv.org/abs/1405.4772

Pylkkanen, P. (2016). Can Bohmian quantum information help us to understand consciousness? In *Quantum interaction: 9th International Conference, QI 2015, Filzbach, Switzerland, July 15-17, 2015, Revised Selected Papers 9* (pp. 76–87). Springer International Publishing.

12 The Benevolent Potential of Spiritual Experiences That Encompass a Variety of States of Consciousness

Ofra Mayseless

Personal Reflections

About twenty years ago, my son Ouri, who was twenty-two at the time, was killed in a car accident in the beautiful, faraway landscapes of Alaska. He was there on a trip with a childhood friend shortly before graduating from the Technion, the Israel Institute of Technology. A driver making a reckless overtake on a winding road collided with them head-on at 100 miles per hour. Both Ouri and his friend were killed, as was the driver.

And me—my whole world came crashing down around me, like a sword ripping through me, from my stomach and uterus to my chest.

For many years I walked around with the feeling that there is something beyond our existence as we know it, some sort of energy, something infinite and eternal . . . But I didn't give it much thought and frankly wasn't really interested in it. But following Ouri's death, I had to know—is there something, or isn't there? And if there is, what is the nature of this energy, and did any part of my sweet, smiling, smart, and beloved Ouri remain? Because if not, I have no life.

At that time, I was wrapped up in extensive academic work—research, teaching, and a major executive position. I left it all behind, troubled and anguished, and told my friends at the university that I would not return to work. I looked for answers—in traditions of wisdom and mysticism, in research on spirituality and parapsychology, in dozens of conversations with other bereaved parents, in dreams, in messages I've received from near and far, and in learning to connect with the spiritual realm. Along the way, I have met many dear friends and soul mates, here and in other dimensions, who have become my companions on this journey, and all of whom have my deepest gratitude. During this process, I have received answers to many of my questions, experienced the strength, joy, and beauty to be found in the spirit, met my son and other spiritual entities, and realized what was important and right for me to do. During this process, I understood that part of my calling in this world involves my taking part in both of these worlds—academia and spirituality, and the integration that I am welcome to promote as a professor and scholar between academia and research, which was my professional home until that time, and the spiritual realms, which have also become my home.

This process led me to a path that wasn't always easy, with struggles and stumbles but also with love and friendships, toward the establishment of the Center for the Study of Human Spirit in 2017. The vision of the Center is to expand and significantly increase our knowledge and understanding of the spiritual realm in human existence. Furthermore, we wish for the spiritual realm to become legitimate and accessible for all individuals and societies, to be fully integrated into academia and every other aspect of life, and for its increased contribution to the fulfillment of the potential within every human being to thrive and grow.

In my personal life and professional life, I currently reside in both realms (academic and spiritual); they are integrated, and each reflects a distinct parallel dimension.

In the Center, there were many integrative activities. For example, together with colleagues, we organized the first Israeli Conference for the Study of Contemporary Spiritualities in 2009, and we have already had the ninth conference. This has helped us create a space where scholars, academic researchers, and field practitioners can meet and exchange ideas, get to know each other, and form alliances.

I have also been engaged, together with many others, in applying in a variety of arenas the insights we have in academia regarding spirituality. For example, we conducted a large-scale study (Teaching with the Heart and Mind) as well as applications of contemplative pedagogy in the education field. In addition, I chaired the first national conference on contemplative education held in 2018, which reflected the great work already taking place in academic circles and the education field, and have also designed a free online course (a MOOC course) promoting a caring, benevolent environment in educational settings. I have also been involved in and working toward the integration of spirituality into psychotherapy through the Gateways to Spirituality in Psychotherapy and Counseling Conference and the establishment of the first Israeli MA specialization in integrative psychotherapy: body-mind-spirit. Other projects revolve around spirituality in businesses and organizations, spirituality as a means for intergroup reconciliation, nature spirituality, and more. These activities, conducted in collaboration with colleagues from the Center and outside it, are just the beginning of our vision to connect and build a bridge—light, wide, and boundless—between academia and the spiritual realms, and between researching the human spirit and working in the community, with the Center acting as a beacon lighting the way to outline the possibilities and welcome a connection between different incredible people, each with their own passions, visions, and wishes to do good in the world.

Overview

Current strands of hardcore science studying consciousness tend to shy away from spiritual experiences and spiritual questions, perhaps because they view it as contrary to scientific ways of knowing. However, the capacity to connect to the spiritual realm and experience spirituality in its diverse forms and states of consciousness relates to the essence of life. It is the basis of the meaning of life for me. Such connection brings joy and meaning and many other benevolent outcomes, including insights, creativity, and love (Park, 2013). Based on my phenomenological conviction, while being a critical scientist, I wondered whether we could study this elusive and "mystical" realm using scientific methods. Can the scientific study of spirituality-based altered states of consciousness contribute to our advancement in deciphering some of the mysteries of life? The answer to this is clear.

Research does not provide the qualia (the essence of the phenomena) as experience does. However, it can shed light on many spiritual states of consciousness issues, including peak experiences, self-transcendence, awe, near-death experiences, and more. We currently have evidence from a variety of scientific approaches that a diverse and distinct group of spiritual experiences, which involve distinct states of consciousness, is associated with benevolent outcomes (e.g., Alexander et al., 1991; Asakawa, 2010; Roseman et al., 2018; Van Lommel et al., 2001; Woods & Barušs, 2004). I believe that the future of consciousness research and a possible breakthrough in this realm relates to our attempts to understand, from a scientific perspective, what is unique in such states of consciousness and how, and in what way, they have such benevolent outcomes.

In this chapter, I discuss spirituality and several known types of spiritual experience and briefly review the literature demonstrating their association with benevolent outcomes. I then outline several core questions that future research needs to tackle. I further add a brief experiential part, suggesting a way for the readers to recall and connect to spiritual experiences that they may have had and their significance. I end with a box delineating my own experience, which brought me to study spirituality.

What Do I Mean by Spirituality?

Spirituality refers to mysterious and elusive realms, and researchers, religious teachers, and scholars have provided many definitions throughout the years and the variety of wisdom traditions (see summary in Pargament et al. [2013]). Nevertheless, several common themes can be identified. First, these include the realization that "there exists a broader paradigm for understanding existence that transcends the immediacy of our consciousness and that binds all things into a more unitive harmony" (Piedmont, 1999, p. 988). This realm is often termed the sacred, the divine, a higher power, God, heaven. It may be transcendent (outside the cosmos or humans) and/or intrinsic (core part of who we are). The second common theme is that we as humans can connect with this realm, perceive, experience, and contemplate it. Such connection provides us with our ultimate purpose and meaning in life. Some researchers also add that spirituality involves transcending oneself and developing a commitment to contributing to others beyond the self (Lerner et al., 2006). Spiritual experiences are conceived as part of spirituality. Spirituality is expressed or manifested in our beliefs, attitudes and worldviews, experiences, feelings and emotions, and behavior in the world. Each of these aspects can be assessed in research.

Spiritual Experiences

In his famous book *The Varieties of Religious Experience*, William James, the initiator of the research on religious and spiritual experiences as part of modern psychology, provided several descriptions of his own experiences. Here is one that describes some form of self-transcendence: "I felt myself one with the grass, the trees, birds, insects, everything in nature. I exalted in the mere fact of existence, of being part of it all. . . . I knew so well the satisfaction of losing self in a perception of supreme power and love (James 1985/1902, pp. 364–365). He further suggested that "we can experience union with something larger than ourselves and in that union find our greatest peace" (James, p. 525).

Abraham Maslow, most well-known for his theory regarding the hierarchy of needs, a former president of the American Psychological Association, and one of the initiators of humanistic psychology, suggested and discussed two other constructs related to spiritual experiences. These included peak experiences, described as rare and exciting and as mystical and magical, including emotions such as wonder, awe, reverence, elevation, humility, and surrender (Maslow 1962a). Maslow described them in different ways: "Think of the most wonderful experience of your life: the happiest moments, ecstatic moments, moments of rapture, perhaps from being in love, or from listening to music or suddenly 'being hit' by a book or painting, or from some creative moment"(Maslow 1962b, p. 67). Maslow further suggested a related but distinct spiritual state, the plateau experience, and described it as "serene and calm rather than a poignantly emotional, climactic, autonomic response to the miraculous, the awesome, the sacralized, the unitive, the B-values. . . . It is far more voluntary than peak experiences are. One can learn to see in this unitive way almost at will. It then becomes a witnessing and appreciating, what one might call a serene, cognitive blissfulness" (Maslow 1970, preface).

Examples of psychological constructs that contain a self-transcendent aspect might be mindfulness, flow (Bakker, 2005), peak experiences, self-transcendence, and specific positive emotions (e.g., love, awe, bliss, elevation). In all of them, it appears that there are some levels of decreased self-salience and increased feelings of connectedness. However, each of these two aspects can occur to varying degrees. The variety of spiritual experiences has other distinct qualities such as vitality, mystical or magical fascination, and equanimity.

There are other spiritual experiences, perhaps less conventional in academic circles than the former, such as channeling, receiving information from "other" realms, near-death experiences, regressions to past life, mystical experiences following the use of mind-altering substances (e.g., psychedelic drugs). Some of these experiences involve other aspects besides self-transcendence or interconnectedness, such as meeting spiritual entities or energies.

There are a variety of contexts/gateways to these varied spiritual experiences, such as praying, fasting, music, dancing, being in nature, reading scriptures, meditation, hypnosis, channeling, dreaming, using mind-altering substances, and so on.

Spiritual Experiences and Benevolent Outcomes

We currently have evidence from various studies that a diverse group of spiritual experiences is associated with benevolent outcomes (e.g., Alexander et al., 1991), some even transformational (Naor & Mayseless, 2020). These outcomes include becoming more personally integrated and in control, showing more minor post-traumatic stress disorder symptoms, developing a higher sense of peace and calm, happiness, and well-being. Outcomes also included having higher levels of meaning of life and a higher tendency to be compassionate and prosocial as well as better physical health (e.g., Bianco et al., 2019; Groarke &

Hogan, 2020; Pargament et al., 1990; Privette 1983; Roseman et al., 2018; Van Lommel et al., 2001).

Such outcomes were found in research on near-death experience phenomena (e.g., Van Lommel et al., 2001); following mindfulness training (Eberth & Sedlmeier, 2012); after experiencing peak experience in nature (Naor & Mayseless, 2020); in psychotherapy that integrates regression to previous lives (Woods & Barušs, 2004); and following the use of mind-altering substances, such as psilocybin or MDMA in psychotherapy (e.g., Palhano-Fontes et al., 2019), showing that the quality of the spiritual experience was the mediating factor between taking the drug and its benevolent effects (Majić et al., 2015; Roseman et al., 2018).

Research Agenda

The phenomenological descriptions of the experiences are varied. However, several recurrent themes emerge and appear to point to numerous kinds of spiritual experiences. These experiences reflect or involve a variety of distinct consciousness states, such as self-transcendence, self-dissolution, encounters with other realms of reality, or bliss. The phenomenological literature has already developed in describing some of these different states (Boniface, 2007; Millière et al., 2018; Privette, 1983; Van Lommel et al., 2001). However, research has only started to examine what is so unique in such states of consciousness and how and in what way they have benevolent outcomes (e.g., Barrett & Griffiths, 2017).

From a neurophenomenological perspective, this involves specific questions: What are the conditions that arouse peak experiences or other spiritual experiences? What are the prerequisites of consciousness to enable such states? What characterizes such states in terms of brain processes? The major underlying questions are as follows: Why and how do such states have benevolent and often far-reaching positive transformations? Regardless of the content, what is different in these states (in the psyche and/or in physiology) compared to "regular" consciousness?

These are significant questions for researchers who study consciousness, and especially for those that combine phenomenology and brain research (Berkovich-Ohana et al., 2013). In light of the growing evidence for the benevolent and unique effects of spiritual experiences, new and creative scientific methods need to be developed to decipher these questions.

Experiential Part

I know that we have many channels to learn about things and arrive at something truthful for us. These include discussion with others, reading, doing research, personal reflection, innate insights, and of course, experience. For most people, personal experience is central in understanding a large variety of constructs, such as love, floating while swimming, the taste of a certain food, the beauty of a piece of art. Experience is crucial when we discuss a variety of states of mind and, in particular spiritual experiences. For example, you can talk about romantic love but it is not until you fall in love that you understand what poets and researchers talked about. Similarly, understanding spiritual experiences and their benevolent outcomes is flawed unless one has also had similar personal experiences.

Indeed, individuals, and most often scientists, who were socialized in a strict positivistic tradition are not sure what I (and others) mean by spiritual experiences. However, in encounters with a large number of people, individually and in groups, I have learned that many of them have had spiritual experiences (as described by researchers of the psychology of religious and spiritual), but they may put them aside because they do not fit their general world view, or they do not term them this way. For example, they may have experienced awe in nature in view of its beauty and grandeur, elevation during or after the birth of their son or daughter, general gratitude not directed to anyone, internal blissful peace and the expansion of their heart, coincidences that they could not have explained, strong intuition that they did not know where it came from but that they felt is deeply true, a dream or a sudden knowledge that foresaw the future (e.g., "I need to contact my mom, NOW"), or directed you to something that needed your attention that you had no previous knowledge about, and the desire to pray for something (e.g., the health of a dear one) and experiencing that the prayer was answered. To help others connect to such experiences or different ones, I suggest the following experience. Here are the guiding words I use:

In a minute, I will ask you to close your eyes and meditate, and I will then provide short guided imagery to recall a spiritual experience that you have had. The spiritual experiences do not necessarily be overwhelming or include God or angels. It could include a sense of awe and wonder in nature, a coincidence that puzzled you, a sense of elevation, or any other experience that comes to your mind. There is no specific outcome that you need to achieve. Whatever comes or does not come is OK and perfect. Try to adopt an open and playful state of mind because this is first and foremost for you.

Please sit comfortably and close your eyes or look away not at another person and try to follow my suggestions but also let yourselves go with what will come. Please take a deep breath and open up or enlarge the upper area of your lungs and open your heart. Take two other deep breaths and then breathe normally. We are going to do a body scan. Pay attention to your body and its position (on the floor or a chair) and go over and pay attention to your bodily experiences. Pay attention to your legs and what they touch, to your thighs, your lower back, upper back, shoulders, your hands, your chest, your breathing in and out, your heart. Pay attention to your throat and your face, your mouth, your ears, your eyes, and your third eye in your forehead between your brows. Pay attention to the experiences in your body and pay attention to your breath in and out as it is naturally happening. [This can take five to seven minutes. I then give the following brief guided meditation.]

Try to recall a spiritual experience, the first one coming to your mind. See what comes. Try to relive it and to be really in that moment when you had experienced it. Where was it? Who else, if at all, was with you? What did you see? Or what happened. Pay attention to your body at that time, your feelings, your emotions, and any other things that happened at the time. Go with what comes and allow yourself to surrender to it. [I allow ten minutes for this and stay quiet. At nine minutes I give a gentle reminder: "We are going to open our eyes in a minute." One minute later I call everyone back.]

Please come back gently. Please, take your time and write up your experience. [I give them five to seven minutes for this.]

Please now answer by writing what is the most significant insight that you gained from this experience.

[If it's done in a group, I invite the members of the group to share.] Please find someone you feel comfortable with and share with each other what happened. [I give twenty-five minutes for this.]

Let's convene again in a circle. Does anyone like to share their own experiences during the meditation and afterward when you had a dyadic sharing?

[To end] Thank you for participating in this session and opening your hearts. I hope that you had a benevolent and meaningful experience. I also want to thank all the entities and energies that were here together with us during this session and helped us, and to my colleagues and friends in this world and in parallel ones, THANK YOU!

References

Alexander, C. N., Rainforth, M. V., & Gelderloos, P. (1991). Transcendental meditation, self-actualization, and psychological health: A conceptual overview and statistical meta-analysis. *Journal of Social Behavior and Personality*, *6*(5), 189.

Asakawa, K. (2010). Flow experience, culture, and well-being: How do autotelic Japanese college students feel, behave, and think in their daily lives? *Journal of Happiness Studies*, *11*(2), 205–223.

Bakker, A. B. (2005). Flow among music teachers and their students: The crossover of peak experiences. *Journal of Vocational Behavior*, *66*(1), 26–44.

Barrett, F. S., & Griffiths, R. R. (2017). Classic hallucinogens and mystical experiences: Phenomenology and neural correlates. In *Behavioral neurobiology of psychedelic drugs* (pp. 393–430). Springer.

Berkovich-Ohana, A., Dor-Ziderman, Y., Glicksohn, J., & Goldstein, A. (2013). Alterations in the sense of time, space, and body in the mindfulness-trained brain: A neurophenomenologically-guided MEG study. *Frontiers in Psychology*, *4*, 912.

Bianco, S., Testoni, I., Palmieri, A., Solomon, S., & Hart, J. (2019). The psychological correlates of decreased death anxiety after a near-death experience: The role of self-esteem, mindfulness, and death representations. *Journal of Humanistic Psychology*. https://doi.org/10.1177/0022167819892107

Boniface, M. R. (2007). Towards an understanding of flow and other positive experience phenomena within outdoor and adventurous activities. *Journal of Adventure Education and Outdoor Learning*, *1*, 55–68.

Eberth, J., & Sedlmeier, P. (2012). The effects of mindfulness meditation: A meta-analysis. *Mindfulness*, *3*(3), 174–189.

Groarke, J. M., & Hogan, M. J. (2020). The eudaimonic functions of music listening scale: An instrument to measure transcendence, flow and peak experience in music. *Frontiers in Psychology*, *11*, 2344.

James, W. (1902/1985). *The varieties of religious experience*. Harvard University Press.

Lerner, R. M., Alberts, A. E., Anderson, P. M., & Dowling, E. M. (2006). On making humans human: Spirituality and the promotion of positive youth development. In E. C. Roehlkepartain, P. E. King, L. M. Wagener, & P. L. Benson (Eds.), *The handbook of spiritual development in childhood and adolescence* (pp. 60–72). Sage.

Majić, T., Schmidt, T. T., & Gallinat, J. (2015). Peak experiences and the afterglow phenomenon: When and how do therapeutic effects of hallucinogens depend on psychedelic experiences? *Journal of Psychopharmacology*, *29*(3), 241–253.

Maslow, A. H. (1962a). Lessons from the peak experiences. *Journal of Humanistic Psychology*, *1*, 9–18.

Maslow, A. H. (1962b). *Toward a psychology of being*. Van Nostrand-Reinhold.

Maslow, A. H. (1970). *Religion, values, and peak experiences*. Viking.

Millière, R., Carhart-Harris, R. L., Roseman, L., Trautwein, F. M., & Berkovich-Ohana, A. (2018). Psychedelics, meditation, and self-consciousness. *Frontiers in Psychology*, *9*, 1475.

Naor, L., & Mayseless, O. (2020). How personal transformation occurs following a single peak experience in Nature: A phenomenological account. *Journal of Humanistic Psychology*, *60*(6), 865–888.

Palhano-Fontes, F., Barreto, D., Onias, H., Andrade, K. C., Novaes, M. M., Pessoa, J. A., Mota-Rolim, S. A., Osório, F. L., Sanches, R., dos Santos, R. G., Tófoli, L. F., Silveira, G. de O., Yonamine, M., Riba, J., Santos, F. R., Silva-Junior, A. A., Alchieri, J. C., Galvão-Coelho, N. L., Lobão-Soares, B., . . . Araújo, D. B. (2019). Rapid antidepressant effects of the psychedelic ayahuasca in treatment-resistant depression: a randomized placebo-controlled trial. *Psychological Medicine*, *49*(4), 655–663.

Pargament, K. I., Ensing, D. S., Falgout, K., Olsen, H., Reilly, B., Van Haitsma, K., & Warren, R. (1990). God help me:(I): Religious coping efforts as predictors of the outcomes to significant negative life events. *American Journal of Community Psychology*, *18*(6), 793–824.

Pargament, K. I., Mahoney, A., Exline, J. J., Jones, J. W., & Shafranske, E. P. (2013). Envisioning an integrative paradigm for the psychology of religion and spirituality. In K. I. Pargament, J. J. Exline, & J. W. Jones (Eds.), *APA handbook of psychology, religion, and spirituality* (Vol. 1: Context, theory, and research; pp. 1–19). American Psychological Association.

Park, C. L. (2013). The meaning making model: A framework for understanding meaning, spirituality, and stress-related growth in health psychology. *European Health Psychologist*, *15*(2), 40–47.

Piedmont, R. L. (1999). Does spirituality represent the sixth factor of personality? Spiritual transcendence and the five-factor model. *Journal of Personality*, *67*, 985–1013.

Privette, G. (1983). Peak experience, peak performance and flow: A comparative analysis of positive human experiences. *Journal of Personality and Social Psychology*, *45*, 1361–1368.

Roseman, L., Nutt, D. J., & Carhart-Harris, R. L. (2018). Quality of acute psychedelic experience predicts therapeutic efficacy of psilocybin for treatment-resistant depression. *Frontiers in Pharmacology*, *8*, 974.

Van Lommel, P., Van Wees, R., Meyers, V., & Elfferich, I. (2001). Near-death experience in survivors of cardiac arrest: A prospective study in the Netherlands. *The Lancet*, *358*(9298), 2039–2045.

Woods, K., & Barušs, I. (2004). Experimental test of possible psychological benefits of past-life regression. *Journal of Scientific Exploration*, *18*(4), 597.

13 Meditation: From Self-Regulation to Ego-Dissolution

Ulrich Ott

The spirit becomes free only when it ceases to be a support.
—Franz Kafka, 1917–1918

Personal Reflections

In my study of meditation, I rely on the approach of psychophysiology, which assigns equal value to subjective experiences, physiological measurements, and behavioral observations. Only by combining these different perspectives and levels of description, can one yield a full picture and study their interaction—for example, when searching for neural correlates of certain experiences. However, meditation is not the only object of my research; at the same time, meditative introspection is also one of my primary research tools. Unexpected, profound experiences during meditation aroused my interest and motivated my study in psychology and my scientific research on altered states of consciousness (see ***glossary****).*

During the workshops in Jerusalem, I found that contemplative pauses helped to widen the intellectual horizon and to facilitate openness to the fascinating new ideas presented by the speakers, which were sometimes not only conveyed verbally but also by the way they presented their insights, guided through meditations, or embodied in an awakened presence while keeping silent.

Overview

Scientific research has conceptualized meditation in different ways ranging from a relaxation method, mental training, and clinical intervention to spiritual practice used to induce mystical states of consciousness. These different concepts are not mutually exclusive or contradictory but rather complementary and reflect the variety of psychological domains affected by meditation: reduction of autonomic arousal, enhanced well-being, focusing of attention, emotion regulation, and, finally, fundamental changes in the perception of self and world. In the past, most research was concerned with studying improvements of self-regulatory abilities through meditation. In contrast, only a few studies investigated deeper meditation states, characterized by experiences of non-duality, overwhelming positive feelings, and transcendence of space and time. These latter states of consciousness have been the focus of my scientific and personal interest for the last three decades. As a yoga practitioner and teacher, I wanted to understand the effects of meditation on mind and brain scientifically. Therefore, I was delighted to observe the developmental trajectory of meditation research, which started with concepts like meditation as a "relaxation response" and now ventures to elucidate the brain dynamics of non-ordinary experiences like "non-duality" and "ego-dissolution" by combining phenomenological assessment with modern brain imaging technology. In this chapter, I describe (1) typical features of mystical experiences (see **glossary**); (2) situations, personality traits, and methods that facilitate their occurrence; and (3) recent findings regarding their possible neural correlates. Neuroimaging and electroencephalographic (EEG) studies with advanced meditators and experimental research using psychedelic substances provide new insights into patterns of brain activity that accompany those experiences, in which the ego seems to dissolve. A final section of the article addresses benefits and possible risks of mystical experiences for mental health based on recent studies with advanced meditation practitioners.

Introduction

Meditation has attracted increasing scientific interest in the last five decades. Research has shown effects of meditation on a variety of bodily and mental processes, ranging from changes in autonomic arousal—relaxation as well as activation—to attentional and emotional self-regulation, and changes in fundamental dimensions of ordinary consciousness, like time, space, and separation of the self from the external world. This variety of effects goes hand in hand with a diversity of applications of meditation techniques. A large part of research has focused on self-regulatory abilities enhanced by meditation and conceptualized it accordingly as a relaxation method, a form of mental training, or a clinical intervention. Meta-analyses have confirmed the effectiveness of meditation for healthy practitioners (Sedlmeier et al., 2018) as well as for patients (Goldberg et al., 2018), and neuroimaging studies found corresponding changes in the activity and structure of the brain (Fox et al., 2014, 2016).

The accumulated evidence on improved self-regulation by meditation is impressive and valuable. However, from my personal perspective, relaxation and enhanced skills of attention and emotion regulation represent important prerequisites for the ultimate goal of meditation training: to gain deep insights into the fundamental nature of reality. Within spiritual traditions, meditation practice primarily aims at inducing altered states of consciousness, designated with a variety of terms like mystical experience, enlightenment, illumination, awakening, samadhi, satori, nirvana, and liberation. Studies on the phenomenology of these non-ordinary states of consciousness led to the development of questionnaires that allow assessment of their characteristic features. In the following, I focus on the phenomenology of mystical experiences, on factors that influence their occurrence, and on possible neural correlates.

Phenomenology of Mystical Experiences

A keen interest in mystical states of consciousness existed already in the early formative years of psychology when William James (1902) gave his famous lecture on the "The Varieties of Religious Experience." Later on, philosophers like Walter Terrence Stace (1960) and Paul Marshall (2005) provided detailed phenomenological analyses of reports, taken from different epochs and cultural areas, with the aim to identify a set of common features. Today, questionnaires allow the assessment of phenomenal core features that characterize mystical experiences. For instance, the items of the Mystical Experience Questionnaire (MacLean et al., 2012) form the following four major factors:

1. Unity with oneself/the world, insightful knowledge of an ultimate reality, sacredness;
2. Positive mood in the form of amazement, tenderness, tranquility, peace, ecstasy, joy, awe;
3. Transcendence of time and space; and
4. Ineffability, or difficulty describing the experience in words.

The experience of unity, also emphasized in the Latin term *unio mystica*, signifies that the boundaries between self (see **glossary**) and world seem to dissolve. In stark contrast to the ordinary wakeful state, the separation into an experiencing subject (ego) and external objects, constituting the surrounding world, breaks down. This change seems to happen rather at a basic perceptual than an intellectual or conceptual level and points to the constructive nature of the models of self and world we create in our minds and brains (for a detailed analysis of the self-model, see Metzinger [2009]; a hypothesis on how this could be related to global perceptual binding through gamma EEG oscillations is presented in Ott [2013]). The experience of time and space, similar fundamental dimensions of our everyday consciousness, can also change profoundly: sense of time and space are lost, and one feels to exist in a realm beyond the confines of time and space (timelessness, limitless). While mystical experiences often entail strong and sometimes overwhelming positive feelings, ego-dissolution can also provoke strong fear and even panic of completely losing control (see section below on risks for mental health).

Meditation practice is one psychological approach to the attainment of mystical states of consciousness and related insights into the nature of mind and reality. However, other methods, conditions, and personality traits can facilitate the occurrence of mystical experiences too. The following section introduces some of these induction methods and influencing factors, because they are important for experimental research and contribute to the effectiveness of meditation practice.

Triggers and Influencing Factors

A variety of naturally occurring situations can trigger mystical experiences. Some people have profound experiences of oneness during encounters with nature, like while hiking in beautiful scenery, or during the reception of art—for example, when listening to music, contemplating poetry, or other artworks (for exemplary reports, see Marshall [2005]). Rhythmical body movements during running, dancing, drumming, and sexual activity have also the potential to induce trance-like flow states and ecstatic feelings of unity with ones actions or the partner. Furthermore, life-threatening events, like cardiac arrest during surgery, traffic accidents, or drowning, sometimes trigger near-death experiences, which share several features with mystical experiences (Vaitl et al., 2005). However, the strongest and most reliable trigger for mystical experiences seem to be psychedelic substances, like LSD, psilocybin, mescaline, and ayahuasca, which have been used by natural religions, individuals for self-exploration, as well as in clinical and basic research (see next section).

Spiritual traditions all over the world use certain conditions conducive for mystical states in monasteries, ashrams, and retreat centers. Located in quiet surroundings, a regular daily routine, with light meals or even fasting, reduces distractions and thereby supports mental absorption (see **glossary**). Meditation, when practiced in isolation, motionless, in silence with closed eyes, represents a mild form of sensory deprivation known to facilitate enhanced wakefulness and the occurrence of unusual visual phenomena up to vivid visions (Britton et al., 2014; Lindahl et al., 2014). Focusing of attention on meditation objects—for example, a mantra, mandala, or "energy center" in the body (Sanskrit *chakra*, i.e., the heart or "third eye" located between the eyebrows)—for long durations can lead to states of deep mental absorption. Other meditation methods use intensive inquiry—for example, koan practice in Zen ("What is the sound of one hand clapping?") or self-reflection ("Who am I?")—or widen awareness to the whole field of sensations, feelings, and thoughts, observing them appearing and vanishing, without clinging to them.

Besides the influencing factors related to the technique practiced and the situational context ("setting"), personality, motivation, and expectations of the practitioner ("set") play a crucial role. In my own research on altered states of consciousness, I found the personality trait of absorption to be of pivotal importance. Persons scoring high on the Tellegen absorption scale are more open to new experiences and are able to become completely absorbed in an attended object (for a detailed description of all aspects of absorption and its biological foundations, see Ott [2007, 2016]). Several studies with psychedelics showed that the trait of absorption allows prediction of the intensity of induced mystical experiences in participants (Russ et al., 2019; Studerus et al., 2012). In our own study (Hölzel & Ott, 2006), we found that the absorption trait was significantly correlated with a measure of meditation depth (non-duality, representing the deepest meditation state). This implies that people with a disposition to experience states of mental absorption easily and frequently in everyday life situations—that is, scoring high on the absorption scale—are likely to have deeper meditation experiences earlier during their practice.

Can science explain the occurrence of these experiences? My answer is "yes, we can." The main reason for my affirmative response is the progress in our understanding of the functioning of the brain within the last decades.

Possible Neural Correlates

Functional neuroimaging has demonstrated the close connection between mental and neural processes. Thus, the fundamental subjective changes during mystical experiences are very likely to go along with similar marked changes in brain activity patterns. Before I describe the plausible candidates for neural correlates that research identified so far, I want to provide some background for readers not familiar with neuroscientific concepts that can serve as a basis for understanding the dynamic and shapeable nature of our mental life.

During my scientific research and teaching at Giessen University (since 1998), I observed the steadily increasing impact of neuroscience on psychology. The Bender Institute of Neuroimaging (BION) was one of the first brain imaging centers in Europe dedicated exclusively to psychological research. Findings from neuroimaging studies were omnipresent in the lectures of almost all psychological disciplines, including clinical psychology. The current director of the BION (Professor Dr. Rudolf Stark) holds the chair for psychotherapy and systems neuroscience, and many of my colleagues doing research as doctoral students and postdocs at the BION are at the same time working as behavioral psychotherapists at the health care center of the university. Thereby, I have witnessed at first-hand the explanatory power of neuroscientific concepts for understanding, explaining, and modifying human experience and behavior.

Today's knowledge of how the self develops and stabilizes itself by feeling the body and acting in the world is one piece of the puzzle needed to understand the changes that can be provoked by meditation. Buzzwords like "embodiment" or "embodied cognition" point to the acknowledgment of the significance of the body in the cognitive sciences. Affective and cognitive neurosciences learned that many mental processes and, most notably, the fundamental feeling of being a self are deeply rooted in the body. Different facets of the self—core self, autobiographical self, social self, self-agency (see **glossary**)—relate to different "bodies" (neuronal body representations) that develop during childhood. This development requires the inner and outer perception of our body including social mirroring. Today, a clear picture of the complex self-model has emerged that integrates a matrix of bodily representations and allows us to successfully navigate through time and space (for an excellent review, see Riva [2018]).

In addition, the theory of perceptual feature binding by EEG gamma oscillations (see **glossary**), proposed by Professor Wolf Singer, director of the Max Planck Institute for Brain Research, provided another important puzzle piece. Visually perceived objects evoke gamma oscillations in different brain regions coding for features like form, color, and movement. Nevertheless, we experience this feature combination as one single object. Separate objects, like moving dots, perceived subjectively as belonging together according to the principles of Gestalt psychology (e.g., common movement trajectory, spatial proximity, similarity), show gamma oscillations more coherent than those evoked by objects seen as clearly separate (e.g., dots moving in different directions; for review, see Singer et al. [1999]). In my doctoral thesis, I proposed that the perceived unity of self and surroundings could result from a transition to a state of global gamma coherence (Ott, 2013). Indeed, EEG studies on ayahuasca reported a marked increase in synchronized gamma activity during the induced mystical experiences (Schenberg et al., 2015; Stuckey et al., 2005), and also studies with advanced meditators found exceptional patterns of widespread gamma EEG activity (e.g., Lutz et al., 2004; Thomas et al., 2014).

These different strands of research clearly demonstrate (1) that our perception of the world and of our self strongly depends on the functioning of brain networks that generate the corresponding neural representations of external objects and of our body; and (2) that, at the same time, brain dynamics are enormously flexible and shapeable by many means. The spiritual traditions have explored and tested over centuries psychological (and sometimes pharmacological) methods to induce certain altered states of consciousness, and thus have accumulated a lot of knowledge in their teachings and practices. Modern science can now examine these traditional methods against the background of what we know today about mind and brain and provide new explanatory models of why and how these methods work.

In my understanding, meditation aims at inducing states of consciousness characterized by enhanced wakefulness ("awakening," see Britton et al. [2014]), unity of oneself with the world ("all is one"), and transcendence of time and space. Based on our knowledge regarding the brain networks that are involved in wakefulness and the generation of stable representations of self, world, time, and space, we can now have a new look at what meditators actually do and how this affects the brain. What happens if one sits motionless in silence—that is, with minimized external stimulation? This condition of sensory deprivation leads to a lowering of thresholds for neural firing that can result in a variety of unusual visual phenomena (see Lindahl et al., 2014). The neural representations of the body fade when not refreshed by somatosensory input or motor action ("sensory deprivation," see Britton et al. [2014]). Thus, the sensed body as an important constituent of our ordinary self-representation drops out. Under these unusual boundary conditions, it is easily conceivable that transitions to other states of consciousness can occur.

While it is rather difficult to investigate these states in meditators during disturbing laboratory conditions, recently, functional magnetic resonance imaging (fMRI) studies with psychedelics allowed observing changes in brain activity and connectivity during induced mystical experiences.

In a seminal review article, Barrett and Griffiths (2018) summarized a variety of findings suggesting the following:

- "Decreased activity and functional connectivity in the PCC [posterior cingulate cortex] and MPFC [medial prefrontal cortex] (medial nodes of the DMN [default-mode network]) mediate aspects of the experience of introvertive unity (including decreased self-referential processing and a loss of sense of self)."
- "Decreased activity and functional connectivity in the IPL [inferior parietal lobule] (a lateral node of the DMN) mediates the experience of timelessness and spacelessness . . . be that occasioned by meditation or by classic hallucinogens."

The activity of the default mode network seems to play a key role for our ordinary self-perception in the everyday waking state. During mystical states of unity and transcendence, the transient suspension of activity and connectivity patterns that underlie the subject first-person perspective (ego), with which we identify, could extend our understanding of the constructive nature of the self. However, as already mentioned above, it sometimes can also provoke strong fears of losing control and becoming crazy. What are possible benefits and risks when trying to attain mystical states of consciousness?

Benefits and Risks for Mental Health

When practiced as relaxation method, mental training, or clinical intervention, meditation seems to be a beneficial tool for improving well-being and enhancing self-regulation and mental health. The spiritual traditions, however, are well aware of possible dangers and unwanted adverse effects of contemplative practice and emphasize the need for proper preparation and skillful supervision by qualified teachers. During intensive meditation, a range of difficult and challenging experiences might occur. In an ongoing research project on "The Variety of Contemplative Experiences," Buddhist meditators reported a wide spectrum of phenomena in different psychological domains (Lindahl et al., 2017, table 4):

- Somatic: Somatic energy phenomena, sleep changes, pain, involuntary movements;
- Affective: Fear, panic, positive feelings, depression, re-experiencing of traumatic memories;
- Perceptual: Hallucination, visions, light perceptions, hypersensitivity, time distortions;
- Cognitive: Change in worldview, delusional beliefs, mental stillness, vivid imagery;
- Motivation: Changes in goals, effort, striving and volition, anhedonia;
- Social: Impairment or integration, changes in relationships, occupational impairment; and
- Sense of self: Changes in boundaries between self and others/world, loss or change of different aspects of self-experience (agency, basic self, embodiment, narrative self, ownership).

Obviously, meditation goes far beyond relaxation and concentration. The contemplative path to a liberating breakthrough and mystical experiences could sometimes result in a breakdown that requires psychotherapeutic or psychiatric treatment. Therefore, an important task for scientific research is to investigate the best fit between characteristics of a practitioner and meditation techniques as well as optimal environmental conditions to reduce the risks for mental health.

Concluding Reflections

The interdisciplinary collaboration of psychologists and neuroscientists, the combination of first-person reports and physiological measurements, opens the avenue for studying specific effects of meditation techniques. In addition, the close cooperation with meditation practitioners is crucial to assess the dynamic interplay between applied technique and changes in mental states during a session and across sessions. Ideally, scientists could study the phenomenological effects of meditation themselves in their own mind while measuring and monitoring their brain activity in real time (for an exemplary neurofeedback setup, see Prestel et al. [2019]).

Self, time, and space provide structure. Usually, humans identify with the mental representation of their body. Self and world model provide a stable basis for navigating the social and spatially extended external world. Yet, closer inspection during meditative inquiry reveals that the mind is less stable than assumed and rather a matrix of mental representations adapting to situations using past experiences to predict the outcomes of possible actions. By reflecting and questioning the identification with mental representations and realizing their fluidity, the human spirit can free itself as stated in the quotation by Kafka at the beginning. Today, science can examine, understand, and evaluate the techniques developed within spiritual traditions as tools for self-realization and as methods for scientific research on the wide range of operation modes of mind and brain.

In my personal view, this kind of research could support humankind to reflect and reconsider materialistic values and further the mental and behavioral changes urgently required to cope successfully with the current global crises.

Glossary

Absorption: Either denotes a mental *state* during which attention is focused completely on one object, or it denotes the disposition to experience such states easily—for example, during reading, listening to music, or watching beautiful scenery in nature. The Tellegen absorption scale allows quantifying this absorption *trait*.

Altered states of consciousness: Ludwig (1966) introduced this umbrella term to denote a wide spectrum of states experienced subjectively as clearly different from the "normal" state of waking consciousness; altered states can be induced by illnesses (e.g., fever), chemical substances (e.g., psychedelics), or psychological means (e.g., hypnosis or meditation). Alterations could affect thinking, time sense, self-control, perception, and emotions (for review, see Vaitl et al. [2005]).

Gamma oscillations: In electroencephalography, the electrical waves, recorded with electrodes placed at the scalp, are classified based on their frequency. Electrical brain waves around 10 Hertz (eight to twelve oscillations per second) are denoted as *alpha* activity. Slower brain waves are denoted as *theta* (4 to 7 Hertz) or *delta* (1 to 3 Hertz) activity; faster brain waves as *beta* (13 to 30 Hertz) or *gamma* (above 30 Hertz) activity. Gamma oscillations around 40 Hertz seem to play an important role for perception and cognitive processing (see Ott, 2013; Singer, 1999).

Mystical experiences/states: In many spiritual traditions, mystical states of consciousness are the ultimate goal of spiritual practice. However, these state can also occur spontaneously (e.g., during encounters with nature or art) or be induced by psychedelic substances (e.g., LSD). Key features of mystical experiences include a profound sense of unity (*unio mystica*), strong positive feeling (love, joy, bliss, sometimes also awe), and transcendence of space and time (see questionnaire by Maclean et al. [2012]).

Self: While growing up, every human develops an image of herself or himself ("self-model") that integrates perceptions from the body (sensations and feelings, constituting the "core self"), from personal experiences stored in memory ("autobiographical self"), and from feedback provided by the social environment ("social self" or "image"). In addition, we experience ourselves as somebody acting willfully according to our own intentions ("self-agency"). Neuroscientific research helps to differentiate the various aspects of the self and to understand how they enable successful interaction with the world.

References

Barrett, F. S., & Griffiths, R. R. (2018). Classic hallucinogens and mystical experiences: Phenomenology and neural correlates. *Current Topics in Behavioral Neurosciences*, *36*, 393–430. https://doi.org/10.1007/7854_2017_474

Britton, W. B., Lindahl, J. R., Cahn, B. R., Davis, J. H., & Goldman, R. E. (2014). Awakening is not a metaphor: The effects of Buddhist meditation practices on basic wakefulness. *Annals of the New York Academy of Sciences*, *1307*, 64–81. https://doi.org/10.1111/nyas.12279

Fox, K. C. R., Dixon, M. L., Nijeboer, S., Girn, M., Floman, J. L., Lifshitz, M., Ellamil, M., Sedlmeier, P., & Christoff, K. (2016). Functional neuroanatomy of meditation: A review and meta-analysis of 78 functional neuroimaging investigations. *Neuroscience and Biobehavioral Reviews*, *65*, 208–228. https://doi.org/10.1016/j.neubiorev.2016.03.021

Fox, K. C. R., Nijeboer, S., Dixon, M. L., Floman, J. L., Ellamil, M., Rumak, S. P., Sedlmeier, P., & Christoff, K. (2014). Is meditation associated with altered brain structure? A systematic review and meta-analysis of morphometric neuroimaging in meditation practitioners. *Neuroscience and Biobehavioral Reviews*, *43*, 48–73. https://doi.org/10.1016/j.neubiorev.2014.03.016

Goldberg, S. B., Tucker, R. P., Greene, P. A., Davidson, R. J., Wampold, B. E., Kearney, D. J., & Simpson, T. L. (2018). Mindfulness-based interventions for psychiatric disorders: A systematic review and meta-analysis. *Clinical Psychology Review*, *59*, 52–60. https://doi.org/10.1016/j.cpr.2017.10.011

Hölzel, B., & Ott, U. (2006). Relationships between meditation depth, absorption, meditation practice, and mindfulness: A latent variable approach. *Journal of Transpersonal Psychology*, *38*(2), 179–199.

James, W. (1902). *The varieties of religious experience: A study in human nature, being the Gifford lectures on natural religion delivered at Edinburgh in 1901–1902*. Longmans, Green & Co.

Lindahl, J. R., Fisher, N. E., Cooper, D. J., Rosen, R. K., & Britton, W. B. (2017). The varieties of contemplative experience: A mixed-methods study of meditation-related challenges in Western Buddhists. *PloS One*, *12*(5), e0176239. https://doi.org/10.1371/journal.pone.0176239

Lindahl, J. R., Kaplan, C. T., Winget, E. M., & Britton, W. B. (2014). A phenomenology of meditation-induced light experiences: Traditional Buddhist and neurobiological perspectives. *Frontiers in Consciousness Research*, *4*, 973. https://doi.org/10.3389/fpsyg.2013.00973

Ludwig, A. M. (1966). Altered states of consciousness. *Archives of General Psychiatry*, *15*(3), 225–234. https://doi.org/10.1001/archpsyc.1966.01730150001001

Lutz, A., Greischar, L. L., Rawlings, N. B., Ricard, M., & Davidson, R. J. (2004). Long-term meditators self-induce high-amplitude gamma synchrony during mental practice. *Proceedings of the National Academy of Sciences of the United States of America*, *101*(46), 16369–16373. https://doi.org/10.1073/pnas.0407401101

Maclean, K. A., Leoutsakos, J.-M. S., Johnson, M. W., & Griffiths, R. R. (2012). Factor analysis of the Mystical Experience Questionnaire: A study of experiences occasioned by the hallucinogen psilocybin. *Journal for the Scientific Study of Religion*, *51*(4), 721–737. https://doi.org/10.1111/j.1468-5906.2012.01685.x

Marshall, P. (2005). *Mystical encounters with the natural world: Experiences and explanations*. Oxford University Press.

Metzinger, T. (2009). *The ego tunnel: The science of the mind and the myth of the self.* Basic Books.

Ott, U. (2007). States of absorption: In search of neurobiological foundations. In G. A. Jamieson (Ed.), *Hypnosis and consciousness states: The cognitive-neuroscience perspective* (pp. 257–270). Oxford University Press.

Ott, U. (2013). Time experience during mystical states. In A. Nicolaidis & W. Achtner (Eds.), *The evolution of time: Studies of time in science, anthropology, theology* (pp. 104–116). Bentam Science ebook. https://doi.org/10.2174/97816080544421130100010

Ott, U. (2016). Absorption in hypnotic trance and meditation. In A. Raz & M. Lifshitz (Eds.). *Hypnosis and meditation. Towards an integrative science of consciousness planes* (pp. 269–278). Oxford University Press.

Prestel, M., Riedl, R., Stark, R., & Ott, U. (2019). Enhancing mindfulness by combining neurofeedback with meditation. *Journal of Consciousness Studies*, *26*(7–8), 268–293.

Riva, G. (2018). The neuroscience of body memory: From the self through the space to the others. *Cortex*, *104*, 241–260. https://doi.org/10.1016/j.cortex.2017.07.013

Russ, S. L., Carhart-Harris, R. L., Maruyama, G., & Elliott, M. S. (2019). Replication and extension of a model predicting response to psilocybin. *Psychopharmacology*, *236*(11), 3221–3230. https://doi.org/10.1007/s00213-019-05279-z

Schenberg, E. E., Alexandre, J. F. M., Filev, R., Cravo, A. M., Sato, J. R., Muthukumaraswamy, S. D., Yonamine, M., Waguespack, M., Lomnicka, I., Barker, S. A., & da Silveira, D. X. (2015). Acute biphasic effects of ayahuasca. *PloS One*, *10*(9), e0137202. https://doi.org/10.1371/journal.pone.0137202

Sedlmeier, P., Loße, C., & Quasten, L. C. (2018). Psychological effects of meditation for healthy practitioners: An update. *Mindfulness*, *9*(2), 371–387. https://doi.org/10.1007/s12671-017-0780-4

Singer, W. (1999). Neuronal synchrony: A versatile code for the definition of relations? *Neuron*, *24*(1), 49–65. https://doi.org/10.1016/S0896-6273(00)80821-1

Stace, W. T. (1960). *Mysticism and philosophy*. Macmillan.

Stuckey, D. E., Lawson, R., & Luna, L. E. (2005). EEG gamma coherence and other correlates of subjective reports during ayahuasca experiences. *Journal of Psychoactive Drugs*, *37*(2), 163–178. https://doi.org/10.1080/02791072.2005.10399798

Studerus, E., Gamma, A., Kometer, M., & Vollenweider, F. X. (2012). Prediction of psilocybin response in healthy volunteers. *PloS One*, *7*(2), e30800. https://doi.org/10.1371/journal.pone.0030800

Thomas, J., Jamieson, G., & Cohen, M. (2014). Low and then high frequency oscillations of distinct right cortical networks are progressively enhanced by medium and long term Satyananda yoga meditation practice. *Frontiers in Human Neuroscience*, *8*, 197. https://doi.org/10.3389/fnhum.2014.00197

Vaitl, D., Birbaumer, N., Gruzelier, J., Jamieson, G. A., Kotchoubey, B., Kübler, A., Lehmann, D., Miltner, W. H. R., Ott, U., Pütz, P., Sammer, G., Strauch, I., Strehl, U., Wackermann, J., & Weiss, T. (2005). Psychobiology of altered states of consciousness. *Psychological Bulletin*, *131*(1), 98–127. https://doi.org/10.1037/0033-2909.131.1.98

14 Meditative In-Action: An Endogenous Epistemic Venture

Giuseppe Pagnoni and Fausto Taiten Guareschi

Ma sedendo e mirando, interminati
spazi di là da quella, e sovrumani
silenzi, e profondissima quiete
io nel pensier mi fingo
But sitting here in a daydream, I picture
The boundless spaces away out there, silences
Deeper than human silence, an unfathomable hush
—Giacomo Leopardi, *L'Infinito*
(transl. Eamon Grennan)

Personal Reflections

Giuseppe Pagnoni: *Subjective experience and science are hardly independent, especially and obviously in consciousness studies. With his neurophenomenology program, Francisco Varela aimed to formalize explicitly their relationship as a dynamic web of mutual constraints—an unavoidable perspective, he argued, for the progress of consciousness research. In fact, how the world appears to us subjectively, the specific makeup of our personality, and our life experiences have always influenced—if implicitly so—the scientific enterprise. Conversely, what we learn through scientific research and practice also shapes our subjective experience and our personal outlook on the world, although perhaps with a weaker capacity to reach the deepest and more spontaneous functions of our consciousness.*

More personally, I feel that contemplative practice, by fostering a greater intimacy with one's own subjective experience, may facilitate access to an intuitive space that is often the shadowy breeding ground of novel perspectives. In this sense, I found the contemplative pauses that punctuated the workshop very helpful in maintaining a good balance, during the lectures and the ensuing discussions, between an open attitude and constructive criticism.

Overview

The theoretical framework of active inference proposed by Karl J. Friston is currently one of the more actively developed research areas in neuroscience. According to this theory the brain adapts its synaptic activity and architecture in such a way that it de facto comes to mirror the causal structure of events that the organism both encounters and actively induces in its environment. In this contribution we show how active inference can provide a useful perspective to better understand the processes engaged by contemplative practices. More specifically, and focusing on the practice of *shikantaza* ("just sitting") in the Japanese Zen Sōtō tradition, we argue that meditation enacts a peculiar policy with high endogenous epistemic value, whereby the practitioner accrues an intimate, but not necessarily explicit, knowledge about herself. This superordinate policy entails the embodied, active suspension of our habitual pleasure-seeking and discomfort-avoiding behavior, an attitude epitomized by the traditional Japanese notion of *mushotoku* (translated as "nothing to be attained"). From this perspective, we also critically examine some popular claims about meditation that we believe are likely to be misconstrued, and may not be without personal, social, and even political consequences.

A Minimal Introduction to Active Inference

The term *active inference* refers to a theoretical framework aiming to understand how self-organizing systems, such as biological organisms, move, perceive, and generally behave in their environment. It is based on the principle of free energy minimization proposed by Friston (Friston, 2010). While its origins lie in the field of neuroscience, it appears to possess a remarkably ample explanatory and applicative scope, and is thus being very rapidly developed (Friston, 2019). (For a hands-on, modeling-based tutorial on the core notions of active inference, see Smith et al. [2021]).

The central notion of active inference is that biological systems keep their identity and viability by minimizing the likelihood of finding themselves in surprising states over the long-term. The role of surprise minimization as a fundamental operating principle may seem bizarre at first, but it becomes easier to understand if we consider the physiological concept of homeostasis. An organism can maintain its vital status and identity only by keeping its internal, physiological parameters (e.g., glucose and oxygen blood levels, pH, temperature) within a narrow range of variation. Trespassing such bounds represents an abrupt surprise for the organism,[1] who will go to great lengths to avoid it.[2] In more general terms, active inference and the principle of free energy minimization represent how living organisms are able to shield themselves temporarily from the increasing entropy described by the fluctuation theorem of statistical mechanics. It is intriguing to note that long before modern physics, the archetypal notions of order and disorder (i.e., cosmos and chaos) were also recognized as fundamental, dynamically opposing forces (but generatively so) in many ancient cultures, and thus placed at the core of their respective creational myths. It is also worth mentioning that active inference's foundational ideas are indeed closely related to the pioneering work of Maturana and Varela on biological autonomy and sensorimotor coupling, leading to the development of the concepts of autopoiesis and enactive cognition (Maturana & Varela, 1980; Varela et al., 1974, 1991; but see also Di Paolo et al. [2021] for a critical perspective on this matter).

Surprise minimization rests necessarily upon the concept of a *generative model*—that is, a model of how hidden (worldly or bodily) causes generate observable quantities (i.e., sensations). The inversion of the generative model allows the organism to *infer* the hidden causes of the current sensations, in turn generating further *expected* sensations. Surprise refers to the mismatch between expected and actual sensations.

But where does such generative model come from, or, in other words, what kind of process *generates* the generative model? The basic idea here is that the history of interaction between an organism and its environmental niche literally *in-forms* the structure and causal dynamics of both organism and environment. In other words, the continuous "friction" between the living agent and its world gives rise in time (and along multiple time scales) to a modification of the internal structure of the agent, so that its dynamics comes to reflect, *re-present* or *re-enact* that of the causal chains of events that the organism habitually encounters in its environment: the agent comes to *embody* a generative model of its world. The environmental niche gets also similarly sculpted by the organisms inhabiting it, although this is an aspect that concerns us only marginally here (the interested reader may want to read Rubin et al. [2020]). On the other hand, it is not irrelevant that for a social species like ours the interaction with other individuals is of utmost importance, and thus that the relationship with the other is, at a very fundamental level, constitutive of our own identity (i.e., of the generative model that we *are*). In fact, it has been argued that it was precisely during the so-called Cambrian explosion, when most of the animal phyla appeared (around six hundred million years ago), that inferring the identity, behavior, and intention of other animals became the cardinal issue for survival and reproduction, and therefore the leading force for the evolution of the nervous (and sensorimotor) system: ultimately, of the mind *in response to other minds* (Smith, 2017). Furthermore, as the gamut of sensorimotor interactions that an organism could entertain with its niche expanded along its evolutionary trajectory, spanning a greater range of spatiotemporal scales, the architecture of its internal generative model had necessarily to become progressively more hierarchical, to be able to reflect the richer and deeper structure of causal dependencies and statistical regularities encountered in its world.

Notably, from the perspective we have been describing, a living system can be seen as if continuously anticipating its sensory input (whence the term *predictive processing* or *predictive coding*), under the single imperative to minimize the discrepancy (*prediction error*) between the expected and the actual sensory signals. This process of prediction error minimization is thought to underlie not only perception and learning, but also decision-making and action selection. It is indeed agency that takes active inference beyond vanilla predictive coding or Bayesian brain formulations: the ability of sentient beings to *actively* solicit from the environment the data they need to confirm or disconfirm their hypotheses, which on longer time scales also underwrites the process of niche construction.

An important aspect of the theory is that the generative model must necessarily include the agent itself in its set of representations in order to be able to anticipate the consequences of its own actions and thus choose which sequences of actions will likely bring about the preferred sensory data. This requirement entails the extension of the predictive horizon far beyond the immediate contingency of "what will happen now," so that the selection of a given course of action by the agent is in fact determined by the *expected* reduction of free energy associated with that policy in the future. This temporal aspect is also particularly interesting for our purposes, because the scope of consciousness of a living system has been linked to the temporal "depth" or "thickness" of its generative model (Friston, 2018). In brief, it has been argued that the longer the temporal arcs the system can cast in its predictive and postdictive processing,[3] the greater its conscious capacity. The terms "depth" or "thickness" refers to the hierarchical architecture of the generative model, where higher layers encode representations of progressively coarser spatiotemporal grain—in other words, in moving up the hierarchy of the model, percepts gradually give way to concepts. Notably, the brain seems to be an organ exquisitely specialized for this, due to its extremely rich connectivity that allows the implementation of the required web of hierarchically structured and recursive informational message-passing.

Now, while perception, learning, and movement can be formulated as processes that unfold based only on the minimization of the *current* prediction error, the higher faculties of decision-making and policy selection—where the term *policy* denotes a purposeful sequence of actions—require also the minimization of the prediction error of the *future* (i.e., the expected prediction error) because the sensory observations on which the computation of prediction error is based have obviously not occurred yet. It can be shown that mathematically this entails the joint minimization of two informational quantities that can be called *risk* and *ambiguity*. Risk scores the probability of reaping sensory data, under the selected policy, that differ considerably from the ones that define the agent's phenotype: in other words, the probability of running into non-preferred states. Ambiguity measures how uncertain (imprecise) the observations are likely to be when the system is in the states aimed at by a policy; a reduction of ambiguity is associated with an improved predictive efficacy of the model because the ensuing observations will be more precise, providing more reliable teaching signals for the generative model.

In summary, minimization of *current* free energy—linked to current prediction error—balances the accuracy of the model (how closely it aligns with the world) with its complexity (how much it needs to change to accommodate the new observations). On the other hand, minimization of *expected* free energy—linked to expected prediction error—expresses the tendency of the organism to avoid nonpreferred sensations and ambiguous information. Perhaps even more intuitively, following an alternative mathematical decomposition of expected free energy, policy selection can be shown to be based on the relative likelihood of available policies to secure (1) the agent's preferred outcomes (the *pragmatic* value of a policy), and (2) novel sensory data that will reduce the model's uncertainty and thus improve future predictions (the *epistemic* value of a policy). Note that while epistemic foraging may look like surprise-seeking, it is in fact *curious* behavior—that is, sensation-seeking in the service of information gathering for future resolution of uncertainty. Also, it is worthwhile to emphasize here that minimization of *expected* free energy is what really distinguishes sentient beings capable of long-term planning from more elementary agents, such as viruses and bacteria, for which "living in the present" (and thus minimization of *current* free energy) is quite sufficient.

Complexity and Accuracy of the Generative Model

Let us delve a little deeper into the issue of model complexity. Statistically, a model that matches too tightly the data that it tries to explain is suboptimal for inferential purposes. This situation, known as overfitting, may result from an increase in the complexity of the model because a larger number of parameters is necessary to accurately describe every minute detail of the data at hand, where a portion of the variance in the data may actually arise from random fluctuations or represent generally inessential features. Thus, an overly complex model is doubly disadvantageous: economically (metabolically, in the case of neural systems), because of its high maintenance costs, and performatively, because it generalizes poorly to situations that have the same core features of the previously learned ones but are furnished with minor variations.

The active inference framework deals very efficiently with this issue because—as briefly mentioned above—the general imperative of free energy (or prediction error) minimization can be mathematically recast in terms of the simultaneous drive to increase accuracy *and* reduce model complexity (see Smith et al. [2021] for an accessible introduction to the mathematical derivations). In other words, free energy minimization is an expression of the principle known as Occam's razor: to seek the most parsimonious explanation that represents with sufficient accuracy the data at hand. Interestingly, REM sleep dreaming has been proposed to have developed phylogenetically to enlist the inactive hours of the night to the task of reducing model complexity: by reenacting and recombining the statistical regularities collected during the waking hours—which could have overly increased model complexity—their essential features can be summarized and consolidated by a process of synaptic pruning (Hobson & Friston, 2012).

On the other hand, when model simplification becomes excessive, it turns maladaptive. As a case in point, the progressive synaptic loss that we face as we age may also correspond to a decrease in complexity of our generative model. Beyond the general age-related deterioration of physiological processes, this is likely a consequence of our agential capacity: barring some drastic environmental changes outside the purview of our control, we are liable to get progressively better in fashioning our niche according to our expectations in a cycle of self-fulfilling prophesy, thus reducing the complexity demands of our generative model. As older adults, we then become experts both at inferring our particular environment and at attaining our preferred sensations, but less capable of adapting to volatile situations: our generative model tends not only to suffer from overgeneralization (i.e., it struggles to encode novel contingencies with sufficient accuracy), but also to become refractory to revision. Children, on the other hand, are extremely adept at epistemic foraging and much less hindered by uncertainty—which makes perfect sense, as in their stage of life the generative model needs to accrue large amounts of information before being able to navigate the world without adult assistance and supervision.

A Work in Regress

Keeping the complexity of the generative model as low as possible, while maintaining good accuracy, allows for an optimally adaptive strategy characterized not only by fitting responses to habitual situations, but also by adequate pliancy when facing the unexpected. While REM sleep has been argued to represent a general mechanism to preserve this balance across the animal kingdom, one may wonder whether the peculiar features of the human species—its highly complex agential capacity, the ability to array its environment with an ever-expanding and dynamically evolving multitude of technical and cultural artifacts, the lengthening of the lifetime, and its increased meta-awareness—have also produced more specifically suited strategies to counteract the progressive loss of cognitive flexibility and adaptability that often comes with aging.

Interestingly, many spiritual traditions point to the necessity of an unlearning process to escape the limitations imposed by a well-established but ossified cognitive structure. Just to quote a few examples:

> In the beginner's mind there are many possibilities, but in the expert's mind there are few.[4] (Shunryū Suzuki, *Zen Mind, Beginner's Mind*)

> In pursuit of learning, every day there is more. In the pursuit of the Dao, every day there is less. (Laozi, *Dao De Jing*, ch. 48)

> Truly I tell you, unless you change and become like little children, you will never enter the kingdom of heaven. (Matthew, 18, pp. 2–4, Holy Bible, NIV)

In a similar vein, a number of important artists, writers, and philosophers have argued that the survival into adult life of the playfulness, openness, immediacy, and sureness of judgment of infancy is an essential requisite for creative work, as epitomized by Pablo Picasso's famous aphorism: "At eight, I was Raphael [. . .] It took me a whole lifetime to paint like a child" (Picasso, 2010).

What could the peculiar *praxis* enabling this kind of fertile backward step be, and what neurocognitive mechanisms could underpin it? To fruitfully approach the question from the point of view we have been espousing here, it is important to introduce a further notion, that of *precision*, and highlight the role it plays in the active inference scheme. In general terms, the outcome of the inferential process depends on the relative strength—that is, precision of, or confidence in, predictions (beliefs) and prediction errors. A high-precision sensory prediction error coupled with a low-precision prediction makes the state of the generative model very sensitive to prediction error: credibility among competing predictions is swiftly redistributed in order to minimize prediction error. In the case of perception, this corresponds to making its content more attuned to the incoming sensory signal by selecting the most fitting hypothesis among the available ones about the cause of the current sensation. On the other hand, a low-precision sensory prediction error coupled with a high-precision prediction will secure the original prediction while the system works toward its fulfillment via appropriate actions: the agent will move and reorient its sensory surfaces so that collecting the very sensory input it is *expecting* becomes more likely. In short, prediction error is always precision-weighted and the variable deployment on precision across the hierarchy of the generative model is a key element in driving the dynamics of the system in perception, learning, and behavior.

In this perspective, the *work in regress* outlined above would be aimed at relaxing the precision of ingrained beliefs, "smoothing out" their probability distributions and thus allowing greater latitude in exploring alternative beliefs. From a different angle, this would amount to increasing the entropy of the system and thus the likelihood of selecting other, usually latent, a priori hypotheses. Incidentally, such a mechanism has also been proposed to explain the effects of psychedelic drugs like psilocybin, LSD, DMT, and MDMA (Carhart-Harris & Friston, 2019), which have recently shown promising long-term efficacy both in the treatment of resistant depression, obsessive-compulsive disorder and post-traumatic stress disorder (Carhart-Harris et al., 2021; Mitchell et al., 2021; Reiff et al., 2020), and in changing the general outlook on life and expectations of terminally-ill patients toward a more accepting and positive stance (Griffiths et al., 2016; Ross et al., 2016). More specifically, the above conditions have been argued to be similarly characterized by an abnormally reduced level of neurocognitive entropy and by overly precise and rigidified priors that have become refractory to revision. Typically, the mentioned psychedelic drugs act as neuromodulators—that is, they change the magnitude of the effect that presynaptic inputs have on postsynaptic firing, a function that matches exactly their hypothesized role of adjusting precision in active inference (Parr & Friston, 2017).

Contemplative Practice as (In)active Inference

Active inference may offer a distinct vantage point for understanding the mechanistic underpinnings of contemplative practice as the kind of *work in regress* discussed in the previous section, while also providing an appropriate language for developing a fruitful dialogue between meditative traditions and cognitive science (see also Farb et al., 2015; Laukkonen & Slagter, 2021; Lutz et al., 2019; Pagnoni, 2019; Pagnoni & Guareschi, 2017). Since the term "meditation" can frequently refer to practices that are quite different in their formal aspects, underlying conceptual framing, and engaged mental processes (see Dahl et al. [2015] for a proposed taxonomy), we will limit our discussion here to the practice of "just sitting" (Japanese *shikantaza*, also called *zazen*) as transmitted within the Japanese Zen Sōtō tradition.[5]

We propose that the practice of *zazen* can be seen as having two thoroughly intertwined aspects: (1) a pragmatic one, based on the priors installed by the meditative instructions, which aim to enact a suitable bodily posture and a mental state of nonjudgmental wakefulness; and (2) an epistemic one, related to the cognizant faculty of realizing the spontaneous appearance of mental content, the effects that the latter triggers in the body, and the instances of failure of the meditative prescriptions (e.g., episodes of absorption in thoughts or sensations, dozing off, or slips of the posture), which is intimately tied to attentional regulation.

Paradoxically—but only so at a cursory examination—the apparent inaction featured by meditation is quite an active process (hence the pun "meditative in-action" in this article's title), and a peculiar one at that. Indeed, *zazen* can be seen as the practice of skillfully orchestrating the suspension of all the usual policies associated with our tendency to move toward what we like and away from what we dislike:[6] a wedge thrown into the mechanics of our fundamental, *constitutive appetite* for self-evidencing (Hohwy, 2014), all the while keeping an open attitude toward any spontaneously arising mental phenomenon. The legs are folded with the feet turned upside down in the lotus position[7]—an embodied symbolic representation of the cessation of the intentionality of walking motion. The hands are placed with the fingers of one hand resting on those of the other hand and the thumbs lightly touching in a horizontal line, a gesture (*hokkai-jōin*) that enacts and symbolizes refraining from both grasping and pushing away. Similarly, the eyes do not chase around—visual saccades are our chief means of environmental epistemic foraging—but are kept slightly downcast and half-open to eschew the seductive pull of dreamlike imagery that tends to manifest when closing our eyes and the associated drowsiness. The mouth is kept naturally closed, temporarily shelving our speaking habits as well. Finally, an attitude of affective equanimity that discourages habitual value judgments about spontaneous thoughts and feelings extends the meditative poise into the cognitive and emotional realms. In summary, contemplative in-action can be viewed as a posture of *radical unavailability* to compulsive reactive patterns, to chasing after what we like (including enlightenment) and avoiding what we don't like, well expressed by the Japanese term *mushotoku* ("nothing to be attained"). In this sense, the embodiment of heedful presence in the sitting posture corresponds to a superordinate policy that effectively decreases the precision (and thus the compelling power) of ingrained policies with a high pragmatic value and a short-term resolution dynamic, by abstaining from following them through with confirmatory (i.e., precision-boosting) actions. Quite literally, the meditator could then be viewed as performing a "sacrifice" (from the Latin, *sacrum facere*, or "sacred-making"), enacting the demarcation of a sacred, ritual space *within him/herself* rather than on the ground (Calasso, 1999), where the usual *busi-ness* of worldly cares and motivations is temporarily suspended. Note that this *praxis* is fundamentally different from an effortful training to control mind and body to reach a purported "enlightened" state based on personal beliefs and ideas, which is denoted in Japanese by the term *shuzen.*[8] Dōgen Zenji, founder of the Zen Sōtō school in thirteenth century Japan, repeatedly warned that "*zazen* is not *shuzen.*" He further clarified that, "The *zazen* I speak of is not learning meditation. It is simply the Dharma gate of repose and bliss, the practice-realization of totally culminated enlightenment" (Waddell & Masao, 1973).

Based on the above considerations, we propose that under the virtual boundary staged by contemplative in-action the process of active inference effectively takes an inward-directed epistemic turn,[9] providing a unique opportunity for the meditator's generative model to learn *about itself*[10]—"Learn the backward step that turns the light inwardly to illuminate your self," we read again in Dōgen's *Fukanzazengi* (Waddell & Masao, 1973). This learning of the generative model about itself is important because, as we previously mentioned, agents capable of a vast spectrum of actions like us must necessarily possess an accurate model of themselves to be able to accurately predict and infer the changes in their environment engendered by their own actions. Evidently, such self-knowledge or representation can only be partial, as the generative model would need to be computationally "bigger" than itself to possess a complete self-representation—an obvious logical impasse. Because of the specific features of the meditative *praxis* engaging body and mind as a whole, we speculate that the auto-epistemic outcome of *zazen* may especially concern the relationship between mental activity and its bodily consequences (e.g., autonomic activity, postural changes). It may be useful to remark here that, while the notion that meditation practice helps us become more "selfless" seems to have become popular lore these days, we believe it is also prone to be misunderstood. Kodo Sawaki, perhaps the most prominent teacher of the Zen Sōto school of the twentieth century, used to say that "*Zazen* is the self selving the self by the self" (Uchiyama & Okumura, 2014), begging the question of what is really this self that we refer to when we talk about selflessness.

In addition, we further argue that the mind-wandering episodes that inevitably punctuate a meditative session are not to be seen as nuisances that we hope will eventually disappear from our practice, but rather—when crucially combined with meditative body-mind presence—as vital opportunities for the generative model *that we are* to learn about itself. In fact, it is well-known that mind wandering increases as the demands placed on us by the external world decrease (Smallwood & Schooler 2015). If the aim of meditation were simply to abolish mind wandering, it would not make much sense to put ourselves in those very conditions that make it flourish—that is, sitting "at rest" in a quiet environment. On the contrary, we propose that meditative in-action be regarded as a skilled optimization of the "default" auto-epistemic modality that characterizes spontaneous mindful rest,[11] when the content of our stream of consciousness provides valuable opportunities for reflection, creative problem solving, and planning (Smallwood & Schooler, 2015).

What are the mechanisms by which such optimization could take place? Our working hypothesis is that the vigilant attitude prescribed and fostered in meditation,[12] coupled with immobility, enhances the (meta-)awareness of spontaneous thoughts, as well as of the interoceptive signals associated with their bodily reverberations. More specifically, not yielding to habitual reactive patterns triggered by spontaneous thoughts during meditation has the dual effect of (1) making them more available to consciousness; and (2) not reinforcing them, in turn redistributing precision more equitably among the existing mental and behavioral policy priors and thus compelling the system to find alternative paths of free energy minimization (e.g., letting the body/mind renormalize *endogenously* a state of increased arousal set off by an anxiogenic thought via autonomic regulation, rather than *exogenously* via compensatory/compulsive overt actions). It is tempting to metaphorically relate the epistemic dimension of the meditative exercise to the Homeric myth of Ulysses having himself tied to the ship's mast by his men, in order to be able to hear the sirens' song without being drawn fatally to the rocks. The picture can be useful for reminding us that meditation is not simply a strategy to assuage anxiety by reducing uncertainty, but rather a way of constraining dynamics against habitual reactive patterns so that tolerance to uncertainty can be learned. This may increase—in time and outside of the meditative session—the likelihood of selecting less-frequently adopted policies whose expected reward lies further in the future. In other words, meditative practice may also work as an antidote to the trend in modern societies toward a steep delay-discounting bias—that is, the increasing preference for attaining immediate rewards and quickly avoiding uncomfortable situations.[13] It is in this way that the epistemically rich dynamic enacted by the very alternation of episodes of mind wandering and meditative presence can beget pragmatically valuable repercussions (Hasenkamp et al., 2012). In fact, mind wandering has also been experimentally linked to reduced delay discounting (Smallwood et al., 2012), supporting the hypothesis that opening up to the spontaneous occurrence of mental events during meditation, rather than trying to suppress them, may increase the probability of selection of policies with higher long-term pragmatic value. Ironically, the acritical apperception of a notion like "living in the present moment", associated with meditation in the popular culture, may be one of the causes of the remarkable diffusion and success of mindfulness practices in the West in the last decade, by virtue of its alignment with the imperatives of a consumer-based capitalistic society (see, e.g., the concept of Buddhist modernism in Sharf [2015] and Thompson [2020], and the associate critiques of Purser [2019] and Purser and Milillo [2014]). Such cultural biases may easily generate perverse effects, and recent experimental evidence suggests that mindfulness practices can actually *decrease* prosocial behavior in individuals with preexisting self-centered personality traits (Poulin et al., 2021). Moreover, it is not difficult to see how the capillary diffusion of digital connectivity and social networks, along its undeniable advantages, has also the potential to entangle us in cognitive and emotional processes of shorter and shorter expectation-confirmation arcs, making vulnerable individuals more liable of being, so to say, held hostage by the present moment.

A few caveats are in order concerning the previous points. First, the meta-awareness fostered by *zazen* is not necessarily explicit or readily amenable to analysis and conceptualization; rather, it is more akin to a process of becoming familiar, intimate with the dynamics of the mind—its pushes and pulls, the peaks and troughs of its clarity, its bodily correlates—and thus is more of an intuitive nature. Second, it is conceivable that what is learned in *zazen* goes beyond mere representational content, concerning more crucially being acquainted with the capacity for *mental action.* In the active inference framework, this is viewed as the ability to deploy different degrees of precision across the hierarchical web of message-passing in the generative model—that is, amplifying specific prediction error streams (especially endogenous ones, in this case) through neuromodulatory transmission, a mechanism that has been linked to the cognitive function of selective attention (Limanowski & Friston, 2018; Sandved-Smith et al., 2021). Finally, it should be pointed out that, according to traditional Zen teachings, the very compulsion to know is also relinquished in *zazen*, in the sense that knowing must be left free to appear, rather than actively pursued according to our own necessarily limited and limiting interests. This suggests again a condition where mental dynamics are somewhat less constrained by our prior beliefs than usual and more exposed—but from the safer vantage point of meditative in-action—to the epistemic vastness of the unexpected.

In closing, it is worthwhile to mention that the approach to the scientific study of contemplative practices proposed here is still very much in the making. Thus, it may contain some internal inconsistencies or theoretical difficulties that further work should be able to elucidate. Computational modeling and experimental testing with brain imaging and behavioral paradigms will be crucial in verifying its viability and usefulness in future studies.

Acknowledgments

Special thanks to Antoine Lutz and the Lyon research group (Oussama Abdoun, Arnaud Poublan, and Jérémie Mattout) for the ongoing collaboration on the topic of active inference and meditation; to Ryan Smith for providing valuable feedback on the manuscript; and to Leonida Gianfagna, Daniele Miracapillo, Giovanni Nubile, and Tiziana Verde for lending their ear to the arguments discussed here and suggesting further exciting lines of speculative inquiry that, for reasons of space and time, will have to find their proper home in a future treatment.

Notes

1. “Surprise” here does not necessarily imply psychological surprise. It refers in fact to an information-theoretical quantity and, to avoid confusion, the term “surprisal” is often used instead in the active inference literature.

2. While the concept of homeostasis is used here for the sake of simplicity, active inference aligns conceptually better to the notion of *allostasis*, which is the maintenance of vital parameters within the range appropriate for the current or expected task requirements (e.g., think heart rate and strenuous exercise).

3. *Postdiction* is intended here as the capacity to infer or locate the causes of the current state of affairs in the past or, in other words, to interpret the present on the basis of past events.

4. To avoid a potential misunderstanding of this quote, here Suzuki was speaking in favor of recovering the openness of the beginner’s mind, which we tend to lose as we become “expert.”

5. The practice of *shikantaza* has some characteristics—the emphasis on the integration of body and mind in the sitting posture while generally forgoing common meditative techniques such as “labeling” of mental content, “decentering,” or breath counting—that license a specific treatment here. However, with suitable adjustments, the active inference framework can certainly be applied to other varieties of contemplative practice.

6. For a more detailed discussion, see Reverend Issho Fujita’s piece “My *zazen sankyu* notebook” in the October 2000 issue of the Zen Sōto School newsletter “Dharma Eye” (https://www.sotozen.com/eng/dharma/pdf/07e.pdf).

7. Alternative sitting postures are recommended for beginners and people who are unable to sit in the full lotus position.

8. See Reverend Issho Fujita’s article “My footnotes on Zazen 1. Zazen is not Shuzen” in the November 2011 issue of the Zen Sōto School newsletter “Dharma Eye” (https://www.sotozen.com/eng/dharma/pdf/28e.pdf).

9. The construction of such a (ritual) boundary can perhaps be usefully formulated in the framework of active inference as a temporary reduction in the degrees of freedom of the active states of the meditator’s “Markov blanket.” For an introduction to the notion of Markov blankets, see Clark (2017).

10. Readers may see an apparent contradiction between the unlearning aspect of meditation and the notion of learning about oneself introduced here. In fact, learning and unlearning are usually interwoven processes because the acquisition of novel representations (or the relationships between them) often requires the modification or unlearning of the old ones. Also, “learning about oneself” includes developing a higher degree of phenomenal opacity of one’s own mental life—that is, of meta-awareness, as discussed further below (Limanowski & Friston, 2018).

11. The use of the term “default” here is not casual: it points to the established association of mind wandering with the activity of a specific set of brain regions that exhibit high metabolism at rest, known as “default-mode network” (Raichle et al. 2001; Smallwood et al. 2021).

12. Maintaining this quality of wakefulness and liveliness in *zazen* is considered crucial: “Like a dragon that reaches water, like a tiger entering the mountain,” as traditional expressions indicate. For example, see Reverend Issho Fujita’s article “*Shobogenzo zazenshin*—A free translation” in the March 2010 issue of the Zen Sōto School newsletter “Dharma Eye” (https://www.sotozen.com/eng/dharma/pdf/25eF.pdf).

13. For reviews on delayed reward discounting, see Frederick et al. (2002) and Kalenscher and Pennartz (2008).

References

Calasso, R. (1999). *Ka: Stories of the mind and gods of India*. Vintage. https://www.ebook.de/de/product/1991558/roberto_calasso_ka_stories_of_the_mind_and_gods_of_india.html

Carhart-Harris, R. L., & Friston, K. J. (2019). REBUS and the anarchic brain: Toward a unified model of the brain action of psychedelics. *Pharmacological Reviews*, *71*(3), 316–344. https://doi.org/10.1124/pr.118.017160

Carhart-Harris, R., Giribaldi, B., Watts, R., Baker-Jones, M., Murphy-Beiner, A., Murphy, R., Martell, J., Blemings, A., Erritzoe, D., & Nutt, D. J. (2021). Trial of psilocybin versus escitalopram for depression. *New England Journal of Medicine*, *384*, 1402–1411. https://doi.org/10.1056/nejmoa2032994

Clark, A. (2017). How to knit your own Markov blanket. In T. K. Metzinger, & W. Wiese (Eds.), *Philosophy and predictive processing*. MIND Group. https://doi.org/10.15502/9783958573031

Dahl, C. J., Lutz, A., & Davidson, R. J. (2015). Reconstructing and deconstructing the self: Cognitive mechanisms in meditation practice. *Trends in Cognitive Sciences*, *19*, 515–523. https://doi.org/10.1016/j.tics.2015.07.001

Di Paolo, E., Thompson, E., & Beer, R. D. (2021). *Laying down a forking path: Incompatibilities between enaction and the free-energy principle*. https://doi.org/10.31234/osf.io/d9v8f

Farb, N., Daubenmier, J., Price, C. J., Gard, T., Kerr, C., Dunn, B. D., Klein, A. C., Paulus, M. P., & Mehling, W. E. (2015). Interoception, contemplative practice, and health. *Frontiers in Psychology*, *6*, 763. https://doi.org/10.3389/fpsyg.2015.00763

Frederick, S., Loewenstein, G., & O'donoghue, T. (2002). Time discounting and time preference: A critical review. *Journal of Economic Literature*, *40*, 351–401. https://doi.org/10.1257/jel.40.2.351

Friston, K. (2010). The free-energy principle: A unified brain theory? *Nature Reviews Neuroscience*, *11*, 127–138. https://doi.org/10.1038/nrn2787

Friston, K. (2018). Am I self-conscious? (Or does self-organization entail self-consciousness?). *Frontiers in Psychology*, *9*, 579. https://doi.org/10.3389/fpsyg.2018.00579

Friston, K. J. (2019). A free energy principle for a particular physics. *Neurons and Cognition*. https://arxiv.org/abs/1906.10184

Griffiths, R. R., Johnson, M. W., Carducci, M. A., Umbricht, A., Richards, W. A., Richards, B. D., Cosimano, M. P., & Klinedinst, M. A. (2016). Psilocybin produces substantial and sustained decreases in depression and anxiety in patients with life-threatening cancer: A randomized double-blind trial. *Journal of Psychopharmacology*, *30*, 1181–1197. https://doi.org/10.1177/0269881116675513

Hasenkamp, W., Wilson-Mendenhall, C. D., Duncan, E., & Barsalou, L. W. (2012). Mind wandering and attention during focused meditation: A fine-grained temporal analysis of fluctuating cognitive states. *Neuroimage*, *59*, 750–760. https://doi.org/10.1016/j.neuroimage.2011.07.008

Hobson, J. A., & Friston, K. J. (2012). Waking and dreaming consciousness: Neurobiological and functional considerations. *Progress in Neurobiology*, *98*, 82–98. https://doi.org/10.1016/j.pneurobio.2012.05.003

Hohwy, J. (2014). The self-evidencing brain. *Noûs*, *50*, 259–285. https://doi.org/10.1111/nous.12062

Kalenscher, T., & Pennartz, C. M. A. (2008). Is a bird in the hand worth two in the future? The neuroeconomics of intertemporal decision-making. *Progress in Neurobiology*, *84*, 284–315. https://doi.org/10.1016/j.pneurobio.2007.11.004

Laukkonen, R. E., & Slagter, H. A. (2021). From many to (n)one: Meditation and the plasticity of the predictive mind. *Neuroscience & Biobehavioral Reviews*, *128*, 199–217. https://doi.org/10.1016/j.neubiorev.2021.06.021

Limanowski, J., & Friston, K. (2018). "Seeing the dark": Grounding phenomenal transparency and opacity in precision estimation for active inference. *Frontiers in Psychology*, *9*, 643. https://doi.org/10.3389/fpsyg.2018.00643

Lutz, A., Mattout, J., & Pagnoni, G. (2019). The epistemic and pragmatic value of non-action: A predictive coding perspective on meditation. *Current Opinion in Psychology*, *28*, 166–171. https://doi.org/10.1016/j.copsyc.2018.12.019

Maturana, H. R., & Varela, F. J. (1980). *Autopoiesis and cognition: The realization of the living* (Vol. 42). Reidel.

Mitchell, J. M., Bogenschutz, M., Lilienstein, A., Harrison, C., Kleiman, S., Parker-Guilbert, K., G., M. O., Garas, W., Paleos, C., Gorman, I., Nicholas, C., Mithoefer, M., Carlin, S., Poulter, B., Mithoefer, A., Quevedo, S., Wells, G., Klaire, S. S., van der Kolk, B., et al. (2021). MDMA-assisted therapy for severe PTSD: A randomized, double-blind, placebo-controlled phase 3 study. *Nature Medicine*, *27*, 1025–1033. https://doi.org/10.1038/s41591-021-01336-3

Pagnoni, G. (2019). The contemplative exercise through the lenses of predictive processing: A promising approach. *Progress in Brain Research*, *244*, 299–322. https://doi.org/10.1016/bs.pbr.2018.10.022

Pagnoni, G., & Guareschi, F. T. (2017). Remembrance of things to come: A conversation between Zen and neuroscience on the predictive nature of the mind. *Mindfulness*, *8*, 27–37. https://doi.org/10.1007/s12671-015-0438-z

Parr, T., & Friston, K. J. (2017). Uncertainty, epistemics and active inference. *Journal of the Royal Society, Interface*, *14*(136). https://doi.org/10.1098/rsif.2017.0376

Picasso, M. (2010). *Picasso: my grandfather*. Random House.

Poulin, M., Ministero, L., Gabriel, S., Morrison, C., & Naidu, E. (2021). *Minding your own business? Mindfulness decreases prosocial behavior for those with independent self-construals*. https://doi.org/10.31234/osf.io/xhyua

Purser, R. (2019). *McMindfulness*. Watkins Media. https://www.ebook.de/de/product/33316247/ronald_purser_mcmindfulness.html

Purser, R. E., & Milillo, J. (2014). Mindfulness revisited. *Journal of Management Inquiry*, *24*, 3–24. https://doi.org/10.1177/1056492614532315

Raichle, M. E., MacLeod, A. M., Snyder, A. Z., Powers, W. J., Gusnard, D. A., & Shulman, G. L. (2001). A default mode of brain function. *Proceedings of the National Academy of Sciences of the United States of America*, *98*(2), 676–682. https://doi.org/10.1073/pnas.98.2.676

Reiff, C. M., Richman, E. E., Nemeroff, C. B., Carpenter, L. L., Widge, A. S., Rodriguez, C. I., Kalin, N. H., & McDonald, W. M. (2020). Psychedelics and psychedelic-assisted psychotherapy. *American Journal of Psychiatry*, *177*, 391–410.

Ross, S., Bossis, A., Guss, J., Agin-Liebes, G., Malone, T., Cohen, B., Mennenga, S. E., Belser, A., Kalliontzi, K., Babb, J., Su, Z., Corby, P., & Schmidt, B. L. (2016). Rapid and sustained symptom reduction following psilocybin treatment for anxiety and depression in patients with life-threatening cancer: A randomized controlled trial. *Journal of Psychopharmacology, 30,* 1165–1180. https://doi.org/10.1177/0269881116675512

Rubin, S., Parr, T., Costa, L. D., & Friston, K. (2020). Future climates: Markov blankets and active inference in the biosphere. *Journal of the Royal Society Interface, 17,* 20200503. https://doi.org/10.1098/rsif.2020.0503

Sandved-Smith, L., Hesp, C., Mattout, J., Friston, K., Lutz, A., & Ramstead, M. J. D. (2021). Towards a computational phenomenology of mental action: Modelling meta-awareness and attentional control with deep parametric active inference. *Neuroscience of Consciousness,* (2), niab018. https://doi.org/10.1093/nc/niab018

Sharf, R. H. (2015). Is mindfulness Buddhist? (and why it matters). *Transcultural Psychiatry, 52,* 470–484.

Smallwood, J., Ruby, F. J. M., & Singer, T. (2012). Letting go of the present: Mind-wandering is associated with reduced delay discounting. *Consciousness and Cognition, 22,* 1–7. https://doi.org/10.1016/j.concog.2012.10.007

Smallwood, J., & Schooler, J. W. (2015). The science of mind wandering: Empirically navigating the stream of consciousness. *Annual Review of Psychology, 66,* 487–518. https://doi.org/10.1146/annurev-psych-010814-015331

Smallwood, J., Turnbull, A., Wang, H.-t., Ho, N. S. P., Poerio, G. L., Karapanagiotidis, T., Konu, D., Mckeown, B., Zhang, M., Murphy, C., Vatansever, D., Bzdok, D., Konishi, M., Leech, R., Seli, P., Schooler, J. W., Bernhardt, B., Margulies, D. S., & Jefferies, E. (2021). The neural correlates of ongoing conscious thought. *iScience, 24,* 102132. https://doi.org/10.1016/j.isci.2021.102132

Smith, P. (2017). *Other minds: The octopus, the sea and the deep origins of consciousness.* William Collins.

Smith, R., Friston, K., & Whyte, C. (2021). *A step-by-step tutorial on active inference and its application to empirical data.* https://doi.org/10.31234/osf.io/b4jm6

Thompson, E. (2020, April 03). *Why I am not a Buddhist.* Yale University Press. https://doi.org/10.2307/j.ctvt1sgfz

Uchiyama, K., & Okumura, S. (2014). *The zen teaching of homeless kodo.* Wisdom Publications. https://www.ebook.de/de/product/23303257/kosho_nchiyama_shohaku_okumura_the_zen_teaching_of_homeless_kodo.html

Varela, F. G., Maturana, H. R., & Uribe, R. (1974). Autopoiesis: The organization of living systems, its characterization and a model. *Biosystems, 5,* 187–196.

Varela, F. J., Thompson, E., & Rosch, E. (1991). *The embodied mind: Cognitive science and human experience.* MIT Press.

Waddell, N., & Masao, A. (1973). Dōgen's "Fukanzazengi" and "Shōbōgenzō zazengi." *The Eastern Buddhist, 6,* 115–128. http://www.jstor.org/stable/44361371

15 My Peregrinations Through Varela's Neurophenomenological Horizon Toward the Mathematics of Subjective Experience

David Rudrauf

bababadalgharaghtakamminarronnkonnbronntonnerronntuonnthunntrovarrhounawnskawntoohoohoordenenthur-nuk!
—James Joyce, *Finnegans Wake*, 1939

Personal Reflections

The chapter you are about to read is much about how other minds played a role in shaping my views and theories about consciousness. It only seemed natural to me from the outset that the phenomenology of subjective experience should play a pivotal role in a scientific theory of consciousness. But my intuitions found firmer methodological ground in my brief yet meaningful, and retrospectively foundational encounter with Francisco Varela as a mentor.

Much of this chapter explores Varela's influences on my views about consciousness. As I'm invited to begin with a more personal and experiential testimony, I reckoned I would bring forth the real, yet retrospectively amusing struggle for self-imposed introspective routines and experiential discipline I progressively built into my daily life as a means of better understanding consciousness. I spent years practicing introspection, including through practices pertaining to meditation, in a more or less formal manner. At some point in my life, I found myself forced to spend time in a form of sustained isolation, which I used as an opportunity, a bit like a retreat, to tackle the question of the essential invariants characterizing subjective experience. I put my mind to work in a quite disciplined manner, sometimes for days without interruption. It was in such a state that the clear and somewhat obvious intuition of the three-dimensional (3D) projective structure of consciousness came to me.

Ever since, I have kept practicing such exploration of the geometry and, more generally, the structure of consciousness, often during the transition between wakefulness and sleep. My friends would often laugh, recounting how they believed I was asleep during conversations late in the night, when I was actually aware—though in an altered state of consciousness—of what they had been saying and could continue the conversation seamlessly.

I would task myself in those moments to explore and test hypotheses about the structure of my own consciousness. The more my endeavor found an anchor in mathematical—and more specifically geometrical—concepts, the more I tried to investigate their phenomenological implications. As the effort was for me quite demanding, I would end up actually falling asleep; so, it became a standard method for me to both work on solving the problem of consciousness and finding sleep. Lucid dreams would often follow and became a place of investigation of the geometry and topology of consciousness.

I recall that, as a little boy (probably five to six years old), while in the process of falling asleep, I often traveled mentally, trying to find the boundaries of space, and experiencing with awe and distress the paradoxes of infinity—a place one should be able to reach but could never really reach. Those early experiences already featured properties I would discover many years later to be of a projective geometrical nature.

My mind is not as intense and fast as it used to be, but I keep trying to push the boundaries of my understanding of consciousness toward its most fundamental and strange properties, some of which I have yet to discover or pinpoint, beyond vague intuitions, and which I know require such work of introspection to be identified.

But I also understand that an Ariadne's thread toward Varela's so-called generative passages is given by mathematics itself, as he firmly believed himself, and that a valid mathematical model of the phenomenology of consciousness must imply properties my introspective quest alone could not necessarily uncover. The analogy that often comes to my mind is that of the theory of general relativity, which was derived from a few principles, and predicted peculiar objects such as black holes, that could not be foreseen based on those principles, but could much later be indirectly observed as the theory prescribed.

Overview

Neurophenomenology, a research program introduced by Francisco Varela, of whom I was a student, is often understood as the disciplined use of introspection to collect data pertaining to the phenomenology of embodied subjective experience in order to probe the neural correlates of consciousness. But it also is a theoretical endeavor aimed at formally understanding how consciousness relates to the brain, and the converse. By using mutual constraints, scientists should arrive at a point of convergence between phenomenology and the brain, which he called "generative passages." For him, an essential, though often overlooked rationale was that this would be achieved by mathematical models—including of the phenomenology of consciousness itself—as an essential moment of the program. This is something I have been pursuing for over two decades, going back and forth between my active introspective investigations and mathematical formalizations, leading to the formulation of the projective consciousness model. Here, I emphasize how Francisco's views influenced my own research, in particular from a methodological standpoint. I also discuss how the investigation of the structure of subjective experience can have implications for psychology, biological cybernetics, artificial intelligence, and neuroscience.

Introduction

For a long time I saw the problem of consciousness as insufficiently universal and tractable to sacrifice the little time left in my life to pursue its resolution. I would keep some distance from it as something to pay attention to, not to lose track of things completely, including oneself, and to fall back on for lack of a better problem, if one failed at handling something more essential. I did fail in my quest—dreaming of being an astrophysicist who would decipher the ultimate equation of the universe, an explorer of worlds beyond earth, a conductor or a composer of cosmic reach, to serve truth and beauty. I found myself in the somewhat (as I felt it) tragi-comic, generally cumbersome universe of cognitive science and academia. So I fell back on the problem of consciousness; never, however, feeling it was completely worthwhile to pursue.

Not that I was full of myself, but rather that I found my mind, and most of the minds I met through the journey, incapable of nailing the problem down seriously in spite of the devouring ambition that sometimes made us all posture with confidence on the depth and importance of our own perspective. Here too I would most likely fail, I realized, and that was neither really sad nor completely enraging; it was just the truth. How could I not be compelled by that fact, as my would-be scientific mind pulled me toward the universal?

Early on, I had the intuition that building a (truly) scientific theory of consciousness would require mathematics of the utmost complexity, applied to an elusive realm of phenomena, anchored in a universe of relations and mechanisms covering the broadest range of levels of organization and existence. Sadly, for me, life did not lead me to master the mathematics at stake, not by a long shot, nor the many other concepts that revealed themselves as necessary. Nevertheless, I could not help dedicating a big chunk of my life to it, as I found myself on the path (a bit trapped into it), fooled perhaps and fueled for sure by an insatiable curiosity inherited from my folks and their folks before them. I thus spent my time accepting that mathematics was an obvious necessity to truly tackle the problem, having no other solution but walking, knowing the slope would be too steep, and that I would probably lose my health if not my mind in a necessarily utterly painful process—I'm sure it was a bad career move some life coach could have objected to, which I would have dismissed of course. So I walked, recalling the saying "sink or swim." I sometimes felt like cattle being pushed through the passageways of a slaughterhouse. More than once, through days and nights of relentless effort, I banged my head against walls (sometimes almost literally), trying to manage the somatic symptoms of pushing my body beyond its natural limits with caffeine and deep breathing. I felt I was becoming a machine servicing some obscure motive, some sort of factory worker embedded in a process bigger than himself, without the right instrument or Organon (Aristotle, 4th century BCE/2016). But I worked with some perhaps hopeless courage and determination. I knew for sure that I would need to surround myself with minds more capable than mine, and dive into a kaleidoscope of apparently unrelated and distant sources of inspirations, and somehow manage to bring them together.

Here my intention is not to offer a précis of the theory I developed with collaborators such as philosopher Kenneth Williford and mathematicians Daniel Bennequin and Grégoire Sergeant-Perthuis, or neuroscientist Karl Friston, dubbed the projective consciousness model.

That can be found elsewhere (Rudrauf et al., 2017, 2020, 2022, 2023; Williford et al., 2018), including in Williford's excellent chapter herein. Instead I revisit some of visions and concepts, and historical background that played a direct or indirect role in how my mind landed on the principles of that theory, before considering some of its implications. I dedicate much of the chapter to the influence Francisco had on my own directions, in a manner that I believe illuminates aspects of Varela's views that are not always well understood. In particular, I emphasize how Francisco's neurophenomenology program entails, according to his own often-overlooked view, a process of mathematization of both phenomenology and neural processes. This is essential for defining mutual constraints between the two realms, for selecting compatible candidate models, and hopefully uncovering interpretable generative passages through mathematical relationships (see also, Rudrauf et al., 2003; Williford et al., 2018). At bottom, this contribution is of a methodological nature.

Varela's Passageway

Although at the time I did not fully realize it, Francisco's mind and work would become pivotal in the construction of my method. I started a PhD in cognitive science in Paris with him on the neurophenomenology[1] of the subjective experience of emotion. I had convinced myself, notably through my own self-observations and through reading writers such as Marcel Proust, Virginia Woolf, and James Joyce, but also several thinkers like Hegel, Husserl, and Merleau-Ponty, to name a few (who all manage to connect the particular to the universal), that it was essential to capture subjective experience itself for the elaboration of an adequate scientific account of consciousness. Francisco died while I was still in the first year of my PhD. Shortly after, several of his students and colleagues decided to write an article in his memory aimed at covering the arc of his work from autopoiesis in the 1970s to neurophenomenology in the 1990s. I wanted to try to connect the dots between these two periods of his work, and find unifying principles and a continuum that could make sense of his notion of "embodied consciousness" (see Rudrauf et al., 2003). In order to do so, I put myself through the task of reading all of his publications. What you will read in this section is basically no more than a digest of that article. As I often find original material more inspiring than second-hand commentaries, I will proceed by emphasizing his own words, quoting extensively.

Neurophenomenology is a research program introduced by Francisco, which is often understood as the disciplined use of introspection to collect data pertaining to the phenomenology of consciousness, notably as an "embodied" subjective phenomenon, in order to probe its neural correlates in a valid and heuristic manner. But it also was a fundamentally theoretical endeavor aimed at formally understanding how consciousness relates to the brain, and more generally to biology. According to the theory, by using a strategy based on the application of mutual constraints, scientists should arrive at a point of convergence between the two domains (phenomenological and neural), which Francisco called "generative passages." This concept, the theoretical and methodological framework and intuitions shaping Francisco's directions of research are not exactly standard and are sometimes hard to penetrate. But, to be sure, they are more sophisticated than it might at first seem. Let us try to elucidate his mindset and vision.

We find at the root of his notion of embodiment, the principles he developed early on with Humberto Maturana around the concepts of autopoieis, organizational closure, and more generally, (biological) autonomy (Maturana & Varela, 1973). Looking for a systemic account of what makes a biological system or organism a living being, starting from the cell, they put forth the concept of autopoiesis (from "auto" as self-generated, and "poiesis" as creation or production): "An autopoietic system is organized as a bounded network of processes of production, transformation and destruction of components which: (i) through their interactions and transformations continuously regenerate and realize the network of processes that produced them, (ii) constitute the system as a concrete entity in the space in which the components exist by specifying the topological realization of the system as such a network" (Varela et al., 1974; see also Maturana & Varela, 1980). This process of self-production and maintenance would be in and of itself "the mechanism which endows living systems with the property of being autonomous; autopoiesis is an explication of the autonomy of the living" (Varela, 1979). Autonomy, individuation against an environmental background, and the existence of a system as an embodied whole would be fundamentally anchored into such "organizational closure"—that is, the functional mutual dependency of the parts of the process within the boundaries of the whole system: "The wholeness of a system is embodied in its organizational closure. The whole is not the sum of its parts; it is the organizational closure of its parts" (Varela & Goguen, 1977). Francisco had a fundamentally mechanistic view and found in such processes a way of equating mechanisms and the observable phenomenology or behaviors of living beings: "The phenomenology of living systems is then the mechanicist phenomenology of the physical autopoietic machines . . . with purely mechanicist notions, true for every mechanicist phenomenon in any space, one can explain completely this organization and its origin" (Varela, 1979). Thus, he said, "We claim that the notion of autopoiesis is necessary and sufficient to define the organization of the living being" (Maturana & Varela, 1973). In fact, "If a physical system is autopoietic, it is living" (Varela, 1979).

His work on biological autonomy, in my understanding, has to be borne in mind when trying to understand his later developments on neurodynamics and embodied consciousness, culminating in his neurophenomenological program. The brain as embedded in the body, and the body as embedded in its environment are organizationally (partially) closed systems that contribute as such in an essential manner to their particular mode of coupling with their environment. This form of coupling was described by Francisco as "enactive," implying it is conditioned, in a constitutive manner, by a system's unique organizational closure and individuality (including its history of interactions within the world), thus defining the "meaning" of information for the system: "Cognition is not the representation of a pre-given world by a pre-given mind but is rather the enactment of a world and a mind on the basis of a history of the variety of actions that a being in the world performs" (Varela et al., 1991). Of note, the concept of enaction was elaborated by Maturana and Varela (see Maturana & Varela, 1987). Nothing too mysterious here, just perhaps a fancy way of saying that a mechanical system with organizational closure, modulated by its history of interactions (e.g., through emotion and memory), will interact with its environment and respond with a certain autonomy and self-determination that cannot be reduced to a simple function of its inputs.

The notion of a system with mutually dependent parts forming a coherent whole in a dynamical manner was naturally amenable to modeling through standard frameworks in mathematical physics such as dynamical system theory or more generally mutually coupled nonlinear differential equations. Of note, coupled nonlinear differential equations describe how the state of a system, as characterized by several interconnected variables, changes in a nonlinear manner, given initial conditions at time t, and thus offer generative models of the system's dynamics. For Francisco, this type of mathematical framework could potentially account for complex behaviors in the animal kingdom, and for how cognition and consciousness relate to the brain and its dynamics. After all, the brain is an interconnected whole of parts (neurons). Neuroscience research, which was better informed and more tractable than the phenomenology of consciousness from a scientific standpoint, made such mathematical concepts more immediately applicable to concrete, brain-related scientific research. That was at least a starting point that Francisco tried to explore in his work on neurodynamics (see Rudrauf et al., 2003).

Importantly, he was attempting to conceptualize how holistic and reductionist views could be rigorously bridged to provide a better scientific account of embodiment: "It is not that one has to have a holistic view as opposed to reductionist view, or vice versa, but rather that the two views of systems are complementary . . . there is a strong current in contemporary culture advocating 'holistic' views as some sort of cure-all . . . Reductionism implies attention to a lower level while holistic implies attention to higher level. These are intertwined in any satisfactory description: and each entails some loss relative to our cognitive preferences, as well as some gain . . . there is no whole system without an interconnection of its parts and there is no whole system without an environment" (Varela, 1997c).

In this context, an interesting, perhaps less known (and somewhat Aristotelian) perspective of Francisco on the problem of understanding and modeling embodiment scientifically, is the causal role he attributed to "forms" or shapes themselves (as geometrical concepts) in the phenomenology of the living. As I saw it, the notion of forms could perhaps become a bridging formal concepts between phenomenology and biology, since forms can be used to characterize both the subjective manifestation of embodied spatial experiences, and physical processes. In Francisco's view, forms literally imposed formal constraints on the generative mechanisms of the inner and outer phenomenology of autonomous systems and the determination of their dynamics. I was particularly compelled by this intuition, which seemed to me to open a realm of productive mathematization relating geometrical concepts and mechanistic processes. According to Francisco, shapes characterizing the whole body and its parts are an "ongoing medium" and a "specified/specifying space." Namely, "The intuition behind our framework is that space is a constitutive element in the dynamics of living organisms just as much as the solidity of their molecular constituents" (Varela & Frenk, 1987). He called for "a research program in which to understand biological shape," and its "morphodynamical constraints" (Varela & Frenk, 1987). He thought of forms as constituting a type of field as in physics: "As in the notion of a field and its corresponding particles, there is in living shape a dynamic complementarity: the entire global shape of the body affects the local conditions for the ECM[2]/cell relationship, but at the same time the local dynamics conditions how the entire body is actually built" (Varela & Frenk, 1987). In such framework, the "cycles of operation" of the system are somewhat "morphocycles" (Varela & Frenk, 1987). He thus saw a science of forms and of their action as an essential framework to bridge the gap between holism and reductionism: "Thus the study of forms, open or closed, is a ground, on which there is a superation of the dichotomy holism/reductionism" (Varela & Goguen 1977).

This type of mereological questioning—that is, questioning about the relationships between wholes and parts and the search for concepts bridging levels of observation—continued to operate in his later developments of the neurophenomenological program toward a science of consciousness.

In his approach to consciousness, Varela emphasized that any theory of consciousness would need to tackle subjective experience and its phenomenology as a nontrivial component of biological autonomy or embodiment (Varela et al., 1991). When it comes to subjective experience, it is easy to be overwhelmed through sheer introspection by its seemingly infinite nuances and variations, but also its elusive phenomenology, which may seem to offer no tangible handle to grasp onto in order to start the systematizing process of theorizing.

But Francisco adopted a rather standard scientific epistemology: "We are similarly assuming that human experience (mine as well as yours), follows fundamental structural principles which, like space, enforce the nature of what is given to us as contents of experience . . . Do I expect the list of structural invariants relevant to human experience to grow ad infinitum? Certainly not. I surmise that the horizon of fundamental topics can be expected to converge toward a corpus of well-integrated knowledge" (Varela, 1996). Importantly, Francisco's scientific aim and proposal were not to only search for the "neurobiological (brain or bodily) correlates of consciousness," because only studying correlates would "leave in the shadow the precise circulation between them" (Varela, 1997b). It would amount to just "putting on one side a list of items or processes, and on the other seeming equivalencies as phenomenological data and separating the two sides by a 'mystery' line, a no-man's land left unexamined" (Varela, 1997b). Thus neurophenomenology is grounded on a pragmatic yet highly theoretical will to progressively and systematically "reduce the distance between subjective and objective . . . a way of narrowing the gap between the mental and the physical" (Varela, 1997b).

In light of the preceding, neurophenomenology appeared to me as a program entailing the study and understanding of the intertwining between the "shape" of subjective experience and that of embodiment, at some level of abstraction. Interestingly, Francisco promoted as much naturalizing phenomenology as "phenomenologizing" neuroscience (Varela, 1999). Now, of course such an endeavor was fundamentally dependent on our ability to characterize rigorously the essential constituents of subjective experience. Thus the program itself entailed devising methodological solutions to create adequate conditions for such a characterization: "My proposal implies that every good student of cognitive science who is also interested in issues at the level of mental experience, must inescapably attain a level of mastery in phenomenological examination in order to work seriously with first-person accounts" (Varela & Shear, 1999).

Neurophenomenology was then conceived as a cross-process of model selection between models of the subjective (models of the phenomenology of consciousness) and models of the objective (models of the brain), under the filter of mutual or reciprocal constraints: "We need to advance a cognitive science where there is a true circulation between lived experience and the biological mechanisms in a seamless and mutually illuminating manner" (Varela, 1996). On the one hand, biophysical or neurodynamical investigations should contribute to illuminate the domain of subjective experience by constraining it: "It is an axiom that we can only experience what corresponds to our organization" (Varela, 1976). And yet, on the other hand, "The novelty of my proposal is that disciplined first-person accounts should be an integral element of the validation of a neurobiological proposal, and not merely coincidental or heuristic information" (Varela, 1996). At the end of such a process of convergence there should appear "generative passages" where "the isomorphic idea [between subjective and objective levels] is taken one step forward to provide the passage where the mutual constraints not only share logical and epistemic accountability, but they are further required to be operationally generative—that is, where there is a mutual circulation and illumination between these domains proper to the entire phenomenal domain" (Varela, 1997b).

But of what would these "generative passages" consist of? This is perhaps, in my opinion, one of the most misunderstood aspects of Francisco's epistemology. An essential, though often overlooked point in his rationale, was that such generative passages would be mediated by mathematical models. This implied modeling the phenomenology of consciousness itself, as an essential moment of the neurophenomenology program. In some sense, that is a natural extension of the search for experiential "invariants": "Putting into proper relevance the structural invariants of experience opens the door for the right domain of formality" (Varela, 1997b). Of course, phenomenological descriptions call for "a dynamical picture" and invariants may be dynamical invariants (Varela, 1997b). Thus, such a "formal level" would constitute a non-reductionist, non-eliminativist bridge between the subjective and the objective (Varela & Shear, 1999). According to him, there is a "strong" parallel between phenomenology and mathematics (Varela, 1997b). After all, a central feature of mathematics is its ability to capture without any metaphysical or ontological assumptions core aspects of the phenomenology of the things themselves, from the behavior of celestial bodies in space-time to the generation of complex forms in nature. This should also apply to the phenomenology of consciousness itself: "The critical role of formal tools is, in short, based on the Janus-faced nature of mathematical ideals: they provide eidetic invariants which can, in turn, be immediately linked to a naturalistic embodiment or implementation" (Varela, 1997b). Therefore, the formal is the level of representation and translation that can "make the image of a global workspace [of consciousness] into an explicit mechanism" (Varela, 1997a; see Baars, 1997).

It is through the mathematization of both the objective and the subjective that, under mutual constraints, generative passages should be found. Mathematics, a form of abstraction, constitutes the unifying conceptual level through which a science of consciousness could be achieved: "We seek to produce epistemological and ontological shifts whereby the two domains of natural objects and phenomenological descriptions can provide a three-dimensional view of mind and experience altogether" (Varela, 1997b).

My Own Spinoff

When I began my own path of research, much of the emerging tree of knowledge growing on the ground of neurophenomenology was attracted to the star of neuroscience, combining qualitative expressions of phenomenological descriptions and neuroimaging methods in search for what amounted to neural correlates. Methods of analyses of brain data were inspired from computational models of the brain, and took inspiration from dynamical system perspectives featuring integrative processes—for example, large-scale synchronization of brain oscillations (Cosmelli et al., 2004; Lutz et al., 2002; Rodriguez et al., 1999; Varela, 1999). I myself spent years of research in neuroscience but came, not surprisingly, to realize that the brain, let alone its relations to subjective experience, was much more complicated and misunderstood than what our optimistic generation of researchers was hoping for (Rudrauf, 2014).

Taking seriously the approach entailed by neurophenomenology, I also became more and more convinced that not enough energy was put into the formalization of the phenomenological side, without which the neurophenomenological program could never be achieved. Likewise, I felt that the empiricism that characterized the domain in practice was becoming a hindrance to breakthroughs; it was an epistemological obstacle of sorts, as it tended in practice to do basically no more than what the standard program of searching for the neural correlates of consciousness had laid out (Crick and Koch, 2003), which Francisco had himself criticized. Indeed, the best way to look for the brain underpinning of consciousness was to probe the brain based on rigorously formulated, formal models of consciousness itself. But neurophenomenology and consciousness science more generally were still looking for their "eidetic" (formal) foundations.

So, I moved away from neuroscience to dive into pure phenomenology, the search for fundamental invariants, and mathematical formalizations that would yield generative models constitutive of the phenomenology itself. I also thought that it could be a productive approach that could be applied not only to neuroscience and behavioral science, but also to the problem of artificial consciousness, which had always been on my horizon. (As a fervent reader of Isaac Asimov in my youth, I could not help but wish to contribute to making real what science fiction had filled my imagination with). I should add that my mind had been shaped in that direction well before my encounter with Francisco's work, since my grandfather, a philologist and art historian, was himself trying to illuminate art production, and more generally aesthetics, based on mathematical, and more specifically geometrical foundations (Rudrauf, 1954, 1961). Likewise, following in the footsteps of his father, my father, a psychiatrist and psychoanalyst, made me acquainted early on (he died when I was 8) with the work of thinkers like Matila Ghyka (1977) and Douglas Hofstadter (1979).

I thus started with this intuition that the subjective perspective of consciousness was a form of aesthetics, combining a peculiar experience of space and affective motives, with an attractive or repulsive power capable of driving our actions. I searched for a possible characterization of the geometry of consciousness itself, and how it could play a central role in driving cognition and behavior in a causal manner.

One "invariant," which after discovering it became obvious to me, was that the subjective perspective of consciousness frames embodied, multimodal sensory, and affective information within a 3D space in perspective, featuring inner and outer dynamical shapes, from a given point of view: the shape of a piano I can hear or see or imagine, the shape of my lived body, of my thorax that expands and contracts as I breathe, the impression that space contracts when I'm scared and that my heart as a volume inside contracts faster, the smell of a nearby flower in a given direction. That space structures the manifestation of its contents between an irreducibly elusive point of view—the location of which seems impossible to pinpoint, despite its apparent generative role—and a horizon of vanishing points at infinity, where all parallel lines in a given direction and objects on those lines appear to converge (such as the railroad tracks I see in front of me or the sailboats I imagine sailing away from a harbor where I could stand). This experience of space could not be reduced to a visual experience but instead subsumes all senses into an integrated experience. It is a space of "representation" and action that operates both in perception and imagination, notably through our capacity for perspective-taking, in both usual and altered states of consciousness (see Rudrauf et al., 2017; Williford et al., 2018; see also Williford, this book). This intuition led me to identify the geometry at stake—that is, 3D projective geometry (two-dimensional projective geometry is familiar as it underlies the generation of perspective in art and architecture). It entails the action of a mathematical group of transformations—that is, a set of transformations that can be composed to yield other transformations of the same kind, in a reversible manner. The application of 3D projective transformations on a 3D projective space—for example, that of subjective experience—constitutes changes of perspective, which are central for the guidance of imagination and action in consciousness (Rudrauf et al., 2017, 2023; Williford et al., 2018).

The question then was how such a geometry could be understood as a central part of biological cybernetics—that is, as a component of the mechanical principles governing autonomous systems like us. The challenge was to formulate a theory of consciousness that would provide us with principles for unifying psychology computationally, with the broadest possible explanatory power concerning a multiplicity of phenomena and behaviors (perception, imagination, appraisal, emotion, social cognition, motivation, and action) and a strong phenomenological validity with respect to subjective experience.

This overarching question found the beginning of a practical answer in the general framework developed by Karl Friston, that of active inference, as based on the free energy principle (Friston, 2010; Friston et al., 2017). A good introduction to Friston's concepts can be found in the chapter by Giuseppe Pagnoni and Fausto Taiten Guareschi in this book. Generally speaking, active inference is a method of inference and optimization regarding the outcomes of actions an organism, or more generally an autonomous system, could and should perform. The system attempts to minimize a quantity that reflects both the divergence of sensory data from prior beliefs (including preferences encoded as expectations) and uncertainty—that is, free energy—in order to guide decision-making and action programming in a manner that balances policies of exploitation and exploration.

Projective geometry brings to the process an internal "representation" and method of transformation that allows an embodied system to perceive itself and the world in perspective from a point of view, anticipate the consequences of actions under various perspectives, highlighting different aspects of outcomes, and more generally plan global bodily motions—for example, approach and avoidance—as dual to changes of point of view corresponding to different perspectives. Importantly, projective geometry preserves incidence relations. For example, if an object, such as a car, appears to move toward your body, then, no matter which perspective you imagine, that object will still appear to be on a collision course with your body—unless you actually move to a different location that is not in the path of the car. Thus the use of projective geometry for the building and transformations of a subjective space offers a global operational solution for an autonomous system to evaluate the adaptive value of perspectives under different points of view. We could then derive a formal concept of a field of consciousness, which is at the heart of the projective consciousness model as an account of subjective experience, and which explicitly models points of view and perspectives on an enacted world, integrating 3D projective geometry and the free energy principle.

Based on these principles, we could account for a wealth of perceptual illusions, including the famous moon illusion, whereby the moon appears bigger when low on the horizon and smaller when high in the sky (Rudrauf et al., 2017, 2020). We could also account for aspects of conscious appraisal and drives, social perspective-taking and theory of mind (the inference of others' beliefs and preferences, and the prediction of their behaviors), and independently documented psychophysical laws relating the intensity of the subjective experience of emotion and the physical distance of an emotionally-competent stimulus, which were implied by the model under straightforward embodiment assumptions (Rudrauf et al., 2022, 2023).

Interestingly to me and particularly relevant to human experience, under the model, infinity as a projective horizon and a place of high sensory uncertainty appeared to act as a driving force and an attractive factor. According to the model, embodied systems attempt to minimize uncertainty (which is equivalent to maximizing information gain) by minimizing free energy, through actual or imagined actions. Therefore, systems governed by the projective consciousness model fundamentally entertain a form of curiosity toward infinity (or more generally toward what is farther), and are intrinsically attracted to it. At the same time, objects associated with positive (or negative) rewards become all the more attractive (or repulsive) as they appear closer and thus bigger, through actual or imagined actions (Rudrauf et al., 2022, 2023). In other words, the action of projective geometry within active inference induces a form of lensing effect of consciousness on information: the closer objects are the bigger they appear subjectively; which entails both epistemic (curiosity) and affective drives under free energy minimization. Through this approach, we could implement virtual and robotic agents, featuring complex adaptive and maladaptive collective behaviors, relevant to psychopathology (see also Rudrauf & Debbané, 2018).

Furthermore, the projective model includes formal properties that appear relevant to prereflective self-consciousness (see Williford's chapter, this book, and Williford et al. [2022]). For instance, it explains the pervasive yet elusive experience we may have when looking at something, where that thing seems to look back at us in a reciprocal manner, reinforcing our awareness that we are both observers and subjects of observation. This phenomenon is expected due to the reciprocity and duality that are unique to projective geometry. In other words, projective geometry entails a fundamental intersubjective structure within subjectivity.

Much remains to be explored and understood, including regarding how such subjective structure and mechanisms relate to other aspects of experience and cognition—for example, language and abstract thinking. Many fundamental theoretical and empirical questions need to be addressed. But at least, we have a framework and model that can serve as guides to further investigate phenomena and derive hypotheses, and contribute to the neurophenomenological endeavor.

Reflections

Francisco's ideas about the essential role of mathematics in characterizing the phenomenology of consciousness and the relationships between the "subjective" and the "objective" encouraged my pursuit of capturing and formalizing core aspects of the phenomenology of consciousness, resonating with intuitions I had since childhood. Projective geometry appeared as a central aspect of consciousness and its subjective perspective, generating new ideas and hypotheses.

A good model should contain more than what we put into it, and, indeed, following the principles of the projective consciousness model, we could understand and model a wealth of apparently heterogeneous phenomena that are central to consciousness and its relation to behavior.

The approach yields hypotheses about the neural implementation of consciousness, and guidelines for searching for its neural correlates. For instance, it predicts that brain processes related to consciousness should reflect the actions of the projective group (perspective-taking according to the laws of that group) on ongoing sensory information. This prediction, in turn, could be quantitatively operationalized thanks to the mathematical definition of the model, and experimentally investigated.

Likewise, it opens promising venues for the development of artificial consciousness and the implementation of virtual and robotic agents operating according to core principles of consciousness (Rudrauf et al., 2022, 2023). This is not to say that those agents would automatically be conscious as we are, just because they would be governed by principles that govern subjective experience. But this is consistent with a scientific endeavor, which can mostly characterize how phenomena relate to each other, sometimes through hidden variables, but not offer an ultimate explanation of why certain relations exist.

The question remains open of how much the gap between subjective experience and more objectively tractable processes—for example, neuronal activity, bodily processes, and observable behaviors—can be reduced. But, to be sure, the passageways of mathematics, including geometry, can lead us in attempting to address that question. Thus, the future of phenomenology may be in the hands of true mathematicians—which I am definitely not.

My own trajectory has been shaped by Francisco's perspective on the problem, sometimes through intuitions he discussed but had not operationalized fully, perhaps for lack of time as his life ended too soon. These intuitions, and his peculiar approach to their conceptualization, resonated with my own intuitions and encouraged me to keep thinking about them as a scientist of sorts. These became like puzzles in my mind, vanishing points of concepts yet to be more closely apprehended, driving my curiosity, and my desire for an elegant and compelling perspective on the issue.

Finally, I would like to emphasize that in the series of workshops on "Perspectives on Consciousness" that motivated this book and chapter, I found both a community and an environment ideal for pursuing the multidisciplinary and experiential inquiry necessary to tackle such questions and, more generally, the mystery of consciousness.

Notes

1. See below for an explanation of that term.
2. "In biology, the extracellular matrix (ECM) is a three-dimensional network consisting of extracellular macromolecules and minerals, such as collagen, enzymes, glycoproteins and hydroxyapatite that provide structural and biochemical support to surrounding cells" (https://en.wikipedia.org/wiki/Extracellular_matrix).

References

Aristotle. (4th century BCE/2016). *The organon, or logical treatises, of Aristotle: With the introduction of porphyry* (literally translated, with notes, syllogistic examples, analysis, and introduction by Octavius Freire Owen). Wentworth Press.

Baars, B. J. (1997). In the theatre of consciousness. Global workspace theory, a rigorous scientific theory of consciousness. *Journal of Consciousness Studies*, *4*(4), 292–309.Cosmelli, D., David, O., Lachaux, J. P.,

Martinerie, J., Garnero, L., Renault, B., & Varela, F. (2004). Waves of consciousness: Ongoing cortical patterns during binocular rivalry. *Neuroimage, 23*(1), 128–140.

Crick, F., & Koch, C. (2003). A framework for consciousness. *Nature Neuroscience, 6*(2), 119–126.

Friston, K. (2010). The free-energy principle: A unified brain theory? *Nature Reviews Neuroscience, 11*(2), 127–138.

Friston, K., FitzGerald, T., Rigoli, F., Schwartenbeck, P., & Pezzulo, G. (2017). Active inference: A process theory. *Neural Computation, 29*(1), 1–49.

Ghyka, M. C. (1977). *The geometry of art and life*. Courier Corporation.

Hofstadter, D. R. (1979). *Gödel, escher, bach* (pp. 582–583). Basic books.

Lutz, A., Lachaux, J. P., Martinerie, J., Varela, F. J. (2002). Guiding the study of brain dynamics by using first-person data: Synchrony patterns correlate with ongoing conscious states during a simple visual task. *PNAS, 99*(3), 1586–1591.

Maturana, H., & Varela, F. J. (1973). *De máquinas y seres vivos: Una teoría sobre la organización biológica* [Of machines and living beings: A theory on biological organization]. Editorial Universitaria.

Maturana, H., & Varela, F. J. (1980). *Autopoiesis and cognition: The realization of the living*. D Reidel.

Maturana, H. R., & Varela, F. J. (1987). *The tree of knowledge: The biological roots of human understanding*. New Science Library/Shambhala Publications.

Rodriguez, E., George, N., Lachaux, J. P., Martinerie, J., Renault, B., & Varela, F. J. (1999) Perception's shadow: Long-distance synchronization in the human brain. *Nature, 397*, 340–343.

Rudrauf, D. (2014). Structure-function relationships behind the phenomenon of cognitive resilience in neurology: Insights for neuroscience and medicine. *Advances in Neuroscience, 2014*(1), 462765.

Rudrauf, D., & Debbané, M. (2018). Building a cybernetic model of psychopathology: Beyond the metaphor. *Psychological Inquiry, 29*(3), 156–164.

Rudrauf, D., Bennequin, D., & Williford, K. (2020). The moon illusion explained by the projective consciousness model. *Journal of Theoretical Biology, 507*, 110455.

Rudrauf, D., Bennequin, D., Granic, I., Landini, G., Friston, K., & Williford, K. (2017). A mathematical model of embodied consciousness. *Journal of Theoretical Biology, 428*, 106–131.

Rudrauf, D., Lutz, A., Cosmelli, D., Lachaux, J. P., & Le Van Quyen, M. (2003). From autopoiesis to neurophenomenology: Francisco Varela's exploration of the biophysics of being. *Biological Research, 36*(1), 27–65.

Rudrauf, D., Sergeant-Perthuis, G., Belli, O., Tisserand, Y., & Serugendo, G. D. M. (2022). Modeling the subjective perspective of consciousness and its role in the control of behaviors. *Journal of Theoretical Biology, 534*, 110957.

Rudrauf, D., Sergeant-Perthuis, G., Tisserand, Y., Poloudenny, G., Williford, K., & Amorim, M. A. (2023). The projective consciousness model: Projective geometry at the core of consciousness and the integration of perception, imagination, motivation, emotion, social cognition and action. *Brain Sciences, 13*(10), 1435.

Rudrauf, L. (1954). The morphology of art and the psychology of the artist. *The Journal of Aesthetics and Art Criticism, 13*(1), 18–36.

Rudrauf, L. (1961). *Les méthodes d'analyse formelle à base mathématique et leur valeur pour la connaissance esthétique et poétique des oeuvres d'art* [Mathematics-based formal analysis methods and their value for the aesthetic and poetic understanding of works of art]. (self-published).

Varela, F. J. (1976). Not one, not two. *CoEvolution Quart, 12*, 62–67.

Varela, F. J. (1979). *Principles of biological autonomy*. Elsevier/North-Holland.

Varela, F. J., Thompson, E. T., & Rosch, E. (1991). *The embodied mind: Cognitive science and human experience*. MIT Press.

Varela, F. J. (1996). Neurophenomenology: A methodological remedy to the hard problem. *Journal of Consciousness Studies, 3*, 330–350.

Varela, F. J. (1997a). Metaphor to mechanism: Natural to disciplined. *Journal of Consciousness Studies, 4*, 344–346.

Varela, F. J. (1997b). The naturalization of phenomenology as the transcendence of nature: Searching for generative mutual constraints. *Alter: Revue de Phénoménologie, 5*, 355–385.

Varela, F. J. (1997c). Patterns of life: Intertwining identity and cognition. *Brain Cognition, 34*, 72–87.

Varela, F. J. (1999). The specious present: the neurophenomenology of time consciousness. In J. Petitot, F. J. Varela, B. Pachoud, & J.-M. Roy (Eds.), *Naturalizing phenomenology* (pp. 266–314). Stanford University Press.

Varela, F. J., & Frenk, S. (1987). The organ of form: Towards a biological theory of shape *Journal of Social and Biological Structures, 10*, 73–83.

Varela, F. J., & Goguen, J. (1977). The arithmetic of closure. In R. Trappl (Ed.), *Progress in cybernetics and systems research* (Vol. 3, pp. 48–63). Wiley Hemisphere.

Varela, F. J., Maturana, H., & Uribe, R. (1974) Autopoiesis: The organization of living systems, its characterization and a model. *Biosyst, 5*, 187–196.

Varela, F. J., & Shear, J. (Eds.). (1999). The view from within: First-person methodologies in the study of consciousness (Special Issue). *Journal of Consciousness Studies, 6*(2–3).

Williford, K., Bennequin, D., Friston, K., & Rudrauf, D. (2018). The projective consciousness model and phenomenal selfhood. *Frontiers in Psychology, 2571*, 1–18.

Williford, K., Bennequin, D., & Rudrauf, D. (2022). Pre-reflective self-consciousness & projective geometry. *Review of Philosophy and Psychology, 13*(2), 365–396.

Figure 0.5
Michal Rovner. Medba East, 2006. Pure pigment on canvas. Copyright: Michal Rovner.

16 Has Quantum Mechanics Something to Say About Consciousness?

Shantena Augusto Sabbadini

Overview

The question that gives the title to this chapter has been with me for a lifetime, from the time I graduated in Milan, Italy, with a thesis on the measurement problem in quantum mechanics. The problem concerns the role of the observer in quantum measurement and therefore, in my view, the precise spot where matter and mind touch each other. It is the at the heart of quantum mechanics. And yet it still does not have a universally accepted solution a hundred years from the inception of the theory.

The difficulty is the fact that the formalism of quantum mechanics includes two parts that seem to be incompatible: one that describes the time evolution of a general quantum system defined by a wave function, the other that describes the measurement process, in which a probability is assigned to each measurement result according to the amplitude of the wave function squared (a prescription called the Born rule). The time evolution applied to the measurement process produces an entangled superposition, typically represented by Schrödinger's famous cat in a box being in a superposition of *dead and alive*, rather than being *either alive or dead.*

One of the early proposed solutions to this paradox directly involved the observer's consciousness. When an observer opens Schrödinger's box and looks inside, the superposition of states "collapses" into a good old either alive or dead cat. In that perspective (called the "idealist" interpretation), consciousness is an essential ingredient of quantum mechanics. But this solution presents serious paradoxical features and is no longer favored by the majority of physicists.

Modern solutions of the quantum measurement problem—for example, the so-called decoherence approach—take a different route: they assume that quantum mechanics correctly describes the measurement process and therefore the outcome really is an entangled superposition including a simultaneously dead and alive cat. But the entangled superposition can be shown to be in all respects equivalent to a classical alternative and therefore we can treat the cat as being either alive or dead.

In this perspective, so long as we interpret quantum mechanics as an essentially statistical science, decoherence solves the quantum measurement problem and consciousness does not appear to play any significant role in it.

But quantum mechanics has nothing to say about the individual result of a measurement. It does not tell us whether in any given instance we will find Schrödinger's cat alive or dead. The outcome in a specific instance is intrinsically unpredictable: it is what Wolfgang Pauli called "an act of creation."

This chapter proposes that such acts of creation happen in consciousness. The world unfolds through continuous acts of creation. While the wave function of the universe evolves in a perfectly deterministic way, compatible with all possible experiences, the experienced world is constantly generating new unpredictable events, new "acts of creation."

Consciousness cannot be accounted for in physical terms.
For consciousness is absolutely fundamental.
It cannot be accounted for in terms of anything else.
—Erwin Schrödinger, 1935

The Quantum Measurement Problem

The question that gives the title to this chapter (or some variant thereof) has been with me for a lifetime, from the time I graduated in Milan, Italy, with a thesis on the measurement problem in quantum mechanics. Since then the journey of my consciousness unfolding took many forms, most of them not directly related to physics: meditation, community building, psychedelics, translating Eastern wisdom texts, and of course relationships, journeys, the whole phantasmagoria of lived experience.

But from time to time the bug of physics would reawaken and I would fall under the charm the of its detached understanding, looking at the eternal questions from the cool third-person perspective of science.

When that happened, I would inevitably go back to my old quantum measurement problem. For good reasons. Because, if modern science has anything to say about the mind/matter interface, and therefore ultimately about the nature of reality, one of the *loci par excellence* of that interface is no doubt the quantum measurement process.

As it is well-known, this key issue, at the heart of quantum mechanics, is still an open question almost a hundred years from the inception of the theory. We have a wonderful theory, whose predictions, even those that seemed most "spooky," have been consistently confirmed. But what the theory really means is still in question. Therefore, in order to discuss the possible implications of quantum mechanics for our understanding of consciousness, we need to briefly review the quantum measurement problem.

A concise description of the problem is given in the following paragraphs. The reader desiring a more detailed understanding is referred to the appendix and to the definition of some key terms given in the **glossary**.

The essence of the measurement problem is the difference between the empirically observed outcome of a measurement process and the outcome calculated by applying standard time evolution to the global system involved in the measurement (observed system, measuring apparatus, observer, environment—technically called *von Neumann chain*).

In the first one, the various measurement results appear to be alternative objectively realized facts, each with a definite probability, and the outcome of a further observation on the global system can be calculated by summing the contributions of these independent facts.

Figure 16.1
Schrödinger's famous cat in a box (1935).

In the second one, applying the continuous time development as realized through the Schrödinger equation the outcome of the measurement process appears to be an entangled superposition (see the **glossary**) of the various results, and the outcome of a further observation on the global system cannot be simply calculated by summing the contributions of the various results. The calculation involves additional terms, called *interference terms* (see the **glossary**), connecting the various results and preventing them from being regarded as alternative objective facts realized with certain probabilities given by the Born rule.

The problem then can also be stated as why in the actual outcome of a quantum measurement processes the interference terms do not appear to be there.

The problem is elegantly emphasized by Schrödinger's famous cat in a box (1935). The object system in this case is a radioactive atom, whose general state is a superposition of decayed and not decayed. The von Neumann chain includes a cat in a box with a devilish contraption that links the state of the radioactive atom to the life or death of the cat. By applying standard time evolution to the system, the outcome is an entangled superposition of product states containing the alive cat and the dead cat.

Many solutions have been suggested in the course of the last ninety years but none so far has obtained the unanimous consensus of the physics community. I will describe a few interpretations that seem particularly relevant for our subject.

The Copenhagen Interpretation

The Copenhagen interpretation (once also known as the "orthodox interpretation") of quantum mechanics, solves the riddle by introducing an axiom ad hoc. It assumes that the dynamics of the interaction is such that the entangled superposition that is the outcome of a measurement process spontaneously collapses into the classical alternative of objectively realized results that is observed in practice (*collapse of the wave function*; see **glossary**). This axiomatic step reflects the experimental evidence and offers a convenient frame for practical calculations. But it can hardly be considered a solution of the quantum measurement problem. In the Copenhagen interpretation one assumes a separation between a classical world accessible to us humans and a quantum world accessible only through measurement. As such the Copenhagen interpretation does not consider the measuring apparatus (and the brain of the observer, if we choose to include it in the von Neumann chain) to evolve according to the laws of quantum mechanics. The collapse of the wave function is incompatible with the linearity of the time evolution operator, and as such it amounts to postulating two different time evolution prescriptions, one applying only to quantum measurement and the other to a general quantum system.

The Idealist Interpretation

Among the early proposed solutions of the quantum measurement problem, the one that stands out for involving consciousness in an essential way is the "consciousness causes collapse" or "idealist" theory. It was hinted at already in von Neumann's *Mathematical Foundations* and was later developed by London and Bauer in 1939 (London & Bauer, 1939) and by Eugene Wigner in the 1960s (Wigner & Margenau, 1967). An interesting variation of it is the theory proposed in our times by Henry Stapp (Stapp, 2001). This interpretation accepts the notion of the collapse of the wave function and associates it with a conscious subject experiencing a definite result and thus collapsing all probabilities to zero except for the one corresponding to the actual event.

The idealist interpretation, although not strictly confutable, has some rather weird implications, which probably contributed to its losing favor in the physics community. Consider, for example, Schrödinger's cat. According to the idealist interpretation the entangled superposition will collapse into the corresponding classical distribution when an observer opens the box and looks inside. But what do we mean by "observer"? Should the cat count as an observer? And if so, how far down the chain of life forms does consciousness collapse the wave function? What about an insect? Or a bacterium?

But even if we limit ourselves to human observers, it is easy to imagine paradoxical consequences. For example, what happens if the observer opens the box without looking inside, takes a Polaroid photo of the content, puts away the photo in a drawer and a month later takes it out and looks at it? For the whole month the cat will have been a superposition of a dead cat in the box and an alive cat going about her business on the roofs.

Also, what about all the cosmological eras before conscious observers appeared in the universe? Was it all an entangled superposition until we collapsed it into an actual universe?

The Many-Worlds Interpretation

A number of recent approaches to the quantum measurement problem give up the notion of collapse of the wave function altogether and assume that quantum measurement is to be entirely described by standard quantum mechanics. The classic example along these lines is the Everett–DeWitt many-worlds interpretation (1957). It takes the entangled superposition that is the outcome of the measurement process at face value, but assumes that each term of the superposition unfolds in a separate universe. In a quantum measurement therefore the world branches off into different worlds, identical except for the different outcomes of the

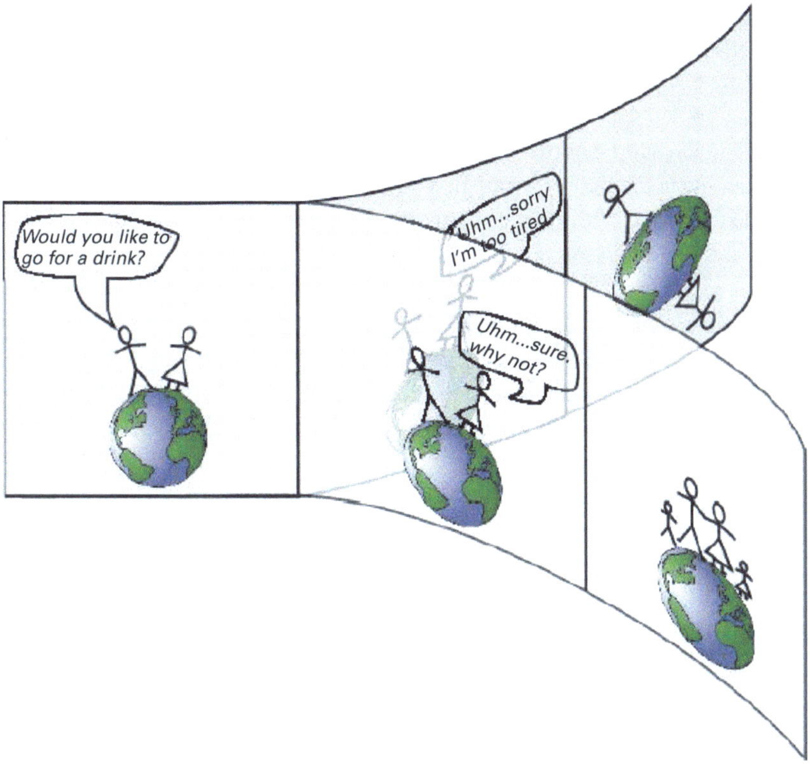

Figure 16.2
A representation of the split that occurs based on the possible outcomes for each action according to Everett's many-worlds interpretation (courtesy of Max Tegmark).

measuring process. The difference involves the whole von Neumann chain, including the observer, who therefore exists in slightly different versions in each universe. Each version has no notion of the existence of other copies of herself or himself and sees the specific outcome of the experiment in her or his world as the only one.

However, one might feel about the idea of multiple universes (an idea that is popping up from a completely different context in cosmology) the main difficulty of Everett's theory is probably a technical one—namely, how to interpret the Born rule in this context. Why does each result appear with just the correct probability in a sequence of experiments conducted in any one world?

Decoherence

Perhaps the most widely accepted interpretation of quantum mechanics today is the one known as *decoherence*. That is the interpretation that I have been involved in, in two stages, forty years apart. The first one was my graduation at the University of Milan in the late 1960s, before the approach was even called decoherence (Lanz L., Prosperi, G. M., & Sabbadini A., 1971). The second was in the years 2000s (Sabbadini S. A., 2006), long after I had left the academia.

The core idea of the decoherence interpretation is that in a macroscopic environment the interference terms decay extremely rapidly. The information they contain diffuses through entanglement with the environment and is lost for all practical purposes.

That means the entangled superposition now extends to the entire macroscopic environment (including the macroscopic apparatus), which we are able to observe only to a very limited extent. When the entangled superposition gets projected onto the observations that we are actually able to make, the difference between entangled superposition and classical alternative vanishes. Only a measurement of the quantum state of the entire macroscopic environment would be able to distinguish them.

The decoherence approach can be formulated in a slightly different way, which I have called *persistence of information.* It is possible to prove that a measurement able to distinguish between the entangled superposition and the classical alternative is necessarily incompatible with all the von Neumann observables. But then such a measurement erases all traces of the previous measurement: it is as if the measurement had not occurred. On the other hand, for all the observations that preserve information about the outcome of the measurement the entangled superposition and the classical alternative are equivalent, they give exactly the same results. If we call "proper measurement" a measurement whose outcome is somehow recorded, then we could state that for all proper measurements the entangled superposition and the classical alternative are equivalent—that is, the evolution of the system can be calculated from the classical alternative as well as from the entangled superposition. Persistence of information, the existence of a trace, is the decisive criterion.

In decoherence there is no collapse of the wave function, but it is as if there were. In other words, the collapse of the wave function is an apparent phenomenon. Its appearance is due to the following:

- The fact that a measurement necessarily involves a macroscopic system (standard decoherence).
- The fact that all proper measurements involve a trace or record (persistence of information).

In all measurements whose outcome we get to know, a human brain is at some point involved. Then the experience gets recorded in the person's brain and the persistence of information criterion is fulfilled. Therefore in all measurements whose outcome we get to know, the outcome can be taken to be a classical distribution. We could extend that argument to perception in general: we perceive the world as classic because all our experiences correspond to trace formation in our brain.

What about "improper measurements," measurements that leave no traces? When no trace is left, the further evolution of the system must be calculated using the full entangled superposition, not the classical distribution. We have a wonderful example of this in the quantum eraser experiments (see appendix 2 in this chapter).

Acts of Creation

Does decoherence solve the quantum measurement problem? It depends on what we require from a solution. If we accept the fact that quantum mechanics is an essentially statistical theory and accept that there is no causal explanation for the single event (no hidden variables!), then yes, decoherence is a complete solution of the quantum measurement problem.

But we may be drawn to investigate further. In decoherence (as in all other nonhidden-variables interpretations) the single event remains unpredictable, indeterminate. You can look at this indeterminacy from two points of view, which are logically equivalent, but have a very different emotional resonance. You can say it is randomness, no meaning attached. Or you can look upon it, as Pauli did, as an "act of creation," or, as Heisenberg said, "nature's choice." All the rest is determined; it is Laplace's clockwork universe ticking away in a perfectly deterministic manner. The quantum state of the universe evolves in a deterministic way (or perhaps does not evolve at all, if we take the Wheeler–DeWitt equation seriously). The only space for creation, for the unpredictable, is the single event that arises at the interface between consciousness and world

We are now ready to discuss the question raised by the title of this paper: Has quantum mechanics something to say about consciousness? On one level the answer is no. The observer's consciousness does not play any role in the solution of the measurement problem. From the point of view of persistence of information just the existence of any record or trace will do. And from the point of view of standard decoherence, only the macroscopic nature of the measuring apparatus is required. The solution of the quantum measurement problem does not require the observer's consciousness at all.

But quantum mechanics suggests where we have to look if we want to see consciousness's footprint in matter: it is there where the predictable ends; it is in the act of creation of the single event. We live in an actual world, we experience actual events: out of the statistical picture one definite actuality emerges in any specific instance. As a metaphor, you can think of the algorithm of quantum mechanics (wave function, time evolution, Born rule for calculating probabilities) as the banks of a river, and the indeterminacy of the single event as the water flowing within the banks. The flow is constrained by the banks, but is otherwise unpredictable, "nature's choice." It is there, I would like to suggest, that we have to look if we want to find the expression of consciousness in matter.

The question that naturally emerges at this point is: Whose consciousness? There is no "who" this consciousness belongs to. It is simply consciousness, the consciousness of the universe, the consciousness of *deus sive natura*, in Baruch Spinoza's terminology. The consciousness that is involved in the emergence of the actual from the womb of the potential is cosmic consciousness, it belongs to the whole.

Consciousness and Life

I would like to propose (this last section is highly speculative) that there is a strict parallelism between consciousness and life and that they are both related to quantum indeterminacy, to the "acts of creation" that are the unpredictable individual events. When we adopt this perspective, the old philosophical conundrum about origins falls away. Where does consciousness arise from unconscious matter? Where does life emerge from the non-living?

There is no "where." Consciousness and life are everywhere. We are invited to extend our notion of life and call life all manifestations of freedom. This includes quantum indeterminacy on the microscopic scale as well as quantum indeterminacy manifesting, in ways that we are still far from understanding, in the macroscopic world, in large-scale coherent phenomena (some recent brain studies point in this direction).

Therefore, there is no sharp divide between life and non-life. I would like to propose that freedom, life and consciousness are different names for the same thing, and the "thing" is everywhere. It appears as quantum indeterminacy on a microscopic scale, as life on a macroscopic scale, as consciousness on a cosmic scale.

This is a very loose definition of both life and consciousness, probably too loose to be practically useful—for example, in biology. But it may well be the only philosophically sound definition, because there is no specific attribute "life" that is not there up to a point in the development of an organism and comes in at that point; there is no specific attribute "consciousness" that is not there up to a point in the development of a nervous system and comes in at that point. All that exists are degrees of complexity. We recognize life and consciousness when they appear in forms close enough to our degree of complexity. We humans are not special—except for the fantastic complexity of the way in which life and consciousness manifest in us. But everything is special. If there is no specific "life" attribute; there is also no specific "non-life" attribute. We are equally justified saying that life does not exist or saying that everything is alive.

Phenomenological Experience and Philosophical Speculation

The Schrödinger quote that opens this chapter captures the essence of what I believe to be a correct rapport between phenomenological experience and philosophical speculation. Phenomenological experience is primary. All the rest is "about" it. All our notions about matter, including our own body and more particularly our brain, are third-person attempts to make sense of our first-person experience of the world.

The fact that such attempts are at all possible is in itself significant (Einstein is quoted to have said that the most incomprehensible fact about the universe is that it is at all comprehensible!). It points to the fact that our first-person experience of the world is structured. The world is a cosmos, not a chaos. Our experience appears to unfold in space and time and to possess that sort of regularity that we call matter. That allows us to construct third-person representations of the world and to speculate about the relation between mind and matter. In those third-person representations consciousness appears to be located "in the world," which leads scientific investigation to ask questions like how does consciousness arise from matter (the "hard problem").

Such questions are not philosophically sound. A better posed question is, how does matter arise from consciousness?

Appendix 1: The Quantum Measurement Problem

The classical analysis of the quantum process of observation is contained in von Neumann's *Mathematical Foundations of Quantum Mechanics* (1932). In the von Neumann description a quantum measurement involves a chain of interacting systems that undergo changes in a correlated way, so that the value of the observable O in the object system S determines the value of an observable M_1 in the next system in the chain, which determines the value of an observable M_2 in the next one, and so on. The whole chain is called the "von Neumann chain," and the observables O, M_1, M_2, . . . are called von Neumann observables. At one end of the chain there is the microscopic system undergoing observation, somewhere along the line there is an amplifier that turns a microscopic signal into a macroscopic one, and at the other end of the chain there is a recording device, or perhaps a human being—we don't need to commit ourselves to a specific choice for now.

For the sake of simplicity in the following I'll consider a von Neumann chain consisting only of the microsystem S being observed and a measuring apparatus M. The observable M of the apparatus whose value gets coordinated with the value of O of the object system will be called "pointer position."

If U_t is the time evolution operator of the composite system $S + M$ during the interaction, the above description of a measurement process implies

$$U_t\phi_n \otimes \Phi_0 = \phi_n \otimes \Phi_n, \tag{16.1}$$

where ϕ_n is an eigenstate of O corresponding to $O = o_n$ and Φ_n is an eigenstate of M corresponding to $M = m_n$. Therefore, by the linearity of U_t, if S is initially in a *superposition* of states corresponding to different values of O

$$U_t \sum_n c_n\phi_n \otimes \Phi_0 = \sum_n c_n\phi_n \otimes \Phi_n, \tag{16.2}$$

The composite system $S + M$ ends up in a superposition of states corresponding to the various possible results. In the language of statistical mechanics, the system is represented by the density matrix (or statistical operator)

$$W_t = P_{\sum_n c_n\phi_n \otimes \Phi_n}. \tag{16.3}$$

On the other hand, the usual axiomatic formulation of quantum mechanics requires the outcome of the measurement process to be such that in a fraction $|c_n|^2$ of the cases the object system is in the state ϕ_n and the measuring apparatus is in the state Φ_n, a situation described by the density matrix

$$\overline{W_t} = \sum_n |c_n|^2 \, P_{\phi_n \otimes \Phi_n} \quad \text{(Born rule)} \tag{16.4}$$

The quantum measurement problem has to do with the compatibility of the two prescriptions (16.3) and (16.4), the first a direct consequence of the linearity of time evolution, the second required by the axiomatic structure of the theory in accord with experimental evidence.

The outcome described by (16.3) is an entangled state of object system and measuring apparatus, a state in which the values of the observable O and the pointer positions are correctly correlated, but no definite result is actualized. All the possible states of the object system and the corresponding pointer positions are simultaneously there, and no result is singled out as actual.

No progress is realized if one extends the chain beyond the apparatus and includes the observer's brain as a further interacting quantum system. If we indicate with B_n the state of the observer's brain corresponding to seeing the pointer in the n position, then applying the same reasoning as before we arrive at the final state

$$W_t = P_{\sum_n c_n \varphi_n \otimes \Phi_n \otimes B_n} \tag{16.3'}$$

as opposed to

$$\overline{W}_t = \sum_n |c_n|^2 \, P_{\varphi_n \otimes \Phi_n \otimes B_n} \tag{16.4'}$$

The observer's brain has become part of the entangled totality, but again no definite result stands out.

Equations (16.3) and (16.4) (or [16.3′] and [16.4′]) predict different probabilities for the outcomes of observations performed on any of the systems in the von Neumann chain. While the results calculated with (16.4) neatly decompose into the sum of contributions corresponding to each value of n, the results calculated with (16.3) contain additional terms, called *interference terms*, mixing the states corresponding to different values of n.

The Quantum Eraser

Figure 16.3 illustrates one version of the quantum eraser experiment. A photon gets "split" at the beam splitter BS_1, travels along two paths in a Mach-Zender interferometer, gets recombined with itself at BS_2 and is detected at detector D_1 or D_2. When one inserts a phase shifter (effectively varying the length of one of the paths) an interference pattern appears at detector D_1 or D_2. Then two down-converter crystals, X_1 and X_2, are inserted on the paths. When the primary photon crosses one of the down-converter crystals, a secondary photon is generated and gets detected by detector d_1 or d_2, providing information

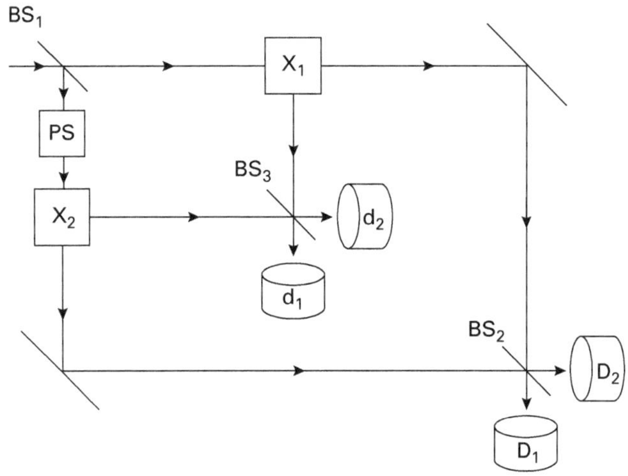

Figure 16.3
The quantum eraser experiment.

on which path the primary photon has traveled. In such configuration the interference pattern disappears. But the clever feature of the experiment is that the secondary photons can be recombined by inserting a third beam splitter BS_3 on their paths, thus erasing the "which" path information. When that is done the interference pattern reappears.

The experiment closely parallels a quantum measurement process. The measurement here concerns which path the primary photon has traveled. The detection of a secondary photon at d_1 or d_2 provides which path information for the primary photon. According to the persistence of information theorem, the existence of such information guarantees the equivalence of the density matrixes (16.3) and (16.4) (see appendix 1) and therefore the vanishing of the interference terms.

But, unlike what happens with a macroscopic measuring device, here the "which" path information can be erased by inserting the third beam splitter BS_3. We are then in an "improper measurement" situation, in which there is no recording of the "which" path measurement. In this case, the density matrixes (16.3) and (16.4) are not equivalent: (16.3), not (16.4), is the outcome and the interference terms reappear.

An interesting side point is that the quantum eraser can also be performed as a *delayed choice* experiment. The erasure by inserting the third beam splitter can take place after the photon has already come through the first beam splitter, and therefore, if we visualize it as a particle, is already traveling along one path or the other. It can also take place when the photon has already interacted with the down-converters X_1 and X_2 and therefore a secondary photon has already been generated.

This aspect of the experiment looks like a *retroactive effect in time*, and it has been the object of much discussion.

Glossary

Born rule: The rule for calculating the probability of obtaining individual results in the measurement of a quantum observable on a system that is in a superposition of states.

Collapse of the wave function: The assumed reduction of a superposition of states to a single component of the superposition when a measurement is performed on the system. This assumption, which was at the base of the Copenhagen interpretation of quantum physics, is highly problematic and still the object of widespread controversy. It is the at the core of the quantum measurement problem.

Density matrix: A generalization of the notion of wave function describing a statistical collection of systems.

Entanglement: A specifically quantum mechanical notion that has no equivalent in classical physics. The states of two quantum systems can be tied together in such a way that a measurement performed on one of them influences the statistical results of measurements performed on the other one.

Entangled superposition: See "entanglement" and "linear superposition."

Hilbert space: A complex vector space endowed with an inner product operation that allows the notions of distance and angle between vectors to be introduced.

Linear operator: An operator that acting on a linear superposition of states produces a linear superposition of its action on the individual states. All quantum mechanical operators are linear.

Linear superposition of states: A specifically quantum mechanical notion that has no equivalent in classical physics. A quantum system can be in a linear superposition of states in which an observable of the system has two or more different values at the same time. A typical example is Schrödinger's cat being alive and dead at the same time.

Measurement (quantum): The measurement of an observable of a quantum system. This generally involves the amplification of a signal from the microscopic (quantum system) to the macroscopic (measuring apparatus) scale. The result of the measurement is in general unpredictable. Only the statistical probability of individual outcomes in a collection of identical measurement can be predicted (see "Born rule").

Observable: A physical quantity of a quantum system that can be measured. In quantum mechanics it is generally represented by an operator in the Hilbert space of the system.

Operator: A mathematical "machine" that when given as input an element of a mathematical space produces as outcome an element of the same (or of a different) mathematical space. In quantum physics, an operator generally acts on the state of a quantum system (a vector in a Hilbert space) and produces a different state of the same system (another vector).

State vector: See "wave function."

Statistical operator: See "density matrix."

Superposition: See "linear superposition."

Time evolution operator: The operator that acting on the state of the system at a given time produces the evolved state of the system at a later time.

Vector: A mathematical entity that can be intuitively represented as an arrow.

Wave function: The mathematical expression describing the state of a quantum system. It is generally represented by a vector in a Hilbert space.

References

Everett, H. (1957). Relative state formulation of quantum mechanics. *Reviews of Modern Physics, 29*(3), 454–462.

Lanz L., Prosperi, G. M., & Sabbadini A. (1971). Time scales and the problem of measurement in quantum mechanics. *Il Nuovo Cimento, 2B*(2), 184–192.

London, F., & Bauer, E. (1939). La théorie de l'observation en mécanique quantique (English translation in *Quantum theory and measurement*, J. A. Wheeler, & W. H. Zurek [Eds.], [1983], pp. 217–259). Princeton University Press.

Sabbadini S. A. (2006). Persistence of information in the quantum measurement problem. *Physics Essays, 19*(1), 135–150.

Schrödinger, E. (1935). Die gegenwärtige Situation in der Quantenmechanik [The present situation in quantum mechanics]. *Naturwissenschaften, 23*(48), 807–812.

Von Neumann, J. (1932). *Mathematische Grundlagen der Quantenmechanik* [*Mathematical foundations of quantum mechanics*]. Julius Springer.

17 The Neurophenomenology of Voluntary Action: The Libet Experiment with an Experienced Meditator

Stefan Schmidt

Truths are as fragile as the people who have formulated them.
—Andreas Steinhöfel, 2005

Overview

Biology, chemistry, and physics are the building blocks of a materialist worldview of consciousness; conversely, consciousness is needed in order to comprehend and investigate these fields. This circularity and the apparent lack of an independent foundation of science are major challenges for consciousness research and are addressed in this chapter. Here, the need for a meaningful combination of first-person and third-person data is explained and emphasized. A fruitful example for this approach is the famous Libet experiment, which presents an apparent paradox about free will. This experiment was expanded by the introduction of the first-person perspective of a very experienced meditator. In a narrative approach with the research team, joint strategies on how they could act differently in the Libet task were developed and subsequently assessed by electroencephalography (EEG) (see **glossary**). While most other researchers try to rule out subjectivity in the Libet task, our approach was driven by the goal of making subjectivity a more reliable source of information. Our participant had the stability of attention and the mental capacity to observe and produce fine-grained differences in the process of initiating a voluntary act in the Libet task. The resulting EEG data provided support for theoretical models and finally resulted in a new interpretation of the Libet paradox termed the slow cortical potential sampling hypothesis.

In this empirical story, the circularity of organism and environment, as described in the model of Varela and Maturana, enabled a new perspective and new insights. It also demonstrates that there is no ultimate starting point or foundation in science, but just "worlds without grounds" (Varela et al., 1991, p. 254).

My Journey into the Study of Consciousness

I am a scientist because I am a curious person. I always wanted to understand how the world works, and why. I aim for an understanding of the big picture. I was particularly interested in the mysteries that science could not solve: experimenter effects, telepathy, shamanic healing, and more. Rupert Sheldrake was an important source on these subjects (e.g., Sheldrake, 1988). Unsurprisingly, there was no academic discipline at any university that covered all of my interests. After starting and failing at physics, I decided to study psychology for the simple reason that it was the least restrictive discipline and encompassed my broad interests.

I was quickly socialized as a normal academic psychologist, thinking and reasoning about human behavior in terms of quantitative data. The bigger picture only became clearer when I had to pick a topic for my master's thesis. I looked for a topic where the outcome was not already known or could be inferred in advance, a real open experiment. I finally conducted an experiment on anomalous perception (identifying a poisonous substance with the sole use of a dowsing rod; see Walach and Schmidt [1997]), which I found very satisfying. I continued with a PhD thesis in parapsychology, studying physiological effects of distant intention (Schmidt, 2002).

While conducting my distant intention experiments, I learned a lot about the foundations of science. These were not as firm as I expected. I was surprised to see that there were quite a lot of empirical results in parapsychology that did not fit into the current scientific worldview, and I was also surprised to see how the scientific mainstream dealt with these issues (Cardeña, 2018; Walach & Schmidt, 2005). In parallel, I started training as a systemic counselor and learned a lot about systems theories. An important moment was my encounter with the systems biology approach by Humberto Maturana and Francisco Varela (1987), and the resulting debates on constructivism (see **glossary**).

A Circular Model and "Worlds without Grounds"

Based on such a constructivist epistemology the best thing one can do in science is to develop a good model, which gives us coherent explanation for a wide array of phenomena. I understood that science cannot reveal eternal truths but only draw maps of the territory we live in. It became clear that our models about the world are not only shaped by the outside world they describe, but also determined by the organisms—that is, humans—that develop these models. Systems theories and cybernetic approaches showed me that there is no firm foundation in science from which to start. They also emphasized that any good scientific model needs to explain its own origination. Heinz von Foerster wrote, "A theory of the brain, that has any aspirations for completeness, has to account for the writing of this theory" (von Foerster, 2003, p. 289).

In 1996, I went to the "Toward a Science of Consciousness" conference in Tucson, Arizona. The science of consciousness was just about to become mainstream and the highly interdisciplinary audience was buzzing. Like everybody else, I was also puzzled by the mind-body problem, or the "hard problem" of consciousness, as it was called by Chalmers (1995). From thereon it took me some years and many discussions before I could see for the first time the *circularity* of the mind-body problem. Initially, I started from a materialistic worldview. I had the view that it all started with the Big Bang, the development of matter and finally, after many steps, with the mysterious emergence of consciousness out of the brain. But then, I realized that consciousness is already needed in the first place to develop these models. From a phenomenological point of view, one can directly experience some phenomena—for example, the structure commonly described as a keyboard, which I am typing on in this moment. But other scientific facts such as electric current, with its electrons and atoms behind the keyboard—they are more theoretical and elusive in nature and one cannot experience them directly (Frank et al., 2019). They already belong to a set of models of the world, or more precisely, to the set of models and constructs that our society has agreed on as a description of the world.

Seeing things in this way, scientific models that describe the world can start either from the Big Bang or from personal experience. There are some good justifications for both approaches but neither of these two very different strands can tell the complete story. This is a classical chicken and egg problem. Most of the problems of this kind can be solved by giving up linear causal reasoning and the search for a first cause. Instead, one can embark on models of circular causality that describe co-dependent causality, a kind of co-evolution. Both parts belong to the same system and in a self-organizing fashion develop certain properties by mutual interaction. Such a circularity of mind and environment is described in the systems biology approach of Maturana and Varela (1987). The epistemological essence of their approach has been further developed in what is now called *enactivism* (see **glossary**) (Stewart et al., 2014). Alongside the biological evolution, the organisms interact with their environment and both shape each other. The models we have from the world are created by our continuous interaction with the environment, and the organisms that develop the capacity to create models are shaped by the environment they live in. Thus, the world we live in is brought forth by us. It is *enacted.* Or as Varela expresses more poetically, we are "laying down a path in walking" (Varela, 1987).

What is the consequence of such an approach to conducting science? The models we can create about the world reflect both the capabilities of our organism and the environment we live in. Consciousness is neither inside nor outside; it encompasses the permanent interaction of the organism with its environment. In such a model, a crucial role is played by an entity that has for a long time been mostly neglected in the study of consciousness: the *body*. Since all interactions with the environment are bodily interactions, consciousness is tied to these processes. This is when we speak about *embodiment* or *embodied cognition.*

To arrive at that understanding and to acquire a taste for a scientific model that is so different from the one I was socialized with, it took some more ingredients. After completing my PhD thesis, I looked for a new topic. I found many, but none of them touched me in a way that mattered in my personal life and gave me a purpose and meaning. In 1998, I undertook a silent Vipassana retreat. This was my first encounter with mindfulness practice and soon I embarked on this topic. The practice of mindfulness and meditation was important to me personally and was also interesting from a scientific perspective. It connected my deep personal interests with my work as a scientist, and this perfect match gave me meaning and motivation. In more than twenty years of mindfulness and meditation research, I always had the feeling that the topic and the questions raised by my work inspired me as a person and vice versa.

By studying meditation and some of the Buddhist background, I understood that the introspective study of consciousness by meditation and the way it works is crucially important for understanding the world. I learned that calming the mind through meditation is an excellent prerequisite for such an introspective approach. On several occasions I was surprised how close Buddhist concepts and theories came to the latest developments in Western science—for example, the notion of self. It felt a bit like there is a hidden path that can be found from both sides.

It was only later on that I discovered that Varela et al. (1991) had already seen these parallels in the foundations of science and Buddhism in 1991. If science is seen in such a self-referential, circular, and evolutionary way as they had described it, this means that there is no firm foundation and no ontological grounding to our scientific models. The same is true for the Buddhist ontology (see **glossary**) and according to Varela, Thompson, and Rosch, our challenge and future task is "building and dwelling in worlds without grounds" (1991, p. 254).

First- and Third-Person Perspectives

My circular understanding of mind and world resulted in ideas about the circularity of methodological approaches concerning first- and third-person perspectives. Since 2006 we had been conducting contemplative research in our EEG laboratory with experienced meditators, and in this research we were troubled by the big gap between what was recorded by an EEG of a meditator's brain and the personal experience that the person had at the same time. The EEG merely displayed changes in the electric potential over the scalp. The meditators reported deep and sometimes rich inner experiences during meditation that were often very difficult to describe with language. How did these two worlds fit together? How could our research advance our understanding of what meditation is and how the brain works if we could not align these two sources of knowledge in a much better way? In psychophysiology, this problem has usually been solved by ignoring experience and looking at the EEG chart only.

In our meditation research, we were not satisfied with this as a solution because only the meditators themselves could tell whether they were at some specific time focused, paying attention, distracted, in a deep state of calm, or agitated and restless. In order to understand the EEG signatures, we needed this information and we tried to obtain it in various ways. For example, we asked meditators to press a button when they lost attention and became aware of it or we asked them to draw a graph indicating different epochs and depth of meditation over a 30-minute period.

This situation changed when we started a very fruitful collaboration with Tilmann Borghardt, also known as Lama Lhündrup.[1] Tilmann had studied medicine; he was also interested in Buddhism, had meditated from a young age, and had finally decided to become a monk in the Tibetan Kagyü tradition (see **glossary**). He spent seventeen years in a monastery in France as a teacher for three-year retreats. Tilmann had a unique capacity to speak about inner states and experiences using very clear and precise language. In addition, being a physician, he also has an understanding of the brain and the processes of science. Listening to Tilmann's experiences during our first EEG session revealed that not only were we able to learn about his inner experiences, but also about how to draft our research questions and what to look for in the respective EEG data. At this stage a circular process started. Tilmann was reporting about his experiences and we simultaneously tried to adapt our EEG experiments and especially our analyses to the information he gave us. This provided new data, which were then discussed together and led to the next set of modifications. This reciprocal combination of first- and third-person perspectives turned out to be a very fruitful process in advancing our understanding of the EEG signals of a meditating person.

From an epistemological perspective, this procedure evolved more or less accidentally in our lab, and it led to some decisive changes. Until this point, meditators had been the passive objects of research that were investigated to gain an understanding of brain processes during meditation due to long-standing experiences in meditation. Tilmann changed this role; he became an active member of the research team that debated the data, reached conclusions, and formulated hypotheses. In some of the publications on this research, Tilmann is the only participant and also the coauthor of the paper (Jo et al., 2013, 2014; Winter et al., 2020).

The role of the meditation practice itself also changed. While we started by investigating meditation as *scientific object*, meditation later became a *scientific method*. Tilmann used his ability to calm his mind and to maintain a continuous attention toward his experience to get introspective insights into mental processes. In this sense, experienced meditators can be employed by science to study phenomena that request stable attention and/or reliable introspective reports (Lutz et al., 2007).

The *neurophenomenological* approach developed in our lab is characterized by the following features. (1) First- and third-person data are aligned by a mutual and iterative process that continuously generates new hypotheses, new experiments, new introspective reports, and new data. (2) Meditation serves as a method to obtain more reliable introspective reports. This is related to the attentional stability of experienced meditators, and to their ability to repeatedly access certain mental states, but it also requires us to find a common language between researchers and meditators. This means developing concepts that can be aligned with the concepts that are used to speak about third-person data. (3) As a consequence of (1) and (2), meditators themselves become an active part of the research. They need to have an understanding of basic scientific procedures involved in this process and serve a double function, which needs to be reflected on carefully.

The most fruitful application of this specific neurophenomenological approach was our research on *voluntary action initiation* in the Libet experiment (see also Schmidt et al., 2024).

The Readiness Potential, Libet, and the Free Will Debate

In 1964, Kornhuber and Deecke discovered the so-called readiness potential (see **glossary**) as an EEG phenomenon (Kornhuber & Deecke, 1965). In their experiment, they asked participants to lift their index finger at a time point of their own choice and recorded the EEG simultaneously. They repeated this procedure many times and averaged the EEG data, which was time-locked to the onset of the movement. In a control condition to this voluntarily initiated movement, the finger of the participants was lifted passively by a string. They found a steadily rising negative potential (see **glossary**) in the EEG over the central area of scalp. This negative potential preceded a voluntary movement but not a passive one. It started approximately 1.5 seconds prior to the movement. They interpreted this signal as being related to the preparation of the upcoming movement and called it the *readiness potential* (RP).

In 1983 Benjamin Libet replicated this experiment and added another feature (Libet et al., 1983). Participants now sat in front of a screen and could see a clock hand rotating with a speed of 2.5 seconds per rotation. Again, they were asked to lift their finger at a time point of their own choice but this time they were also asked to observe the position of the clock hand at the moment they sensed the first *urge to act* or had a subjective experience of *wanting to move*. Libet called this time point "*w*-time." He found a somewhat paradoxical result. While the RP started approximately 1.5 seconds (i.e., 1,500 milliseconds) prior to the movement, the urge and decision to move was reported at approximately 240 milliseconds. Thus, the signal interpreted as movement preparation preceded the conscious decision to move by a substantial time interval (see also Dominik et al. [2024] for an overview on this paradigm).

This was the starting point of a long-lasting debate regarding *free will*, as many argued that this finding challenged the notion of free will (see, e.g., Klemm, 2010). This interpretation is based on the obvious fact that some brain processes already indicate the preparation of a movement *before* the person has the conscious experience of deciding to move. From a neuroscientific perspective, we see in Libet's experiment a paradox of sequence and timing. Here we will address this paradox only within the reference frame of neuroscience. This is, because the debate on free will in combination with the Libet paradigm requires additional presuppositions about the relationship between neural events on one hand and subjective experience on the other. Most discussions linking Libet's data and free will assume a reductionist relationship between experience and brain processes without making this explicit. However, from a non-reductionist perspective the timing paradox in Libet's data and the question of free will cannot be linked so easily.

The Libet Experiment with Experienced Meditators

The Libet experiment raised many intense discussions and still does. What are the weak points in Libet's study? How could the mismatch in timing be explained in a way that aligns the brain processes of preparation as seen in the RP with subjective timing of the participants? Many of the early debates centered on the reliability of the subjective reports of *w*-time. While we are able to measure the onset of changes in the EEG to the millisecond, the process of recording *w*-time is less exact. The participants are asked to observe the clock hand and to align it to an inner process. This raises some questions. How long does it take to realize consciously the position of the clock? Is there a delay? How long does it take to realize a felt urge? How long does it take to align these two observations coming from different sources? We know from research into consciousness that these are not trivial questions. For example, the binding problem refers to the fact that different features of an observed object (e.g., the roundness and the redness of a red ball) are processed in the brain not exactly at the same time but are nevertheless aligned within consciousness into one coherent experience. Therefore, it looks like there are some hidden processes in the brain where the timing of events are shifted and aligned with other events. Thus, some researchers tried to get rid of the subjective report and to replace it by some variable that can be measured unambiguously. However, it turned out that this is close to impossible with respect to the "urge to move" as this is by definition a subjective experience. Other strategies were to make the measurement process more reliable by, for example, replacing the clock hand with series of distinct letters being presented one after another and asking the participant which letter was present when they first realized the urge to move (e.g., Alexander et al., 2016). Nevertheless, the basic problems of the external timing of inner subjective experiences remained.

The idea to conduct the Libet experiment with experienced meditators followed from the idea to make the subjective report more reliable by asking people that have long-standing experience in the meticulous observation of inner states (Bornemann & Singer, 2017; Fox et al., 2012). Our initial hypothesis was that meditators have an earlier *w*-time than people without meditation experience. We assumed that through a stable and calm observation of the mind the urge to move is detected and reported earlier by meditators, perhaps even so early that it corresponds with the timing of the RP and thus solves the paradox. We invited Tilmann and five matched participants without meditation experience to perform the Libet experiment (Jo et al., 2014). The results of this first experiment are shown in figure 17.1. It shows the mean values and the variance (see **glossary**) of the performance of these six participants. Tilmann had a mean *w*-time of 230 milliseconds prior to the button press and came quite close to the often reported mean of 240 milliseconds. It is obvious from the graph that our hypothesis was not confirmed since Tilmann did not display a decisive earlier "urge" than the control participants. He showed a remarkably small variance in the forty trials but so did two other participants from the control group.

The Libet task is not as easy as it seems at first glance. The person is asked to press a button at a moment of their own choice while observing a clock. The person is not allowed to pre-plan the button press. The voluntary action needs to be spontaneous. However, at the same time the person has to observe their own mind for the upcoming urge and also to remember the position of the clock hand at that particular moment. Participants usually need quite some time and have to perform several test trials to get acquainted with the task. Tilmann, with his long-standing experience in observing his own mental activity, needed hardly any training and was immediately able to integrate the various aspects of the Libet task. We later conducted another study in which twenty experienced meditators were compared to twenty non-meditator controls in performing the Libet task. When we performed the Libet task according to the standard protocol as describe above, no difference was found between the groups. Then we repeated this procedure with only one change in the instruction. This time we asked all participants *explicitly* "to pay attention to their inner impulse, intention, or inner feeling that emerges internally in the process of making a voluntary movement" (Jo et al. 2015, p. 151). When looking at the distribution and temporal development of the electrical potential over the scalp with the EEG during this process, we found for meditators more or less the same pattern in both tasks. However, in non-meditators the respective pattern changed markedly when performing the second task where the instructions where more geared toward introspection (Jo et al., 2015). We interpreted this finding as another indication that meditators have improved and more stable introspective capabilities.

At about the same time as we performed the first experiments with Tilmann, we became aware of a different interpretation of the Libet experiment based on the idea of *backward sampling* (see **glossary**). Also at that time, we made another hypothesis, regarding an electrophysiological correlate of the felt urge to act, that we wanted to follow up. To test these

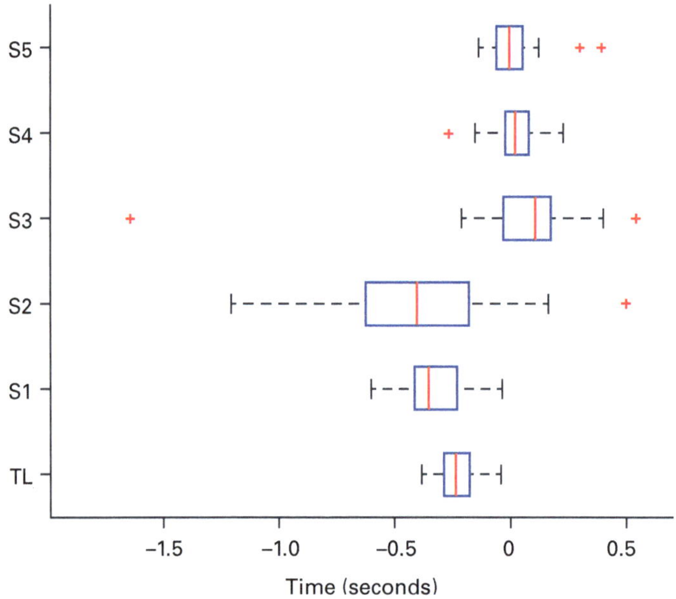

Figure 17.1
Boxplots of subjective reports for the Libet task of an experienced meditator (TL) and five matched controls (S1–S5). The outliers of each participant are marked by a cross (Jo et al., 2014, p. 109).

two new factors, we decided to continue further with Tilmann in performing the Libet experiment.

Interpretations of the Libet Task Based on Backward Sampling

In 1985 John Eccles made an interesting suggestion as to how to solve the Libet paradox (Eccles, 1985). He suggested that there is a continuous background fluctuation of positive and negative electrical potentials in the cerebral cortex and that these fluctuations have an influence on the timing of the participant's initiation of an action. More specifically, a negative potential in these fluctuations will have an excitatory effect because such a potential lowers the threshold for a neuron to fire and thus raises the likelihood for action initiation. The important factor here is that the RP can be only determined by averaging the EEG recordings of many trials (approximately forty to fifty). Otherwise, the negativity of the RP might just reflect the unbalanced averaging of spontaneous fluctuations preceding the intention to act.

Eccles's interpretation refers to the issue of backward sampling. Only when an action takes place do the researchers look at the EEG recordings. This is not only a technical problem but also a logical one. It might be the case that the RP is also there if no action takes place, but such a case is never assessed. If one assumes the RP to be the cause for the upcoming action it is not enough to demonstrate that it is there if the action takes place. One also needs to demonstrate that the RP is *not present* if the action *does not take place*. But this proof was missing in the Libet case.

The idea of explaining the Libet paradox through backward sampling was taken up by Schurger and colleagues in 2012 (Schurger et al., 2012). They suggested a threshold model to initiate action. The idea is that the decision to act (i.e., lifting the finger) takes place if some spontaneous electrical activity in the brain is getting so strong that it crosses a certain threshold. Due to backward sampling only these cases are assessed and the spontaneous activity ramping up to the threshold is averaged and results in the shape of the well-known RP. Schurger and colleagues provided a mathematical and an empirical model that supported their interpretation. Seen that way, the presence of the readiness potential prior to movement can be explained, but this time the RP itself is no longer the *cause* for the movement. Also the RP is no longer interpreted as preparation for the upcoming movement (Neafsey, 2021).

The Role of the Slow Cortical Potentials

Taking up this idea of unbalanced backward sampling, we conducted more experiments with Tilmann. Our interest centered on the question of whether the subjective "urge to move" or "inner intention" as a first-person report can also be seen in the EEG recordings and thus aligned with third-person data. One likely candidate as such an electrophysiological correlate are the *slow cortical potentials* (SCPs) (see **glossary**). These are slow fluctuations in the EEG recordings below a frequency of 1 Hertz. They have no rhythmic activity and are therefore called fluctuations rather than oscillations. It has been demonstrated in several studies that a shift toward a more negative potential within the SCP is related to shorter reaction times and to a lower sensory threshold (Schmidt et al., 2016). Thus, they seem to be associated with excitatory functions, meaning the likelihood for the neurons to fire is increased (compared to a inhibitory state where the likelihood is reduced). Furthermore, a study by Keller and Heckhausen (1990) has supported the idea that a negative shift in the SCP might also be related to a feeling of "wanting to move" or an urge to act.

Indeed our experiments have shown that participants, and especially Tilmann, more often initiated a button press when the SCP were becoming negative and less often when they were becoming positive (Jo et al., 2014, figure 3D). This means he more often pressed the button when the summarized electrical signal from his scalp was moving toward a more excitatory peak than toward an inhibitory one of these fluctuations. We further asked Tilmann about the nature of this "urge" or "impulse" and its role in the process of making the decision to move and initiating the movement. He described them as

> (. . .) non-conceptual movements of the mind. Since they are pre-conceptual one can recognize them not by their content but as slight movements of the mind which, have a certain taste. . . . With a trained calm mind one can observe this pre-conceptual readiness to act. One needs to trust the mind and to relax anxiety and control. . . . Also with a calm mind one has the possibility not to follow these impulses and continue relaxing. (Jo et al., 2014, p. 115)

Based on these statements we asked Tilmann to take different approaches to these "inner impulses" or "movements of the mind." In one series of trials, he tried to press the button whenever such a movement of the mind was present (the less effort condition). In the next series he then tried to press the button in moments where *no* such movement of the mind was present. This demanded a stronger effort according to his report. When we compared the RPs of these two series, we saw an impressive difference (see figure 17.2A). When Tilmann followed the movements of the mind (less effort) there was a pronounced RP present. However, in the other series (strong effort) the RP was very weak especially in the early part. In another series, we asked him *not* to press the button whenever he observed a movement of the mind but to hold this action back as long as possible. He told us that he was able to withhold the button press for approximately one rotation of the clock, which corresponds to 2.5 seconds. Indeed, when assessing the RP we found a second negative peak about 2.5 seconds prior to the button press (see figure 17.2B).

The SCP Sampling Hypothesis

The results of our neurophenomenological approach with Tilmann supported our idea that the RP seen in the Libet experiment can be explained on the basis of the SCP. In 2016 we published the *SCP Sampling Hypothesis* that combined John Eccles' idea of backward sampling with the observation that the first-person experience of an "urge" or "movement of the mind" corresponds to the third-person observation of a negative deflection of SCP in the EEG. The SCP sampling hypothesis entails the following five statements: "(1) there are continuous fluctuations of slow cortical potentials; (2) during the negative phase of these neural fluctuations the excitability of the neuronal system related to action initiation is raised; (3) during the negative phase of these fluctuations the feeling of an urge to act is reported; (4) spontaneous voluntary movements are started more likely during a negative phase or negative peak of these fluctuations; and (5) consequently, the averaging of many trials started during negative slow cortical potentials and time-locked to movement onset results in the curve of the early readiness potential." (Schmidt et al., 2016, p. 642).

This hypothesis solves the paradox yielded in the Libet experiment. The RP is no longer seen as a signal reflecting movement preparation; it is rather an artifact of unbalanced backward sampling. As such, it has no direct causal influence on the initiation of movement. This interpretation of the RP has meanwhile received some confirmation based on other experiments (for a review, see Neafsey [2021]). However, although the RP is no longer seen as a compelling cause for the self-initiated movement, it is of course not entirely unrelated to it. It reflects the idea that the *likelihood* of initiating a voluntary action is higher when the SCP is negative and thus when the excitatory thresholds are low. As such, it can be best considered as a *moderator of action initiation.* Sometimes the continuously fluctuating states of the brain seem to be more prone to action than at other moments. That is why the term chosen at first, *readiness potential*, is still an adequate description of this phenomenon.

Personal Reflections

> Cognitive scientists should not fear that introspective evidence will impugn the scientific credibility of their work. They should fear the Frankenstein science they will create without it.
> —Jack and Roepstorff (2003)

Based on my epistemological approach laid out in the beginning of the chapter in my view it is clear that we will never find unique all-encompassing explanation of consciousness. We are only able to create models about consciousness. The special case that needs to be taken into account here is the observation from second-order cybernetics by Heinz von Foerster (2003). A model about consciousness is created within consciousness. Thus, any such model must be able to explain its own origin.

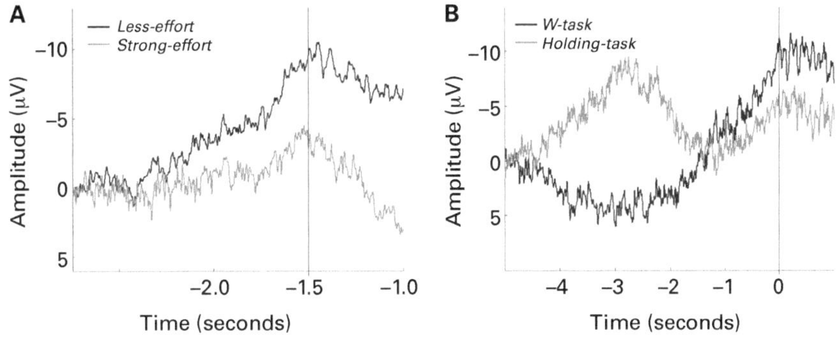

Figure 17.2
Readiness potentials of two series with Tilmann. (A) "Less effort" display of the classical Libet task. Tilmann reacted toward inner movements of the mind. "Strong effort" refers to a series in which he pressed the button when no such inner movement was present. (B) "*W*-task" refers to the classical Libet task. "Holding task" displays results of trials where Tilmann noticed the inner movement but waited for 2.5 seconds before pressing the button. Taken from Jo et al. (2014, p. 111).

Although this is often assumed, there is no position that is external to consciousness. All reasoning, modeling, and theorizing about consciousness is taking place within consciousness. In that sense there is no objectivity. Every observation regarding consciousness is the observation of somebody. Thus, subjectivity is already part of the equation. One can deny this simple fact, but this will not solve the issue.

The consequence of this view is that the subjective part needs to be taken seriously; it needs to be taken as data and has to be aligned with the so-called third-person methods, better termed intersubjective methods. Based on the systemic approach this alignment of observations from different sources should be integrated in a systematic mutual process.

Based on this approach of enactive cognition, all cognition arises from the interaction of the organism with its environment. These interactions are of course bodily interactions and thus the body has a crucial role in the study of cognition and consciousness. Why we are conscious of certain interactions and cognitions and not of others cannot yet be explained by these approaches. This remains a mystery that needs to be addressed. We know that by attending to our own conscious states the experiences and mental events we are aware of can be changed. The investigation of these processes is the realm of first-person methods. As for the fundamental question of why we are conscious at all, I would suggest that we also look at unconventional ideas to understand this mystery, because if there were a conventional solution, I think we would know it by now.

The approach suggested here is based on challenging some of the basic assumptions of science held by many people—for example, the primacy of matter. Basic assumptions are what they are: assumptions—they are unproven. Nevertheless, they are the foundation of all our scientific models. We should understand that science can only build "worlds without ground" (Varela et al., 1991), and that sometimes it is important to get back to the foundation and to find out what consequences a change at the base of the scientific building would have.

The enactivist approach is one such change, as is phenomenology. In these models, consciousness is no longer in the brain. It results from the interaction of the organism with its environment and thus encompasses them both as an inseparable unit. There is no inside and no outside. According to Fuchs (2018), the primary dimension of consciousness is related to this sensorimotor relationship of the living organism and its environment. Only in a secondary dimension can humans decouple from direct experience and attend to imaginations, virtual objects, or memories. In this dimension, they can also attend in a self-reflexive fashion to their own experience and experience themselves as experiencing subjects. And it is only this aspect of our conscious experience that results in the impression of an inner world (Fuchs, 2018).

Glossary

Backward sampling: Analysis technique applied in physiological measurements. If a certain event or behavior takes place, data *preceding* this event are sampled and analyzed. The event or behavior (e.g., participant pressing a button or heart rate reaching a certain threshold) marks the case of analysis from which data is analyzed *backward* in time (i.e., before the event).

Constructivism: View in the philosophy of science. In constructivism, it is assumed that truth cannot be achieved, neither in science nor in daily life. Rather scientists as well as laypersons construct models about the world and these models are also dependent on the human mind creating these models. Thus, there is no knowledge that is independent from the human perspective.

Electroencephalography (EEG): Physiological measurement technique to record electrical activity from the scalp. The recorded potentials reflect synchronous activity of larger populations of neurons in the brain underneath.

EEG potentials: Certain electrical charge measured by EEG at a certain scalp location and a certain point time in comparison to a reference point.

Negative potential: Negative electrical charge found in the EEG in a certain area and during a certain time period compared to a reference point.

Readiness potential (RP): Rising negative potential found over central areas of the scalp prior to a self-initiated movement. The readiness potential starts approx. 2.5 to 1.5 seconds prior to the movement.

Slow cortical potentials (SCPs): Ongoing fluctuating electrical potentials found in the EEG with a frequency below 1 Hertz.

Enactivism: View from the cognitive science stating that cognition arises by the active and dynamic interaction of the organism with its environment. Thus, a living organism actively enacts the world it lives in (Stewart, 2014). The most basic process thereby is the sensorimotor coupling of organism and its environment. This position toward cognition results in the 4E-model of cognition where cognitions are seen as enacted, embodied, embedded and extended (Ward & Stapleton, 2012).

Kagyü tradition: The Kagyü tradition is one of the four schools of Tibetan Buddhism, the others being Gelug, Nyingma, and Sakya. The Kagyü tradition traces back to Indian Buddhist scholars from the eleventh century who brought Buddhist teachings to Tibet and provided translations.

Neurophenomenology: Methodological approach in the study consciousness and cognitive sciences introduced by the Chilean biologist Francisco Varela (1996). It draws from the phenomenological approach toward conscious experience, a method of studying phenomena from a first-person perspective developed by the philosopher Edmund Husserl in the beginning of the twentieth century. These "first-person data" are aligned with data from a third-person perspective, usually obtained in a scientific experiment. Its aim is to bridge the explanatory gap (Chalmers, 1995) between neuroscientific findings on the one hand and subjective experience on the other.

Ontology: Philosophical branch that is part of metaphysics. It is concerned with the question of what are the most basic categories and how they can form a system in order to explain all kinds of entities. For example, a Buddhist ontology describes the most basic entities and principles of a Buddhist worldview.

Variance: Term from statistics. It describes in a set of observations the width of their distribution. Take as an example two sets of five observations (e.g., number of petals in a flower). One set is 4, 5, 6 ,7, 8. The other set is 2, 4, 6, 8, 10. Both sets have a mean value of six, but the latter set has a larger variance than the former one.

Note

1. See also https://retreathouse.ekayana-institut.de/the-project/about-us/about-tilmann-lhundrup/

References

Alexander, P., Schlegel, A., Sinnott-Armstrong, W., Roskies, A. L., Wheatley, T., & Tse, P. U. (2016). Readiness potentials driven by non-motoric processes. *Consciousness and Cognition, 39*, 38–47. https://doi.org/10.1016/j.concog.2015.11.011

Bornemann, B., & Singer, T. (2017). Taking time to feel our body: Steady increases in heartbeat perception accuracy and decreases in alexithymia over 9 months of contemplative mental training: Contemplative training improves interoception. *Psychophysiology, 54*(3), 469–482. https://doi.org/10.1111/psyp.12790

Cardeña, E. (2018). The experimental evidence for parapsychological phenomena: A review. *American Psychologist, 73*(5), 663–677. https://doi.org/10.1037/amp0000236

Chalmers, D. J. (1995). Facing up to the problem of consciousness. *Journal of Consciousness Studies, 2*(3), 200–219.

Dominik, T., Mele, A., Schurger, A., & Maoz, U. (2024). Libet's legacy: A primer to the neuroscience of volition. *Neuroscience & Biobehavioral Reviews, 157*, 105503. https://doi.org/10.1016/j.neubiorev.2023.105503

Eccles, J. C. (1985). Mental summation: The timing of voluntary intentions by cortical activity. *Behavioral and Brain Sciences, 8*(04), 542–543. https://doi.org/10.1017/S0140525X00044952

Fox, K. C. R., Zakarauskas, P., Dixon, M., Ellamil, M., Thompson, E., & Christoff, K. (2012). Meditation experience predicts introspective accuracy. *PLoS ONE, 7*(9), e45370. https://doi.org/10.1371/journal.pone.0045370

Frank, A., Gleiser, M., & Thompson, E. (2019). The blind spot. *Aeon*. https://aeon.co/essays/the-blind-spot-of-science-is-the-neglect-of-lived-experience

Fuchs, T. (2018). *Ecology of the brain: The phenomenology and biology of the embodied mind* (1st edition). Oxford University Press.

Jack, A., & Roepstorff, A. (2003). Why trust the subject? *Journal of Consciousness Studies, 10*(9–10), v–xx.

Jo, H.-G., Hinterberger, T., Wittmann, M., Borghardt, T. L., & Schmidt, S. (2013). Spontaneous EEG fluctuations determine the readiness potential: Is preconscious brain activation a preparation process to move? *Experimental Brain Research, 231*(4), 495–500. https://doi.org/10.1007/s00221-013-3713-z

Jo, H.-G., Hinterberger, T., Wittmann, M., & Schmidt, S. (2015). Do meditators have higher awareness of their intentions to act? *Cortex, 65*, 149–158. https://doi.org/10.1016/j.cortex.2014.12.015

Jo, H.-G., Wittmann, M., Borghardt, T. L., Hinterberger, T., & Schmidt, S. (2014). First-person approaches in neuroscience of consciousness: Brain dynamics correlate with the intention to act. *Consciousness and Cognition, 26*, 105–116. https://doi.org/10.1016/j.concog.2014.03.004

Keller, I., & Heckhausen, H. (1990). Readiness potentials preceding spontaneous motor acts: Voluntary vs. involuntary control. *Electroencephalography and Clinical Neurophysiology, 76*(4), 351–361. https://doi.org/10.1016/0013-4694(90)90036-J

Klemm, W. (2010). Free will debates: Simple experiments are not so simple. *Advances in Cognitive Psychology, 6*(1), 47–65.

Kornhuber, H. H., & Deecke, L. (1965). Hirnpotentialänderungen bei willkürbewegungen und passiven bewegungen des menschen: Bereitschaftspotential und reafferente potentiale [Brain potential changes during voluntary and passive movements in humans: Readiness potential and reafferent potentials. *Pflüger's Archiv Für Die Gesamte Physiologie Des Menschen Und Der Tiere, 284*(1), 1–17.

Libet, B., Gleason, C. A., Wright, E. W., & Pearl, D. K. (1983). Time of conscious intention to act in relation to onset of cerebral activity (readiness-potential) the unconscious initiation of a freely voluntary act. *Brain, 106*(3), 623–642. https://doi.org/10.1093/brain/106.3.623

Lutz, A., Dunne, J. D., & Davidson, R. J. (2007). Meditation and the neuroscience of consciousness: An introduction. In P. Zelazo, M. Moscovitch, & E. Thompson (Eds.), *Cambridge handbook of consciousness* (pp. 499–554). Cambridge University Press.

Maturana, H. R., & Varela, F. J. (1987). *Tree of knowledge: The biological roots of human understanding* (1st ed.). Shambhala.

Neafsey, E. J. (2021). Conscious intention and human action: Review of the rise and fall of the readiness potential and Libet's clock. *Consciousness and Cognition, 94*, 103171. https://doi.org/10.1016/j.concog.2021.103171

Schmidt, S. (2002). *Außergewöhnliche Kommunikation? Eine kritische Evaluation des parapsychologischen Standardexperimentes zur direkten mentalen Interaktion* [Extraordinary communication? A critical evaluation of the parapsychological standard experiment on direct mental interaction]. Bibliotheks- und Informationssystem der Universität.

Schmidt, S., Bauer, P. R., & Trautwein, F.-M. (2024). Neurophenomenology in action: Integrating the first-person perspective into the libet experiment. *Mindfulness*. https://doi.org/10.1007/s12671-024-02368-9

Schmidt, S., Jo, H.-G., Wittmann, M., & Hinterberger, T. (2016). "Catching the waves"—slow cortical potentials as moderator of voluntary action. *Neuroscience & Biobehavioral Reviews*, *68*, 639–650. https://doi.org/10.1016/j.neubiorev.2016.06.023

Schurger, A., Sitt, J. D., & Dehaene, S. (2012). An accumulator model for spontaneous neural activity prior to self-initiated movement. *Proceedings of the National Academy of Sciences*, *109*(42), E2904–E2913. https://doi.org/10.1073/pnas.1210467109

Sheldrake, R. (1988). *The presence of the past. Morphic resonance and habits of nature*. Collins.

Steinhöfel, A. (2005). *The center of the world*. Delacorte Press.

Stewart, J. R. (2014). Foundation issues in enaction as a paradigm for cognitive science: From the origin of life to consciousness and writing. In J. R. Stewart, O. Gapenne, & E. A. Di Paolo (Eds.), *Enaction: Toward a new paradigm for cognitive science* (pp. 1–31). MIT Press.

Stewart, J. R., Gapenne, O., & Di Paolo, E. A. (Eds.). (2014). *Enaction: Toward a new paradigm for cognitive science* (1st paperback ed.). MIT Press.

Varela, F. J. (1987). Laying down a path in walking: A biologist's look at a new biology. *Cybernetic*, *2*, 6–15.

Varela, F. J. (1996). Neurophenomenology. A methodological remedy for the hard problem. *Journal of Consciousness Studies*, *3*(4), 330–349.

Varela, F. J., Thompson, E., & Rosch, E. (1991). *The embodied mind: Cognitive science and human experience*. MIT Press. https://katalog.ub.uni-freiburg.de/link?kid=1604414057

von Foerster, H. (2003). *Understanding understanding: Essays on cybernetics and cognition*. Springer.

Walach, H., & Schmidt, S. (1997). Empirical evidence for a non-classical experimenter effect. An experimental, double-blind investigation of unconventional information transfer. *Journal of Scientific Exploration*, *11*(1), 59–68.

Walach, H., & Schmidt, S. (2005). Repairing Plato's life boat with Ockham's razor. The important function of research in anomalies for mainstream science. *Journal of Consciousness Studies*, *12*, 52–70.

Ward, D., & Stapleton, M. (2012). Es are good. In F. Paglieri (Ed.), *Consciousness in interaction: The role of the natural and social context in shaping consciousness* (pp. 87–104). John Benjamins Publishing Company.

Winter, U., LeVan, P., Borghardt, T. L., Akin, B., Wittmann, M., Leyens, Y., & Schmidt, S. (2020). Content-free awareness: EEG-fcMRI correlates of consciousness as such in an expert meditator. *Frontiers in Psychology*, *10*, 3064. https://doi.org/10.3389/fpsyg.2019.03064

18 Consciousness Is Definition-Dependent

Or Raphael

Personal Reflections

My first mindful encounter with pure consciousness occurred after my first heartbreak. The emotional turmoil of a teenager's aching heart led me to seek relief in the form of inner peace. I intuitively realized that this peace lies in a deeper anchor located somewhere "behind" my emotional world. I stumbled across Herman Hesse's Siddhartha *and was further inspired to search for a stable ground of being. As I read the book, my mind occasionally entered into an empty state that gradually brought a sense of spaciousness and stability, gently softening my emotional suffering. These experiences pulled me toward the mysterious ground of being, expanding my inner world and offering glimpses of detachment from my personal story and identity*

While studying physics at university, I delved deeper into Buddhism and Advaita Vedanta. One day, while waiting for a bus, I encountered the statement: "There is no self"; this triggered a burst of uncontrollable, all-encompassing laughter from deep within my being. In that moment, a very elusive yet absolutely concrete weight was suddenly taken off my shoulders.

During the following years, I found myself feeding on spiritual texts in breaks between physics classes, particularly enjoying The Truth Is *by Sri H. W. L. Poonja. My initial belief that physics would provide critical "glasses" of knowledge that would help unravel the mysteries of the universe led to frustration; I found the subject filled my mind with information and theories, but did not significantly contribute to my understanding of the world, nor myself. Fortunately, my meditative practice and connection to the spacious ground of being helped alleviate the cognitive load and frustration I was experiencing in my studies.*

Over the next decade, the pure ground of being became a central part of my continuous experience and sense of self. To me, this ground of being is consciousness—an infinite, ever-present, unchanging "space" of existence. I perceive this "space" as encompassing not only my personal experience, but also the entirety of perceived and unperceived worlds. As the totality of all that exists, consciousness transcends the thinking mind and is beyond language, equations, and definitions, ever shining as the enduring mystery of life.

Overview

Consciousness is mystery. Not a particular mystery—just mystery. It is a subject of countless scientific debates and theories, widely regarded as one of the greatest unresolved questions in science. In the first, theoretical section of my piece, I draw from my experience with scientific methods and spiritual practices to present the idea that consciousness studies involve multiple "consciousnesses," each equally legitimate and potentially useful for inquiry. In this section I argue that combining subjective inquiry and practices with traditional scientific approaches is essential because consciousness and experience are inseparable and therefore the study of experience is an integral aspect of the study of consciousness. Additionally, self-reflection and contemplation can reveal new aspects of consciousness to scientists and clarify the limits of our knowledge. In the second, experiential section of this piece, I propose two practices that illustrate the value of contemplative methods in the scientific study of consciousness. These practices aim to explore our core ability to know and identify the limits of objective knowledge, and offer an alternative, subjective pathway to objective knowledge.

Flow down and down in always widening rings of being.
—Rumi

Introduction

It's strange how the simple things in life go on while we become difficult.
—Richard Brautigan

Throughout my adult life, I have been continuously engaged in both science and spirituality. My attraction to both stems from a desire to connect with a deeper truth, gain clarity, and uncover the "hidden hands" that are shaping the world and my experience within it. Initially, I believed physics would provide the tools to understand the world more clearly. However, over time, I found that physics offered an overwhelming array of "glasses" of perspectives, each revealing different aspects of reality as portrayed by the well-known phrase "more is different" (Anderson, 1972). Seeking simplicity, I turned to spirituality, which helped me connect with the simplest possible experience that could be named "bare existence." Yet, spirituality also provided a variety of "glasses," presenting multitude of theories and viewpoints through which to see myself and the world. As my awareness gradually cleared up, I increasingly noticed more of the subtleties of my mental life, leading me right back to a multiplicity of perspectives.

My journey through science and spirituality has led me to consider that an 'ultimate truth'—free of any perspective, if it exists at all—may lie beyond the reach of either domain. In science, even if we find equations that encapsulate the essence of how nature operates, they will not simplify the complexities of "the world" that contains human existence. Similarly, in spirituality, even profound experiences of ultimate reality become additional lenses through which we can view the world. Science and spirituality provide endless pairs of glasses, and as we discover more, the scope of viewpoints and perspectives expand. Some perspectives will be developed within existing frameworks, while others might lead to "paradigm shifts" (Kuhn, 1970). The question then becomes, which glasses should we wear, when, and for what purpose? This multiplicity of perspectives, particularly in consciousness studies, highlights that science can offer various definitions and insights into consciousness, but will probably never converge into a singular understanding of what it is. Instead, it will continually diverge as our understanding evolves, offering new ways of seeing and relating to consciousness.

Consciousness Is Definition-Dependent

The first claim I want to present is that consciousness, as a scientific object of investigation, depends on its definition. This may seem both obvious and significant—obvious because science relies on precise definitions to study phenomena, and significant because the definition itself shapes our understanding of the studied phenomena. We should therefore follow Haruki Murakami's way of thought in his book "What I Talk About When I Talk About Running" to clarify 'what we talk about when we talk about consciousness.' However, this fundamental step is not always rigorously applied in the scientific study of consciousness.

A deeper aspect of this claim is that there is no single, universal consciousness to be discovered or studied. Instead, there are different consciousnesses that can be defined and explored using different tools and methods. (Whether or not we should use the same word to describe different consciousnesses is another issue worth discussing, but is beyond the scope of this chapter.) In my view, it is sufficient to clearly define consciousness within a specific study or field, and to discuss the insights derived from that definition. While debates about the nature of consciousness can help refine our definitions and promote interdisciplinary integration, they should not be used as a means to find a singular, ultimate definition or essence of consciousness.

Definitions and Approaches to Consciousness

There are numerous scientific definitions and approaches to consciousness, some of which are explored in this volume (See Sattin et al. [2021] and Wilber [1997] for detailed reviews). One notable definition is that consciousness is integrated information (Tononi, 2008). Many researchers, however, do not define consciousness directly but the conditions for the constitution of consciousness. For example, some see it as an emergent property of brain activity (Crick & Clark, 1994), while others attribute it to complex interactions between brain, body, and environment (Thompson & Varela, 2001), or between private and collective subjective and objective domains (Wilber, 1997). Other scholars focus on a "particular aspect of consciousness," such as sensory awareness (e.g., Engel & Singer, 2001), or on "different concepts of consciousness," such as the stream of consciousness, self-awareness, and intentionality (Zahavi, 2002).

Which of these references to consciousness is "right"? And on what basis can we validate them? In my view, these perspectives are equally valid, as each conceptualizes consciousness in a distinct, complementary way, defining either a unique phenomenon or an aspect of an overarching, indefinable "phenomenon." Each perspective is valuable, provided it is applied consciously and accurately, offering a clear framework for research. Different fields naturally adopt the definitions that best serve their purposes and adapt or neglect them over time.

While we could compare the usefulness of different perspectives, it may not be necessary, as each provides unique insights and benefits. Collectively, these perspectives offer various lenses through which to view consciousness, each as valid and real as the others.

My personal approach to consciousness is rooted not in conceptual definitions but in nonverbal, direct experience. However, to provide a functional definition for this discussion, I draw inspiration from a beautiful poem by Lalleshwari:

How shall the Nameless be defined
A thousand times my Guru I asked:
How shall the Nameless be defined?
I asked and asked but all in vain.
The Nameless Unknown, it seems to me,
Is the source of the something that we see.

In light of these reflections, I define consciousness as the non-material "substrate" of existence—the totality from which all appearances arise, to which they return, and of which they are composed. This concept aligns with well-known mystical traditions, such as the Brahman in Hindu philosophy, the Dao in the Daodejing, and the Self as taught by Sri Ramana Maharshi. In modern philosophical terms, it parallels Bernardo Kastrup's unbound consciousness, described as nature's sole ontological primitive (Kastrup, 2017), and George Berkeley's perceiving, active entity termed "mind," "spirit," "soul," or "myself" (Berkeley & Jessop, 1952, p. 11). This view also resonates with Husserl's idea of consciousness as the medium through which everything objective is manifested (Moran, 2005), and aligns with the concept of autopoiesis (Maturana, 2002).

In some respects, this view of consciousness overlaps with certain interpretations of quantum mechanics, where consciousness is considered a "unified field" (e.g., Hagelin, 1987; Kafatos, 2015). However, my definition places consciousness beyond the realm of objective scientific study, as it is not a strictly definable object. To me, consciousness is the source of the mind, always beyond its reach. It is present in all phenomena and indistinguishable from both the objects of investigation and the investigator. Thus, consciousness cannot be isolated or studied using objective tools. The study of consciousness is a meditative practice, connecting me to the essence of existence beyond objective knowledge. Simultaneously, this study is intertwined with living life as a dynamic, ever-learning organism, encompassing all levels of experience, including cognition, which is the foundation of scientific knowledge.

Science Is in Consciousness

Regardless of how we define consciousness, we can probably agree that conscious thought occurs within consciousness. Modern science, which relies on conscious thought processes such as designing experiments, analyzing data, and formulating models, inherently operates within the realm of consciousness. Scientists are consciously aware of these intentional actions, aligning with Husserl's view that knowledge acquisition cannot be understood without consciousness (Moran, 2005). Conscious thought is just one aspect of the activities occurring in consciousness, such as sensory perceptions and emotional experiences. Therefore, consciousness cannot be fully captured by science because it is "bigger" than science.

As previously outlined, science can define consciousness, describing, modeling, and relating each defined consciousness to different phenomena. In this sense, consciousness is *in* science and can be meaningfully addressed by science. But is the consciousness within which science operates the same as the consciousness studied by science? This relationship between science and consciousness can be likened to an ouroboros or a Möbius strip, illustrating an ungraspable correlation. To me, this seemingly paradoxical relationship highlights the limits of standard logical thinking and hints at deeper aspects of reality.

The Scientific Study of Consciousness

In my view, the scientific study of consciousness should aim to enhance our awareness and understanding of internal, subjective processes and the interconnectivity between internal and external phenomena. A comprehensive approach should include the study of the self (encompassing first-person perspectives, both conscious and nonconscious) and external perspectives (third-person phenomena), along with their correlations. It should also explore the concept of knowledge and its boundaries, not only from a theoretical, philosophical standpoint but also through experiential and contemplative practices. One such practice involves becoming aware of pure, objectless consciousness, often referred to as "void" or "nothingness" that lies beyond cognitive knowledge (see practice 1 in the experiential section below). This awareness could illuminate the nature and limits of cognitive knowledge. Additionally, alternative methods of accessing information and gaining insights, which involve different modalities of knowing, should be explored (see practice 2 below).

It is important to remember that scientists are primarily humans, not just "thinking machines" (for a critique of the narrow focus of modern science, see Strevens [2020, ch. 12]). Contemplative practices hold significant potential to enhance various aspects of human experience and expression. For instance, studies suggest that meditation can improve cognitive abilities, such as working memory and executive functions (Chiesa et al., 2011), and can also positively impact well-being and social interaction (Galante et al., 2014). As we engage in contemplative practices, we transform ourselves and, consequently, our scientific pursuits. The study of consciousness, in particular, encourages us to become more aware of our cognizance. This enhanced awareness can allow us to access new insights and consequently further deepen our understanding of consciousness.

From a societal perspective, the scientific study of consciousness should address important questions about the various consciousnesses it defines. These include a variety of questions: How can we alter our conscious experience internally and externally? What are the relationships between the psyche, brain, and body? How does consciousness relate to nature? Exploring these questions can help society progress in meaningful ways, provided we maintain good intentions, use science for the collective good, and remain humble about the limitations of our fields.

I hope and believe that conferences like the one we held in Israel will enable scientists and the general public to connect more deeply with the incredible mysteries of consciousness, foster the development of new methodologies for conducting science, and encourage the establishment of new ways of seeing, being, and creating—individually and collectively.

Experiential Section

In this section, I offer two short practices that explore our ability to know. The first practice combines simple mindfulness with a "direct path" practice, as commonly taught by Sri Ramana Maharshi's successors. This practice provides a glimpse into the core part of our being that is beyond objective knowledge. The second practice, inspired by shamanic shape-shifting traditions (see Leder [2012] for a theoretical review), aims to facilitate understanding through the experiential embodiment of the phenomena under study.

Practice 1

Close your eyes, take a few deep breaths, and gradually release any tension in your body. Once you feel calm and settled, direct your attention toward the insides of your body. Begin by scanning your body from your feet upward, through your shins, knees, thighs, belly, chest, neck, and head. Sense each area somatically without labeling your sensations. After finishing this scan, relax your attention and let it "spread" evenly throughout your entire body. Try to maintain this state of spread awareness, with a balance of calmness and alertness.

Now, direct your attention to your head and begin noticing your thoughts. Ask yourself repeatedly: Who is observing these thoughts? Notice if any experiential answer arises that is not based on thought (such as a word like "I" or a mental image of your body). Each time a thought appears, try to experience it as a distinct object with clear boundaries, set against a wider, spacious background of awareness. Keep asking who is observing the thought, while maintaining awareness of thoughts as objects. Finally, explore this question: What is the relationship between what you call "I" and the background of knowingness against which all thoughts are observed?

Practice 2

We experience the world through our human bodies, which shape our perceptions via our senses and other capabilities (mental, emotional, physical, etc.). This embodiment influences how the world appears to us. But what degrees of freedom do we have to experience the world differently from our standard mode? Can our consciousness adopt perspectives beyond our default one? Let us explore this possibility with a simple experiment.

Choose an animate or inanimate object and fully embody it—become it with your entire body, mind, and perception. Start by imagining your human body transforming into the body of the chosen object. Feel its size, texture, and movement capabilities. Once you have a clear sense of how this new body feels from within, direct your gaze out and ask yourself: What does the world look like as this object? How do I relate to my surroundings through this form? After a few minutes, compare this experience with your usual human perspective. What differences in perception do you notice? Does one seem more real or objective than the other?

Repeat the exercise with several different objects. You can also choose dynamic phenomena, such as wind or waves from the natural world, or dynamic objects from the human-made world. After completing this practice, reflect on your experiential insights. Consider what these experiences reveal about your own consciousness, or about consciousness at large.

I would like to complement these practices and end this chapter with a beautiful poem by Rainer Maria Rilke:

Widening Circles
I live my life in widening circles
that reach out across the world.
I may not complete this last one
but I give myself to it.
I circle around God, around the primordial tower.
I've been circling for thousands of years
and I still don't know: am I a falcon,
a storm, or a great song?

References

Anderson, P. W. (1972). More is different: Broken symmetry and the nature of the hierarchical structure of science. *Science, 177*(4047), 393–396.

Berkeley, G., & Jessop, T. E. (1952). *The principles of human knowledge* (Vol. 35, pp. 401–444). Encyclopaedia Britannica.

Chiesa, A., Calati, R., & Serretti, A. (2011). Does mindfulness training improve cognitive abilities? A systematic review of neuropsychological findings. *Clinical Psychology Review, 31*(3), 449–464.

Crick, F., & Clark, J. (1994). The astonishing hypothesis. *Journal of Consciousness Studies, 1*(1), 10–16.

Engel, A. K., & Singer, W. (2001). Temporal binding and the neural correlates of sensory awareness. *Trends in Cognitive Sciences, 5*(1), 16–25.

Galante, J., Galante, I., Bekkers, M. J., & Gallacher, J. (2014). Effect of kindness-based meditation on health and well-being: A systematic review and meta-analysis. *Journal of Consulting and Clinical Psychology, 82*(6), 1101.

Hagelin, J. S. (1987). *Is consciousness the unified field?: A field theorist's perspective* (pp. 29–87). Maharishi International University.

Kafatos, M. (2015). Fundamental mathematics of consciousness. *Cosmos and History: The Journal of Natural and Social Philosophy, 11*(2), 175–188.

Kastrup, B. (2017). An ontological solution to the mind-body problem. *Philosophies, 2*(2), 10.

Kuhn, T. S. (1970). *The structure of scientific revolutions* (Vol. 111). University of Chicago Press.

Leder, D. (2012). Embodying otherness: Shape-shifting and the natural world. *Environmental Philosophy, 9*(2), 123–141.

Maturana, H. (2002). Autopoiesis, structural coupling and cognition: a history of these and other notions in the biology of cognition. *Cybernetics & Human Knowing, 9*(3–4), 5–34.

Moran, D. (2005). *Edmund Husserl: Founder of phenomenology*. Polity.

Thompson, E., & Varela, F. J. (2001). Radical embodiment: Neural dynamics and consciousness. *Trends in Cognitive Sciences, 5*(10), 418–425.

Tononi, G. (2008). Consciousness as integrated information: A provisional manifesto. *The Biological Bulletin, 215*(3), 216–242.

Sattin, D., Magnani, F. G., Bartesaghi, L., Caputo, M., Fittipaldo, A. V., Cacciatore, M., Picozzi, M., & Leonardi, M. (2021). Theoretical Models of Consciousness: A Scoping Review. *Brain Sciences, 11*(5), 535.

Strevens, M. (2020). *The knowledge machine: How irrationality created modern science*. Liveright Publishing.

Wilber, K. (1997). An integral theory of consciousness. *Journal of Consciousness Studies, 4*(1), 71–92.

Zahavi, D. (2002). The three concepts of consciousness in Logische Untersuchungen. *Husserl Studies, 18*(1), 51–64.

19 Biophysics of Consciousness, or Can My Brain Explain My Pain?

Idan Segev

Personal Reflections

Why brain: *I was attracted to study the brain because, as a child, I saw around me a lot of emotional suffering. I then thought to myself that, if I could understand the source of such suffering in the brain, I could provide a direct solution to suffering by developing some anti-suffering pill or a focal electrical stimulation to relieve the source of pain. Already, during my army service, I consulted top scientists on how I should go about it. They all strongly recommended that I should study biology in order to become acquainted with the brain and mathematics to uncover the general principles of brain function. So I did. It was excellent advice.*

Naïveté: *Early in my university studies, I quickly recognized my naivety in expecting a swift, miraculous solution to "human suffering" through brain interventions. The brain's complexity became apparent, from its one hundred billion neurons and their numerous types—each often linked to a specific disease—to the intricate, dynamic circuits they form and their constantly changing connections ("plasticity" underpinning learning). This complexity left me wondering how such a delicate system of chemical and electrical elements usually functions as a coherent whole. To grasp this, I realized I needed to focus on studying the brain's fundamental units, the neurons, both experimentally and theoretically.*

Inspirations: *My work was profoundly influenced by two major theoretical advances: the mathematical model of Hodgkin and Huxley (1952) that detailed the initiation and propagation of the spike—that is, the brain's basic electrical "bit" of information—and by the "cable theory for dendrites" by Wilfrid Rall (1959), who was my insightful and compassionate postdoc advisor. Rall's theory clarified how thousands of synaptic inputs to a neuron's dendritic tree produce output spikes transmitted to other neurons. After many years of almost "becoming a neuron" myself, I feel that I have made good progress in understanding how these powerful (and very beautiful) elementary processing units of the nervous system receive, process, and transmit information like sensory information.*

Something is missing: *It is exciting to study mathematically the cardinal elements that (from a materialistic viewpoint) generate my feelings, my sense of identity, my creativity* *(even though this rarely happens). But, something very fundamental was missing in all these scientific explorations. First, emotionally, the scientific process was not experiential enough. Science, by its conventional ethos, is austere and too serious—very normative. It claims that "the data should speak for itself," and therefore personal impressions and sharing emotions when presenting a scientific result are irrelevant, almost "forbidden." Second, intellectually, although I found great satisfaction in "understanding neurons and neuronal circuits," I had this constant nag that the most fundamental aspects of my own being—my conscious experiences, my pain and aspirations, my awareness of my own thoughts—are puzzles that science cannot really say much about. And when it does, it typically provides correlations (but sometime also causation; see below) between the subjective experience (e.g., the feeling of pain, enjoying the smell of a rose) and the activity in the respective brain region and neuronal networks involved. Recent advances in linking brain mechanisms and conscious experiences are indeed extremely interesting and of key clinical importance because they provide data-based tools to intervene and alter human experiences (e.g., reduce pain). But this does not provide an explanation for the big leap "from spikes to consciousness." I was therefore attracted to books and articles about the science of consciousness, about the "easy" and the "hard" problems, and about new technologies to manipulate and measure degrees of consciousness. I particularly enjoyed reading* The Astonishing Hypothesis *by Francis Crick (1995), which puts the reductionistic/physicalist viewpoint on consciousness so powerfully: "You, your joys and sorrows, your memories and your ambitions, your sense of personal identity and your free will are in fact no more than the behavior of a vast assembly of nerve cells and their associated molecules."*

Materialistic mind: *I really like this materialistic phrasing of who I am because it raises a sense of wonder about the properties of the physical material itself (atoms, molecules, DNA) that generates the whole me, including my feelings. It actually says that these "dead" and "boring" physical elements themselves, when they form living cells that interact with each other in our brain, can generate the so-very-beautiful music of Bach; my love for my family; my urge to understand myself—my own brain and body. According to physicalism, life itself emerges out of the properties of physical matter (Nurse, 2021). But finding correlations and even casual relationships between brain activity and my subjective feelings, including the development of a mathematical measure for the information capacity of consciousness (Tononi & Koch, 2015) does not explain this most imposing leap "from spikes to feelings."*

Out of myself: *With this realization of the limits of science in providing an explanation for "the ghost in the machine," and with the understanding that it is of great importance (intellectually, clinically, and ethically) to advance our understanding in the phenomena of consciousness, we have initiated two unique encounters (2019, 2020) entitled "Perspectives on Consciousness." The rationale was that when scientists like me will be engaged with researchers and contemplative practitioners from diverse fields, it will push us all beyond our comfort zones and that this will spur creativity, new ideas. In these two encounters, it was perfectly acceptable to "give up" on complete rationality and articulation. I embraced this freedom, dancing with about thirty other participants, sometimes spinning on my back on the floor. This experience allowed me to temporarily shed my internal self-critique. Remarkably, we could attend these meetings barefoot—a rarity in scientific gatherings. Emotional moments arose, such as when attendees shared their profound sadness, allowing for a genuine expression of feelings.*

This experience has had a lasting impact on me. I now find it easier to share both feelings and failures with my students in the lab and during classes, emphasizing the importance of learning from all outcomes, not just successful ones. I frequently discuss these "Perspectives on Consciousness" experiences, encouraging students to engage in similar unconventional activities. This approach helps them to accept making mistakes, embrace moments of folly, and feel more at ease with peers and myself. I believe fostering this open emotional exchange among scientists and students enhances creativity and often leads to those transformative "aha!" moments, where new connections suddenly illuminate understanding, like realizing that $E = mc^2$.

Overview

Scientists do not typically use words like miraculous, astonishing, or amazing in their scientific work. But they do so when they speak about consciousness. This reflects the persistent puzzle of consciousness—it is a phenomenon we experience very intensely, but find hard to explain scientifically. Despite impressive progress in understanding the biophysical mechanisms operating in the brain during conscious states, the "big leap" between these mechanisms and conscious experiences remains vast, and might remain unexplained by the scientific method as we know it (but who says that science should explain all things?). This, in no way, means that we should not continue very determinedly, and with a sense of urgency, to uncover the neuronal mechanisms and computational principles linked to consciousness. On the contrary, understanding consciousness has far-reaching clinical, ethical, and technological implications.

In this chapter, I discuss recent advancements in identifying correlations and causative links between the biophysical properties of individual neurons and neuronal circuits, and consciousness, facilitated by genetic tools like optogenetics. I argue that progress in understanding consciousness can be significantly boosted by forming a "society of brains" where philosophers and spiritual practitioners who study consciousness "within" collaborate closely with neuroscientists who study consciousness "without." I conclude by reflecting on how interdisciplinary interactions, the inspiration for this book, have enriched my research and broadly benefited science.

And still they gaz'd and still the wonder grew, that one small head could carry all he knew.
—Oliver Goldsmith, "The Deserted Village"

Introduction

This article emerged from my unsettled scientific relationship with the problem of consciousness. Investigating the brain's structure and activity intimately for fifty years, I could only find a variety of physical elements—genes, chemicals, membrane ion channels, and spiking activity in large connected-networks of neurons. Spikes, the brain's fundamental information "bits," act as the basic "letters" of its internal "language," enabling the representation and processing of sensory information across nearly all nervous systems. These spiking activities in large networks correlate strongly with our subjective experiences, both conscious and unconscious (Malach, Chapter 10). Recent studies have revealed direct causative links between specific spiking patterns in certain neuronal groups and subjective experiences like fear (Liu et al., 2012) or pain (Copits et al., 2016). While the biophysical mechanisms of spike generation and transmission are well understood, largely thanks to Hodgkin and Huxley's Nobel Prize-winning research (1952), the critical question remains: How do these neural spikes translate into subjective experiences such as pleasure or pain?

In this chapter, I discuss what neuroscience has to say about the cellular and biophysical mechanisms underpinning consciousness, with a specific focus on pain as a rudimentary archetypical example of conscious experience, common to both "mice and men." I explain how intense pain triggers a series of spikes starting from the periphery, traveling through pain-specific pathways to the spinal cord, and ultimately reaching specific brain regions where this spiking activity is interpreted as pain. I also introduce the use of recently developed "optogenetic" tools and other sophisticated techniques to either block or generate pain by manipulating spiking activity in targeted groups of neurons within specific brain areas. Although there is increasing and undeniable evidence for the tight correlation-causation links between spiking activity in the brain and conscious experience, I argue that the significant leap from these physical processes to subjective experience remains unexplained by current scientific methods.

Indeed, we scientists may look at spatial patterns of neuronal activity that are correlated with a certain conscious experience with, for example, functional magnetic resonance imaging (fMRI), calcium imaging, or multi-electrodes. We may even causally invoke by activating specific groups of neurons (e.g., by using optogenetic tools) neuronal activity that gives rise to a particular conscious experience. But the brain does not have "eyes" to look at this activity. So, how are these spatially and temporally spiking activity decoded to give rise to subjective conscious experience? I believe that, no matter how refined the technology is that we (our brain) invent, it most likely will not give us the "hook" to develop a theoretical framework that explains the puzzling leap "from spikes to feelings." We therefore will stay with the unsolved problem of the "ghost in the machine."

This humbling limitation of the scientific paradigm does not mean in any way that we should "give up" on studying the biophysical and anatomical underpinning of consciousness. On the contrary, the more we understand the biophysical and anatomical mechanisms of consciousness, the more refined correlation we will have between the various brain states and respective states of conscious experience, leading to better interventions for conditions like pain or fear. Progress also hinges on creating a community of diverse brains, where philosophers and spiritual practitioners, who study consciousness from "within," interact with neuroscientists like myself who study the brain from "without." Through introspection, these expert practitioners are likely to provide refined characterization of various states of consciousness that can then be systematically studied using modern neuroscientific tools. My personal experience confirms that such interdisciplinary encounters are likely to spark creative breakthroughs.

Spikes, Spikes, Spikes: The Elementary Signals of Brain States, Consciousness Included

A face (photons arriving at your eyes), the smell of a rose (odorant molecules flowing through your nose), and Bach's cantatas (wave pressure on your ear drum) are all transformed into abstract representations in your brain, carried in all cases by a universal electrical "bit" of activity—the "spike." Each spike, an "all or none" signal, has an amplitude of about 100 millivolts and lasts roughly 1 millisecond. These spikes occur in both peripheral neurons at the sensory organs (e.g., in the retina, the cochlea in the ear, or the nose epithelium) as well as in central neurons in the brain (e.g., in the neocortex or hippocampus). This neuronal spiking is represented as a sequence of binary zeros (no spike) and ones (spike) over time. For example, one neuron might display a firing sequence like 0,0,1,1,0,0,0,1,0, while another neuron, which may be connected via synapses to the first neuron, might exhibit a different pattern of spiking.

When the nervous system, from the periphery to higher brain regions, processes sensory input, large groups of neurons in specific brain areas fire spikes collectively. This collective activity generates global electrical "music" in the brain, produced by neurons in response to external (e.g., a face) or internal (e.g., a memory) stimuli. There are no movies in your brain (no screen), nor a rose, nor music. Just trains of spikes in individual neurons composing large neuronal networks in particular brain regions. Somehow this electrical activity recreates in your brain the various physical stimuli arriving to our senses from the outside world (electromagnetic radiation, mechanical pressure, heat). This electrical activity somehow mirrors the physical properties of the external world—like light, sound, and heat. Occasionally, this collective spiking leads to conscious experiences such as love or pain; often, however, it does not produce consciousness, as was shown in a recent study by Hayat et al. (2022) focusing on the conscious and unconscious perception of music.

In other words, our perception of the world is not a "projection" of the world around us onto the brain (we are not taking a "photocopy" or "screenshot" of the world). Rather, this perception is constructed actively by the brain using its "in-house" mechanisms—spiking activity in individual neurons and neuronal networks in particular brain regions. Thus, our experiences of faces, music, and smells are re-created by our brains responding to sensory inputs. Intriguingly, under certain conditions, when spikes flow through specific networks and regions, you might become consciously aware of these experiences, such as feeling pain (figure 19.1).

One of the landmark achievements in neuroscience was made by Hodgkin and Huxley in 1952 when they elucidated the mechanisms that generate and propagate spikes in neurons. They demonstrated that spikes result from the opening and closing of specific "gates" embedded in the neuron's membrane. Through these gates, ions such as sodium and potassium flow across the neuron's membrane and this generates voltage change across the membrane, and in particular the voltage signals called spikes. Their biophysical-based mathematical model, which explained spike generation and propagation, was a scientific breakthrough that earned them the Nobel Prize.

Building on this foundation, Sakmann and Neher (1984) further advanced our understanding of the spike by showing that the membrane "gates" are actually protein-based ion channels that open and close as a function of the voltage across the neuron's membrane. This discovery, which also led to a Nobel Prize, highlights the fundamental role of these ion channels in both sensory neurons—such as those in our fingertips that generate electrical signals in response to painful stimuli or in the nose that respond to odors—and in central brain neurons, where neurotransmitters like dopamine trigger spiking activity in specific groups of neurons.

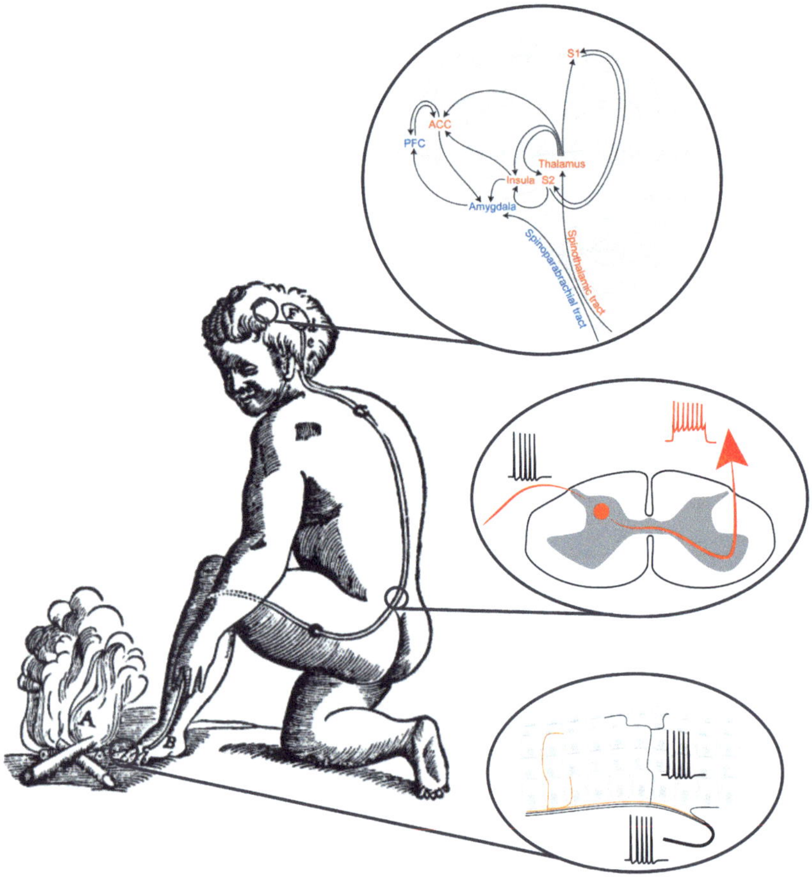

Figure 19.1
Progress in understanding pain in the brain. (Left) René Descartes' illustration of the pain pathway in his 1664 *Traite de l'homme*. According to his theory, when the body encounters a stimulus that causes pain (like heat), it pulls on a long nerve fiber extending to the brain, specifically the pineal gland, which then releases a fluid that prompts a muscle contraction and the sensation of pain. (Right) Present-day depiction of pain pathways: (bottom) Noxious stimuli are detected by the fine branches of specified sensory nerves (nociceptors) ending in the skin (thin branches); (middle) these nociceptors transduce the noxious stimulus to steams of spikes (black and red traces) that propagate through specific nerve fibers (axons) to the spinal cord; and (top) after some processing in local spinal networks, the spiking activity propagates to specific brain regions (such as S1 [the primary somatosensory region in cortex], the amygdala [involved in emotional responses], and the thalamus [a relay station of sensory information]). Consequently, pain is consciously experienced. The withdrawal reflect of the foot from the fire is done automatically, before the involvement of high brain regions, because the sensory impulse that travels to the spinal cord activates local neurons (interneurons) that connect to spinal cord's neurons (motor neurons) that send spikes to the muscles leading to withdrawal from the painful stimulus (not shown). The study of the different receptor types, membrane ion channels and specific fibers and pathways involved in pain generation, enable the development of a variety of drugs and a variety of electrical intervention that operate at different stations along the "pain path" to ameliorate pain. Figure courtesy of Alexander Binshtok.

Understanding spike mechanisms has been pivotal for developing both targeted drugs that act on specific ion channels as well as electrical interventions that modulate spike generation in specific neuronal pathways. These strategies are used to regulate sensory impressions either in the periphery or in progressively higher brain regions, from spinal cord to brain stem to cortex. Indeed, pain perception typically starts with the activation (spiking) of particular peripheral nerve fibers (e.g., at your skin); these spikes propagate via elongated fibers (axons) to the spinal cord and from there to various neuronal networks in the brain (figure 19.1), sometimes giving rise to a subjective feeling of, for example, pain! Let's then take pain as an example and see how conscious experience of pain emerges.

Pain in the Brain

Biophysics of Consciousness: From Spikes in Your Finger to Feeling Pain—Correlations

It is easier to discuss the case of an acute pain experienced from burning a finger. Already in the seventeenth century, René Descartes proposed that when you burn your finger a message is generated at the finger and travels to the brain, which then sends another message back, causing the finger to withdraw from the heat source (figure 19.1). We know today that these messages are carried (both ways) by trains of spikes, and that these spikes are generated at the heat-sensitive receptor cells in the skin (finger). Modern research has unraveled the fine mechanisms underlying pain generation at these receptors, thus refining Descartes' view on the generation of pain signals in the finger (figure 19.1). This also includes the characterization of the different types of pain receptors (known as nociceptors), and the specific fibers (axons) that carry pain information to the spinal cord and subsequently to the brain (Goldstein et al., 2019). Contemporary imaging technologies and direct recordings of spiking activity have provided a detailed map of brain regions involved in the processing of pain signals (figure 19.1).

This understanding is crucial for developing targeted pain treatments. Techniques include blocking the initiation of spikes at the periphery or interrupting the transmission of pain information in the spinal cord before it reaches the brain. Surprisingly, only in the brain (rather than in the spinal cord) can spikes give rise to the conscious experience and it is there that general anesthesia modulates pain. Recent advances also use electrical and optogenetic (light induced) stimulation to prevent specific brain networks from interpreting electrical signals as painful. Importantly, the subjective experience of pain can be adjusted through top-down control, which involves reinterpreting the same spiking signals originating from a painful stimulus in a way that feels less painful. This process, which incorporates practicing and learning, highlights the plasticity of neuronal pathways in the spinal cord and brain. Recent studies, such as Zorn et al. (2020), suggest that mindfulness meditation can modulate how pain is experienced, specifically reducing the emotional impact of pain rather than its intensity.

While general anesthesia has been widely used in hospitals globally for many years, the precise mechanisms by which it blocks consciousness are not fully understood. However, a notable recent study by Aru et al. (2020) sheds light on potential cellular mechanisms of

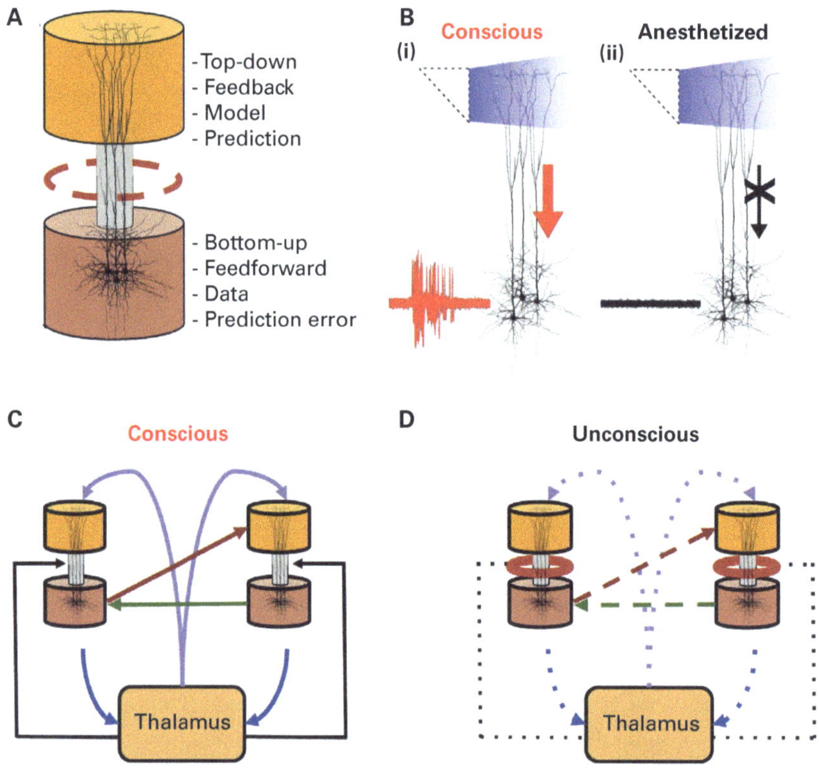

Figure 19.2
Cellular mechanisms supporting consciousness. (A) Layer 5 cortical pyramidal neurons (L5PCs) have a central role in the mechanisms of consciousness. These cells are the key elements through which the cortex exerts its control over behavior. They have a large dendritic tree (in black, the input site of the neuron) and long-range axonal projections within and outside the cortex (the output, not shown). L5PCs have two distinct functional compartments. The basal tree (bottom red cylinder) receives "bottom up" information from the outside environment and the apical tree (top orange cylinder) that integrates contextual information and expectations, semantic knowledge, and attention arising from higher brain regions. The coupling compartment (encircled by dashed red line) enables flexible integration of these two data streams. (B) In conscious states (i) optogenetic stimulation to the apical compartment (schematic top, blue light) leads to high-frequency spike firing of L5PCs (bottom, red trace). Therefore, in this state, "top-down" information affects the output of L5PCs. In contrast, under anesthesia (ii) stimulation to the apical compartment does not generate an output (flat black trace). (C) In the conscious state, the "gate" connecting the basal and apical compartment of L5PCs is open and allows signals to propagate and evolve in the thalamocortical and corticocortical loops. (D) During the unconscious state, basal and apical compartments are decoupled (as indicated by the red ellipses), leading to cessation of the activity propagation in the thalamocortical system. Adapted from Aru et al. (2020).

anesthesia and, by extension, consciousness processing. They propose that consciousness relies on the integration of two distinct streams of data within the dendrites of cortical layer five pyramidal neurons (L5PCs); see figure 19.2. Specifically, the bottom-up stream carries sensory information from external sources (e.g., a face activating the retina) and then travels to the visual cortex where it primarily targets the basal dendrites of L5PCs (figure 19.2A). Concurrently, the top-down stream, which carries internal information such as memory, predictions, and contextual data impacts the apical dendrites of these same neurons (figure 19.2A). Aru et al. (2020) showed that anesthesia disrupts the coupling between these basal and apical dendritic compartments (figure 19.2B), thereby impairing the brain's ability to integrate and reverberate this information across neuronal assemblies—a process essential for generating conscious perception (figures 19.2C–D). Further supporting this model, a study by Hayat et al. (2022) found that while sensory feedforward input remains largely intact during sleep, the crucial feedback from other brain regions is significantly reduced. This highlights the importance of feedback signaling in conscious sensory processing.

Pain is a more complex experience than visual or auditory processing due to its multidimensional and multilevel nature, involving intricate sensory processing influenced by emotional and cognitive factors. The specific brain region where pain sensation originates remains unclear (Baron & Devor, 2022; Damascelli et al., 2022; Mano, H., & Seymour, B., 2015). However, a recent study demonstrated that targeting a deep brain region known as MPTA with a localized anesthetic injection in mice could deactivate conscious experiences, including sensations, movement, and pain. This suggests that the MPTA acts as a critical neural "hub" for pain sensation, which when inactivated blocks pain as observed in mouse behavior (Minert et al., 2017).

Interestingly, there are extensive similarities in the anatomical substrates of painful and pleasant sensations. Recent molecular-imaging and animal studies have demonstrated the important role of the opioid and dopamine systems in modulating both pain and pleasure. Understanding the mutually inhibitory effects that pain and reward processing have on each other, and the neural mechanisms that underpin such modulations, will be key for alleviating unnecessary suffering and improving well-being (see review by Leknes and Tracy., 2008).

Generating Subjective Experience by Igniting Spikes in the Brain—Causality

Stimulating specific regions in the brain to invoke specific conscious (or unconscious) experiences is not new (e.g., Libet [1973], and recently Desmurget et al. [2009]). Recent advances in genetic and optical methods, and fine-scale electrodes implanted in the brain, have made it possible to manipulate specific neurons and neuronal networks, thereby changing/affecting the subjective experience of the animal, typically in mice. Using these new tools, one can detect groups of neurons in the mouse's brain that become activated when the mouse is exposed to a fearful stimulus (e.g., an electric shock to its foot). These group of neurons can then be artificially activated (generating spikes) by shining on them a beam of blue light, resulting in the mouse becoming frozen, similar to their reaction during the initial stressful event. This method effectively establishes a clear causal link between the activity of specific neurons and the resultant fear response (Ritter et al., 2024; Tonegawa et al., 2018).

Furthermore, the development and application of optogenetic and chemogenetic tools have expanded our capabilities to activate not just neurons but other cell types within the brain, such as astrocytes (Kol et al., 2020). This advancement provides compelling evidence of the direct connection (establishing causation) between the activity of specific groups of neurons and the emergence of respective conscious experiences, thus dramatically enriching our understanding of brain function in general and conscious experience in particular.

We will soon uncover more refined links between particular neuron types and conscious states. Our ongoing mission involves developing novel technologies, such as *in vivo* characterization of various types of human neurons and synapses using molecular techniques (Lein et al. 2021), alongside tools like optogenetics, chemogenetics, and voltage-dependent sensors for activating specific neuron types (e.g., only layer 5 or layer 3 pyramidal cells in the cortex). To explore how certain cell types influence brain states or specific emotions like fear, we will leverage fine-scale simulations of selected groups of cells together with physical emulations of extensive brain circuits, exemplified by projects like the Blue Brain Project (Markram et al., 2015). This will bring us closer to determine how activating particular subgroups of neurons, embedded in large networks, results in distinct conscious experiences rather than other experiences.

Discussion

What more can we do in order to provide a scientific explanation and theory to conscious experiences that we may have, for example, of pain? Science has made enormous progress in understanding the very basic mechanisms used by the brain to transform sensory inputs into trains of spikes in the periphery. We profoundly understand the fundamental mechanisms of the spike and have a compact mathematical description of it (the Hodgkin and Huxley [1952] equations). We know that these spikes are the building blocks, the units of information, used by the brain to represent the external and our internal world. We may therefore block these spikes (e.g., by drugs or by electrical/optical stimuli) either in the periphery or upstream (in the spinal cord or in the brain itself) and thereby affect conscious (e.g., of pain) as well as unconscious experiences. This provides a convincing correlation between the specific conscious experience and the spiking activity at various neuronal pathways. But we can do more. We can causally generate conscious experiences (e.g., of pain) via direct activation of specific neurons/neuronal circuits (using optogenetic, chemogenetic, or by specific electrical stimulation). This causal relationship between a particular spiking activity (that we may trigger experimentally in specific neuronal networks and brain regions) and a specific conscious experience gives scientists a very strong feeling that we "understand" consciousness, as we have demonstrated both correlation and causation between specific spiking activity and specific conscious experience.

Despite the advancements I have described, the fundamental puzzle remains unresolved: How do my spikes translate into my experience of pain? This "big leap" in understanding persists, and it is unclear how science will fully address this enigma. Indeed, this intriguing issue was recently highlighted in a thought-provoking experiment by Gidon et al. (2022) that challenged the direct causal links between neural spikes and consciousness.

One way to go about the puzzle of consciousness is to claim that when we better understand consciousness at the molecular and biophysical levels, we will no longer need to explain it as having some kind of special essence. This mirrors how we no longer attribute life to a "vital force" or explain burning through the outdated "phlogiston theory." Alternatively, we may claim that everything (living and non-living things, including electrons and photons) has consciousness ("panpsychism"). This idea remains speculative, and its scientific examination is challenging. Furthermore, although thought provoking and highly interesting, I cannot see how the leading theories of consciousness, such as the global neuronal workspace (Dehaene et al., 2011) and the integrated information theory (Tononi & Koch, 2015) and recent additional contributions (Mudrik et al. 2023) help in explaining the persisting "big leap" from spikes running in my brain to the emergence of conscious experience. So, perhaps this "hard problem" of consciousness is beyond the scope of scientific inquiry? And indeed, must science explain everything?

Let me conclude with a brief anecdote from about twenty years ago, during an interesting scientific conference in Asilomar, California. In one of the sessions, a colleague seated a row ahead of me posed me a question about the talk in progress. As I leaned forward to respond, I was suddenly struck by a severe pain in my lower back and fainted; it turned out to be a slipped disk. I woke up a short while later and found myself lying on the grass outside of the auditorium, with many worried faces of colleagues looking at me. I told them that I have intense pain in my back. One of them, a prominent professor and friend, reminded me: "But you know well that this pain is just a stream of spikes running from your spinal cord to your brain." What a relief that was!

Final Reflections

To me, science is a poetic and romantic endeavor. It stems from a unique combination of the working of twenty thousand genes and environmental input that shaped our species' brain around three hundred thousand years ago. This brain, through interaction with others and the world around us, invented the scientific method—a systematic way to explore the wonders of the physical world, including our body and brain. This method has led to incredible technologies to study the cosmos, elementary particles, and our brain and body in ultrafine detail. We can now measure and manipulate individual neurons and synapses during behavior and even edit and repair our genes. Undoubtedly, an amazing fit—a romantic journey (if used appropriately) into ourselves, and therefore into the properties of matter itself (unless shown otherwise in future years)—that gives rise to both life and consciousness.

When science is viewed as a romantic endeavor, one can't help but wonder why scientists refrain from expressing emotions during lectures and rarely collaborate with other disciplines like art and philosophy. Why is it almost "forbidden" for scientists to write in a poetic, experiential language, as seen in Nicolelis' (2001) most interesting article? I believe we should embrace our experiential capacities within science, as our brains are driven largely by emotions and unconscious activities. Allowing these aspects into scientific practice could enhance creativity and enrich interactions with colleagues, students, and the community at large. This book, born from experiential and thought-provoking encounters, aims to broaden readers' minds and hopefully have a profound experiential impact on them.

References

Aru, J., Suzuki, M., & Larkum M. E. (2020). Cellular mechanisms of conscious processing. *Trends in Cognitive Sciences*, *24*(10), 814–825.

Baron, M., & Devor M. (2022). Might pain be experienced in the brainstem rather than in the cerebral cortex? *Behavioral Brain Research*, *427*, 113861

Copits, B. A., Pullen M. Y., Gereau R. W. (2016). Spotlight on pain: Optogenetic approaches for interrogating somatosensory circuits. *Pain*, *157*(11):2424–2433.

Crick, F. (1995). *The astonishing hypothesis: The scientific search for the soul.* Scribner.

Damascelli, M., Woodward, T.S., Sanford, N., Zahid, H.B., Lim, R., Scott, A. and Kramer, J.K. (2022). *Multiple functional brain networks related to pain perception revealed by fMRI.* Neuroinformatics 20, 155–172.

Dehaene, S., Changeux, J. P., & Naccache, L. (2011). The global neuronal workspace model of conscious access: From neuronal architectures to clinical applications. In S. Dehaene, & Y. Christen (Eds.), *Characterizing consciousness: From cognition to the clinic?* Research and Perspectives in Neurosciences. Springer.

Desmurget, M., Reilly, K. T., Richard, N., Szathmari, A., Mottolese, C., & Sirigu, A. (2009). Movement intention after parietal cortex stimulation in humans. *Science*, *324*(5928), 811–813.

Gidon, A., Aru, J., & Larkum, M. E. (2022). Does brain activity cause consciousness? A thought experiment. *PLoS Biol*, *20*(6), e3001651.

Goldstein R. H, Barkai, O., Íñigo-Portugués, A., Katz, B., Lev, S., & Binshtok A. M. (2019). Location and plasticity of the sodium spike initiation zone in nociceptive terminals in vivo. *Neuron*, *102*(4), 801–812.e5.

Hayat, H., Marmelshtein, A., Krom, A. J., Sela, Y., Tankus, A., Strauss, I., Fahoum, F., Fried, I., & Nir, Y. (2022). Reduced neural feedback signaling despite robust neuron and gamma auditory responses during human sleep. *Nature Neuroscience*, *25*, 935–943.

Hodgkin, A. L., & Huxley, A. F. (1952). A quantitative description of membrane current and its application to conduction and excitation in nerve. *Journal of Physiology*, *1 17*, 500–544

Kol, A., Adamsky, A., Groysman, M. et al. (2020). Astrocytes contribute to remote memory formation by modulating hippocampal–cortical communication during learning. *Nature Neuroscience*, *23*, 1229–1239.

Leknes, S. and Tracey, I., 2008. A common neurobiology for pain and pleasure. *Nature Reviews Neuroscience, 9(4)*, pp.314–320.

Libet, B. (1973). Electrical stimulation of cortex in human subjects, and conscious sensory aspects. In A. Iggo (Ed.), *Somatosensory system. Handbook of sensory physiology* (Vol. 2, pp, 1–20. Springer.

Liu, X., Ramirez, S., Pang, P. T., Puryear, C. B., Govindarajan, A., Deisseroth, K., & Tonegawa S. (2012). Optogenetic stimulation of a hippocampal engram activates fear memory recall. *Nature*, *484*(7394), 381–385.

Mano, H., & Seymour, B. (2015). Pain: A distributed brain information network? *PLoS Biol*, *13*(1), 1–14.

Markram, H., Muller, E., Ramaswamy, S., Reimann, M.W., Abdellah, M., Sanchez, C.A., Ailamaki, A., Alonso-Nanclares, L., Antille, N., Arsever, S. and Kahou, G.A.A., Berger, T.K., Bilgili, A., Buncic, N., Chalimourdal, N., Chindemil, G., Courcol, JD., Delalondre, F., Delattre, V., Druckmann, S., Dumusc, R., et al. (2015). Reconstruction and Simulation of Neocortical Microcircuitry. *Cell*, *163*(2), 456–492

Minert, A., Yatziv, S. L., & Devor, M. (2017). Location of the mesopontine neurons responsible for maintenance of anesthetic loss of consciousness. *Journal of Neuroscience*, *37*(38), 9320–9331.

Nicolelis, M. (2001). Actions from thoughts. *Nature*, *409*, 403–407.

Mudrik, L., Mylopoulos, M., Negro, N., & Schurger, A. (2023). Theories of consciousness and a life worth living. *Current Opinion in Behavioral Sciences*, *53*, 101299.

Nurse, P. (2021). *What is life?* David Fickling Books

Rall, W., (1959). Branching dendritic trees and motoneuron membrane resistivity. *Experimental Neurology, 1(5)*, pp.491–527.

Ritter, A., Habusha, S., Givon, L., Edut, S. and Klavir, O., (2024). Prefrontal control of superior colliculus modulates innate escape behavior following adversity. *Nature Communication, 15*, 2158.

Sakmann B., and Neher, E. (1984). Patch clamp techniques for studying ionic channels in excitable membranes. *Annual Review of Physiology*, *46*, 455–472.

Tonegawa, S., Morrissey, M. D., & Kitamura, T. (2018). The role of engram cells in the systems consolidation of memory. *Nature Reviews Neuroscience, 19*, 485–498.

Tononi, G., & Koch, C. (2015). Consciousness: Here, there and everywhere? *Philosophical Transactions of the Royal Society B, 370*, 20140167.

Zorn, J., Abdoun, O., Bouet, R., & Lutz, A. (2020). Mindfulness meditation is related to sensory-affective uncoupling of pain in trained novice and expert practitioners. *European Journal of Pain*, *24*(7), 1301–1313.

Figure 0.6
Michal Rovner. Red Sands, 2021 (detail). LCD screen and video. Copyright: Michal Rovner.

20 Consciousness and the First-Person

Yaron Senderowicz

Personal Reflections

Since my first years as a mature person, the puzzles relating to subjective experience perplexed me. The attempts to describe what subjectivity means and to distinguish it from an objective realm often use the (metaphorical) distinction between an inner sphere and an outer sphere. According to this view, the subjective content of consciousness constitutes an inner sphere that only the conscious subject can access. However, what fascinated me already at a very early stage of my intellectual development, was precisely the reverse feature of conscious experience. My early intuition was that I have special access to my conscious experiences. Nevertheless, this insight was coupled with the awareness that my conscious experiences involve having the whole world in view, the objective world that includes myself as being "out there," in the world, together with other experiencing subjects that share it. To be sure, I am the conscious individual, and I am conscious of the world as being my world. But I realized quite soon that although I have special access to my conscious experiences, the content of all my conscious thoughts, intentions, and activities could be grasped only by attending to the shared world that includes me, the experiencing subject, as one of the things in it.

How can the world that I consciously experience be grasped from my subjective viewpoint, although as an experiencing subject I am only one of the things included in it? What does it mean to consciously have the objective world in view? These questions refer, I believe, to the essence of subjectivity. They constitute the main motivational ground that underlies my philosophical research activities. The methods that I use are similar to the contemplative methods used by phenomenologists and philosophers of mind. They combine phenomenological reflection and conceptual analysis. Reflection on my stream of conscious experience and on the variety of ways of being involved in the world is always my starting point. The meditation sessions in our workshop were particularly helpful for me in this respect. Meditating is similar to phenomenological reflection. It nevertheless differs from it. Since I am not a meditation practitioner, the experience of meditating was novel to me although it was partly familiar. This experience helped me to grasp the differences between the contemplative approaches that I used in the past and the one used by the meditator. I believe that each may reveal different aspects of consciousness. I intend to carry out this activity in the future.

Overview

In this chapter, I claim that the first-person perspective "from within" is the most essential feature of consciousness. It pertains to *how* we are conscious of what we are conscious of. I nevertheless demonstrate that the capacity to be conscious of oneself and one's own states and processes "from within" does not constitute a single capacity but rather several interrelated capacities.

In the first section, I examine the common characteristics of the variety of the first-person perspectives "from within" that distinguish it from the third-person perspective. These characteristics include the use of terms such as "I," "me," "mine," or their mental analogues as essential indexicals. They also include what Sydney Shoemaker characterized as the immunity to error through misidentification with respect to the term "I." Finally, the first-person perspective "from within" involves a unique type of epistemic access to oneself and one's own states that differs from the third-person perceptual access. I then examine three arguments that aim to articulate a real *ontological* distinction between the body and the mind from *epistemic* premises. I clarify that the first-person viewpoint "from within" serves a key role in these arguments.

In the last section, I explain why the first-person view "from within" does not constitute a single cognitive capacity but rather several interrelated capacities.

> Two things fill the mind with ever new and increasing admiration and reverence, the more often and more steadily one reflects on them: the starry heavens above me and the moral law within me. I do not need to search for them and merely conjecture them as though they were veiled in obscurity or in the transcendent region beyond my horizon; I see them before me and connect them immediately with the consciousness of my existence.
> —Immanuel Kant, *Critique of Practical Reason*

Introduction

One of the perplexing puzzles related to consciousness is the difficulty to specify what it is. It is sometimes claimed that consciousness could be identified with what I fail to possess when I am in a state of dreamless sleep and what I regain when I wake up.[1] However, the attempt to single out what consciousness is by reference to my capacity to *identify* some of my own conscious states does not suffice to explain what consciousness is. What precisely do I lose when I am in a state of dreamless sleep and what is the nature of the change that goes on in me when I become conscious again?

We apply the term "conscious" to a variety of kinds of mental states and processes such as sensations, thoughts, judgments and inferences, explanations, intentions, beliefs, desires, emotions, and feelings of pain and of joy. It seems that the answer to the question of what consciousness is, should begin by providing a list of properties that all conscious states and processes share by virtue of which they are characterized as conscious. But is there *one set* of properties that all the varied conscious states and processes share by virtue of which they should all be characterized as conscious? It is sometimes claimed that conscious mental states and processes involve a way it feels like experiencing them.[2] However, although many conscious states and processes exemplify this feature, it is *not* exemplified by *all* conscious states and processes. I *can* consciously understand a mathematical demonstration, a scientific explanation, moral demands, and the rules of chess without involving a way it feels like being in these cognitive states.[3]

The difficulty of singling out the set of properties that the term "conscious" designates is evident when one considers the fact that we also characterize a *bodily activity* as conscious. My walk to the fridge to take out a cool bottle of water is conscious *walking*. Dancing or playing basketball are conscious *bodily activities*. Prima facie, it seems that the *dancing* is conscious, and the *playing* of the game is conscious.[4] But what common feature is shared by these conscious activities and conscious mental states such as thoughts and judgments? Here one might be tempted to claim that bodily activities are *not directly* conscious but only *derivatively* conscious by virtue of the conscious *mental representations* that represent them. According to this objection, the representation of the dancing is conscious, not the dancing itself, the representation of the smile is conscious but not the facial expression of the smile. However, why couldn't bodily activities be directly conscious? Anyone who denies that bodily activities could be directly conscious owes us an explanation of what consciousness is.

The First-Person Perspective "From the Inside"

By a first approximation, it seems that there is at least one feature shared by all conscious states and processes. They all seem to involve the first-person perspective. One of my goals here is to clarify why the term "first-person perspective" does not designate a single cognitive capacity but rather several interrelated capacities. However, let us first try to see what the core features of these capacities are. When I consciously sense something, I am *immediately* aware that the sensation is my own sensation. When I consciously judge that a given proposition p is true, I am directly conscious that *I myself* have judged that p is true. When I am consciously involved in a given bodily activity, I am directly aware that *I myself* have performed it. It seems that the term "subjective" designates precisely this type of self-awareness "from within" that the sensation is mine, that I am the thinker of my thoughts, and that I am the one performing my bodily activities. This type of self-awareness is not always explicit. Conscious self-awareness could be what the phenomenologists characterize as prereflective self-awareness.[5] However, at least where the human kind of consciousness is concerned, if I entertain a conscious state or if I consciously perform an activity, it seems that I must have the disposition to be *explicitly* conscious that the state is my own state and that the activity is my own activity.

The first-person perspective involves a unique type of use of indexical expressions similar to "I" and "mine" or their mental analogue. Indexical expressions are expressions such as "you," "here," "now," "then," and "there." They refer to particular entities (objects, persons, times, and places) only within a particular *context of use*.[6] When I say "I am tired," my use of "I" refers to me, the user of this expression. When you say "I am tired" your use of "I" refers to you. When I consciously think that I think a given thought, my use of "I" refers to myself, the thinker of the thought. As Castañeda, Perry, and others showed, it is impossible to eliminate what indexical terms such as "I," "mine," or their mental analogues express from the content of cognitive states that involve them without canceling their cognitive import. "I" (or its mental analogue) functions as an *essential* indexical.[7] When I consciously sense the sweetness of the chocolate in my mouth and say, "Wow! I like the taste of this piece of chocolate," my assertion expresses the *same proposition* that is expressed by the assertion "Yaron Senderowicz likes the taste of this piece of chocolate." Nevertheless, the use of this sentence has different cognitive properties. Due to some disability, I may forget that my name is Yaron Senderowicz. Based on some relevant third-person kind of evidence, I may even know that the person named Yaron Senderowicz likes the taste of a given piece of chocolate without realizing that Yaron Senderowicz is *my* name. However, if I consciously experience the sweet taste of a given piece of chocolate, I must be, at least tacitly, aware that *I* am tasting it and that *I* like it.[8]

The first-person perspective is not only characterized by the fact that "I," "me," "mine," and similar expressions function as essential indexicals. It has some unique epistemic characteristics. When I point to my reflection in the mirror and utter "I look funny today", my use of "I" functions as an essential indexical. But this type of use of "I" differs from my use of "I" when I say "I am in pain." I may misidentify myself in the mirror; someone else that looks similar to me might be reflected there. But *no possibility of misidentifying myself* is at stake when I consciously think a thought, sense something, or am aware that I am dancing. In these contexts, my use of "I" is immune to error through misidentification with respect to the term "I."[9]

From Epistemic Premises to Ontological Conclusions

The unique epistemic characteristics of the first-person perspective are involved in what became known as the hard problem of consciousness. This problem concerns the nature of the qualitative features of our experiences, the way it feels to experience them. Philosophers distinguish between two kinds of hard problems: an ontological hard problem and an epistemic hard problem. In both cases, it seems that the problem arises due to the fact that conscious awareness involves the first-person perspective. The ontological hard problem regarding consciousness can be stated as follows: Do the semantic and epistemic characteristics of the first-person perspective imply an *ontological distinction* between mental substances and physical substances, or between mental properties and physical properties?[10] The premises of this type of argument include propositions that express self-knowledge of my own existence or my own conscious states "from the inside." The conclusion supposedly determines the *kind of being* that pertains to entities accessible from the first-person viewpoint.

Descartes was the first modern philosopher who presented arguments that entail ontological conclusions from epistemic premises. According to Descartes, the first-person viewpoint "from within" that I have of my existence as a thinker implies a *real ontological distinction* between the thinking being that I am and my body. As he points out, I cannot fail to know that I exist if I consciously think that I do, but I can fail to know that I have a body, because my knowledge that I have a body must be based on a type of a mental representation. And according to Descartes, I can represent myself as an embodied creature even if I do not have a body.[11] Since I must know that I exist if I think that I do but I can represent myself as an embodied creature even if I do not have a body, the individual that thinks differs from the body connected to it.

The epistemic differences between self-knowledge as a thinker and self-knowledge as an embodied creature clearly involve the first-person perspective.[12] In the implication "I think, therefore I exist," I cannot replace the term "I" with the term "Yaron" and preserve the epistemic certainty it expresses. If, for example, I forgot that my name is Yaron, I may fail to form the *belief* that Yaron thinks. In this case, I will fail to *know* that Yaron thinks. In contrast, if I am conscious that *I* think, I cannot fail to know that I think and therefore that *I* exist.

The Knowledge Arguments

Self-knowledge of my existence as a thinker is only one kind of knowledge that motivates arguments that infer ontological conclusions from epistemic premises. The questions regarding the qualitative features of experiences (*qualia*) motivate Cartesian-like arguments that entail ontological (dualistic) conclusions from epistemic premises. The attempts to respond to them are the subject of several current controversies regarding consciousness.[13] Three eminent arguments presented in the last fifty years that address the nature of qualia seem to entail ontological conclusions from epistemic premises: Kripke's modal argument,[14] Jackson's knowledge argument,[15] and Chalmers's zombie argument.[16] As I will demonstrate now, they involve the unique first-person perspective that we have of our conscious mental states. For lack of space, I briefly address only Jackson's and Chalmers's arguments.

Jackson's thought experiment consists of the imagined story of Mary, a brilliant neuroscientist who has been locked in a black and white room all her life.[17] Jackson presumes that Mary knows everything the physical sciences can tell us about the neural and behavioral properties of the perception of colors. Since Mary has never had experiences of red things, she does not know what it feels like to experience red objects. According to Jackson, when Mary is released from the black and white room and consciously perceives a red rose for the first time, she must learn something *new* about the content of visual experiences. She now knows what it *feels like* to experience red objects. Before she was released from the black and white room, Mary knew everything the physical sciences could tell us about the perception of red things. Therefore, her novel knowledge (the way it feels to experience red things) *could not* be knowledge of a physical property. According to Jackson, the qualitative property that constitutes the experience of redness is an *epiphenomenal* property. It is caused by the neurophysiological processes involved in the visual perceptions of red things, but it does not cause anything in the physical world.

Jackson's argument aims to demonstrate that conscious experiences involve properties of which we are conscious, which could not be physical properties. Since these conscious properties are epiphenomenal, they are irrelevant to a comprehensive account of physical reality. The first-person perspective "from within" is tacitly involved in this argument. The argument seems to presuppose that the conscious properties of experience are accessible to Mary only from her first-person perspective "from within." Mary did not know what the experiences of redness felt like, because she did not consciously experience redness directly, "from within." When she was released, she acquired new concepts—phenomenal concepts of colors—that express her novel knowledge. The first-person viewpoint is required in order to form a phenomenal concept of redness.[18] Mary had all the relevant physical/functional scientific concepts before her release from the black and white room, but she lacked the phenomenal concept of redness.

Chalmers's zombie argument also involves a tacit appeal to the first-person perspective. According to Chalmers, the hard problem of consciousness concerns the possibility of providing a reductive physical/functional explanation of the qualitative conscious features of experience. Chalmers supposes that the conceivability of a given state of affairs entails its possibility. He develops this supposition by means of his two-dimensional semantics.[19] His account of reductive explanations is based on this supposition. According to this account, E reductively explains P only if E *a priori entails* P—that is, only if P is *inconceivable* without E. This means that P *logically supervenes* on E: there are no possible worlds in which phenomena of type P exist and are not correlated with phenomena of type E. A reductive explanation of the conscious qualitative features of experience must therefore clarify why the realization of qualia is inconceivable if the correlative neurophysiological states and processes are not realized. Yet, according to Chalmers, qualia cannot be reductively explained; they do not logically supervene on brain states and processes. Zombies are creatures that are physically identical to human beings and nevertheless lack the qualitative conscious properties of human mental states. According to Chalmers, zombies are conceivable. That is, a physical duplicate of me whose cognitive states do not instantiate the variety of conscious phenomenal properties that I entertain is conceivable. Since conceivability entails possibility, if zombies are conceivable, they are possible. But if zombies are possible, the qualitative features of my experiences do not logically supervene on the physical/functional states of my brain. For zombies are physically identical to conscious human persons. We must therefore conclude that there is a real ontological distinction between qualia and physical reality.

Chalmers's reasons for claiming that zombies are conceivable involve a tacit commitment to the first-person perspective. His argument *presupposes* that we can conceive, say, what pain feels like only because we are sometimes conscious of being in pain by experiencing it. We can conceive the possibility of a zombie, of a creature that is physically identical to a human being that nevertheless lacks conscious experiences, by comparing this first-person conception of pain to the description of such a creature. The first-person conception of phenomenal, qualitative states involves phenomenal concepts. The failure to reductively explain qualia is therefore tantamount to the failure to reduce phenomenal concepts to physical/functional concepts. As in Jackson's case, my phenomenal concept of pain differs from my physical/functional concept that refers to the correlative brain state. Phenomenal concepts can be formed only from within the first-person perspective; they depend on it. They express the ways in which we conceive our own mental states "from within." Therefore, in Chalmers's case too, the ontological distinction between body and mind involves the first-person viewpoint.

The Variety of First-Person Viewpoints

Does the failure of reductive explanations of phenomenal concepts imply ontological dualistic conclusions? Several philosophers deny this for a variety of reasons.[20] I will not examine here the controversies regarding this matter.[21] I wish to address a different issue in this context. The excessive attention given to the questions regarding the ontological nature of qualia seems to blur the fact that the puzzles regarding conscious experiences do not primarily pertain to *what* we are conscious of but also to the *cognitive means* by which we are conscious of what we are conscious of. Clearly, any puzzle regarding qualia must also be a puzzle about how they are accessible "from within." Since qualia are supposedly accessible only from the first-person viewpoint "from within," the first-person/third-person asymmetry is in fact reflected in the nature of qualia.

So far, I have used the terms "first-person perspective from within" as if they refer to a *single* cognitive capacity. I wish to make three claims in this context:

1. The first-person perspective "from the inside" is not only limited to the qualitative features of our experiences.
2. There are reasons to suppose that the characteristics of the first-person viewpoint "from within" are not characteristics of a single cognitive capacity but of several cognitive capacities.
3. These cognitive capacities are interrelated.

The qualitative features of my sensations and perceptual experiences is just one kind of entity that is accessible to me in the unique first-person mode of awareness. Other people cannot have a first-person access to what I consciously sense and feel. But a similar difference is also exemplified by my conscious awareness of being bodily present in the world. When I walk, run, or dance, I am conscious "from within" of being bodily present in the world. To be sure, other people can know that I am bodily present. But they cannot know this in the same unique way in which I am conscious of it "from within." Needless to say, the first-person/third-person asymmetry that pertains to being embodied cannot raise the same ontological puzzle that my conscious awareness of the qualitative features of my experiences raises. I am my body, the very same body that others can see and touch. But only I can be conscious of it as *my* body and of its variety of activities as *my* activities.[22]

The differences between being conscious of a sensation "from within" and being first-personally conscious of being embodied "from within" invite the question of whether the first-person viewpoint constitutes a single cognitive capacity. The differences between my conscious awareness of my sensations and my conscious awareness of my thoughts, my judgments, and my intellectual activities provide additional support for the claim that the first-person perspective "from within" constitutes more than one cognitive capacity.

I am conscious of my own sensations from a subjective perspective. Only I can have this kind of access to my sensations. A similar first-person/third-person asymmetry is exemplified by my conscious awareness of my thoughts and judgments. But is the first-person viewpoint that allows me to access my thoughts and judgments "from within" the same cognitive capacity involved in my first-person conscious awareness of the qualitative feature of my experiences and my bodily activities? Let us consider, for example, the differences between being conscious that I sense something and being conscious that I judge something. Conscious sensations seem to function as the intentional objects of reflection. The unique first-person access that I have to my sensations presumably consists of some sort of inner perception or self-monitoring mechanism.[23] It allows the conscious subject to acquire knowledge of the properties of the sensations that they instantiate. However, the conscious characteristics of acts of judging, of drawing inferences, and similar intellectual activities cannot be based on the same kind of cognitive capacity. Say that I consciously judge that $7+5=12$. The fact that I make this judgment, normally involves *my conscious understanding* of what I judge. Otherwise, I would be similar to a parrot that merely seems to say something without saying anything. For that purpose, I must possess the relevant concepts of numbers, of mathematical functions, and I must consciously know how these concepts should be used in order to form propositions. But it is nonsensical to suppose that I can determine that I judge something in the above sense by *perceiving* the judgment that I make as an object of inner perception. How can an object—any object—manifest the characteristics that pertain to understanding what one says and thinks?[24] I cannot be conscious that I judge without being conscious that I understand what I judge, and I cannot be conscious that I understand what I judge by turning my judgment into an *object* of my inner gaze. *Consciously understanding what I judge must be an intrinsic feature of the act of judgment itself.* We seem to be led astray here by the fact that judgments are made by using sentences, which are representational entities that have perceivable physical characteristics. But judgments are not mere tokens of sentences, and they are definitely not sensations.[25] The subjective first-person perspective involved in consciously judging and other intellectual activities differs from the first-person relation that I have to the qualitative features of my experiences.

The different first-person cognitive capacities are also interrelated. Conscious thoughts, judgments, and inferences involve being conscious of types of *objective unities*. The judgments that we make and the variety of our intellectual activities expresses this. I now sit in my room, I see the monitor of my computer. I reflectively distinguish its color from its shape, and its other properties. They are all bound together in the conscious representation of the object that I perceive. My judgment "this monitor is rectangular" expresses my awareness of the unity of the properties of being a monitor and being rectangular, the fact that they together constitute properties of *this* object. When I am consciously aware of the fact that these are different properties of this object, I must also be at least tacitly conscious *of the unity of my own conscious representation* of it.[26] I now focus on the table on which the monitor stands, the wall behind it, this room, and the relation between these objects. Each of these conscious representations of objects constitutes a focal point of similar unity of a multiplicity of objective properties and relations. Any judgment that I make about the properties of these objects expresses my awareness of these unities.

The conscious awareness of the unity of my consciousness involves the subjective first-person perspective "from within." It involves at least a tacit use of the mental analogue of the term "I." It manifests the same first-person/third-person asymmetry that pertains to the first-person perspective.

I am now consciously focusing on this room, how I am related to it and to the objects that are there with me. The conscious awareness of the objective unity of the room and the objects that include me as one of them involves my first-person awareness "from within" of the unity of my consciousness. I am also aware "from within" of being *bodily present* "here." These are two different kinds of first-person self-awareness "from within." I cannot have the whole room in view merely by being first-personally conscious that I am bodily present "here." Similarly, I cannot be conscious in the first-person mode of awareness of being bodily present "here" merely by being conscious of the unity of my consciousness of the room, even if it includes a representation of myself as one of the things in it. But I am both conscious of the unity of the world and of myself as one of the things that is bodily present in it.[27] Bodily first-person awareness "from within" and intellectual first-person awareness "from within" of the unity of the representation of the world must therefore cooperate. Conscious experience must involve both types of first-person awareness "from within."

Concluding Remarks

In this chapter I argued that whatever consciousness is, it involves the first-person perspective "from within" that differs from the third-person viewpoint. The ontological and epistemological puzzles of consciousness are in fact founded on this asymmetry. Nevertheless, I suggested that the supposition that there is just *one* capacity for first-person perspective "from within" responsible for the variety of conscious capacities is mistaken. We possess substantive phenomenological and philosophical reasons to claim that there is more than just one first-person viewpoint. These cognitive capacities are indeed similar to each other and they are interrelated. Nonetheless there are substantive differences between them.

The similarities, interrelatedness, and substantive differences between these cognitive capacities should guide an interdisciplinary study of consciousness. The phenomenological and philosophical study of consciousness from the first-person viewpoint should initially separate between the reflective examination of sensations, of bodily activities, and of executing the intellectual capacities. The philosophical examination of the conscious character of thoughts and judgments cannot replace the first-person examination of movement and bodily activities carried out by other disciplines, and it cannot be replaced by the neuroscientific examination of these processes. The investigation of the nature of intellectual, conscious activities, of body awareness, and of sensations and feelings requires different methods. Conducting these different investigations would make it possible to determine the differences between them as well as the ways in which they complement one another in the search for a comprehensive account of consciousness.

Glossary

Epiphenomenal: A version of mind/body dualism according to which mental causation is an *asymmetric* relation. According to this view, physical states and processes can cause conscious non-physical properties and states of affair, but non-physical properties and states of affairs cannot cause physical states and processes.

Epistemic property: A characteristic of a claim to knowledge.

Epistemology: Theory of knowledge.

Ontological distinction: A distinction that pertains to the existence of entities.

Ontology: The philosophical field that investigates the most basic characteristics of the being of beings.

Qualia: The qualitative properties of conscious states; the way *it feels like* to experience them.

Notes

1. See, for example, Searle (1992, p. 83).

2. See Nagel (1974).

3. Some philosophers argue that this type of cognitive state also has phenomenological characteristics, or a way it feels like to be in these states. See Chudnoff (2015) and Kriegel (2015). Let me stress that I am not claiming that consciously understanding a mathematical demonstration or a scientific explanation is *never* accompanied with a way it feels to be conscious of these processes. My claim is rather that it is *not a necessary feature* of being conscious of these processes and that it therefore cannot determine what consciousness is. The claim that *all* types of conscious states involve a way it feels like experiencing them is an *empirical contention.* However, there are many counterexamples to this claim.

4. The claim that bodily activities are directly conscious is the enactivist's view. For a discussion of this point see Gallagher (2005, 2017), Merleau-Ponty (2012), and Thompson (2007).

5. See Zahavi (2006) for a discussion of this topic.

6. On this topic see Kaplan (1989).

7. See Castañeda (1966, 1967) and Perry (1979). For a similar view, see Chisholm (1981) and Lewis (1979).

8. This is a phenomenological and an epistemic claim. Both features determine the concept of this type of consciousness states. I discuss this issue in what follows.

9. The immunity to error through misidentification with respect to the term "I" is according to Shoemaker (1968) the main feature that distinguishes the use of "I" as a subject and the use of "I" as an object. Shoemaker bases his account on Wittgenstein (1958). For a discussion of various aspects of this distinction see Coliva (2006, 2012), Longuenesse (2012, 2017), Peacocke (2012).

10. This is not always how the hard problem of consciousness is explicitly presented. Nevertheless, it is usually presented by singling out the asymmetry between the subjective, first-person access to one's own conscious mental states and the third-person access to them. See, for example, Nagel (1974), Jackson (1982, 1986), and Chalmers (1996). The way I formulated the hard problem makes explicit what I believe is implicit in the way in which it is stated. I discuss this point in what follows.

11. See Descartes (1984, ch. 2).

12. This is not something that Descartes explicitly says but rather a feature of conscious thought that motivates his argument.

13. For a discussion of this topic, see Senderowicz (2010).

14. Kripke (1980, ch. 3).

15. Jackson (1982, 1986).

16. Chalmers (1996).

17. Jackson (1982, 1986).

18. For an account of phenomenal concepts and of Jackson's knowledge argument as involving the difference between phenomenal concepts and physical concepts see Loar (1997).

19. See Chalmers (1996, ch. 2).

20. Levine claims, for example, that the differences between phenomenal and physical/functional concepts ground an explanatory gap which is an *epistemic* hard problem and not an *ontological* hard problem. See Levine (1983, 2001).

21. On this topic, see Senderowicz (2010).

22. On the first-person/third-person asymmetry regarding our conscious awareness of our bodies see Sartre (2020).

23. High-order theories of consciousness—namely, high-order perception (HOP) and high-order thought (HOT)—provide different accounts of self-monitoring. For the HOP theory see Armstrong (1997); for the HOT theory see Rosenthal (1997).

24. For a similar criticism of the perceptual model of self-knowledge see Shoemaker (1996a, 1996b, 1996c).

25. The differences between sensibility and understanding is the core idea of Kant's theory of knowledge. See Kant (1998). For a discussion of the differences between first-person awareness of being a thinking subject and an embodied person in Kant's theory see Kitcher (2011), Longuenesse (2012, 2017), and Senderowicz (2021).

26. For the significance of consciousness of unity to the awareness of objects and the world see Shoemaker (1996d).

27. Husserl dubbed this phenomenon as "the paradox of subjectivity" (see Husserl 1970, section §53). For a discussion of the paradox of subjectivity see Carr (1999).

References

Armstrong, D. (1997). What is consciousness?. In N. Block, O. Flanagan, and G. Güzeldere (Eds.) *The nature of consciousness* (pp. 721–728). MIT Press.

Carr, D. (1999). *The paradox of subjectivity*. Oxford University Press.

Castañeda, H. N. (1966). "He": A study in the logic of self-consciousness. *Ratio*, *8*, 130–157.

Castañeda, H. N. (1967). Indicators and quasi-indicators. *American Philosophical Quarterly*, *4*(2), 85–100.

Chalmers, D. J. (1996). *The conscious mind*. Oxford University Press.

Chisholm, R. M. (1981). *The first-person*. University of Minnesota Press.

Chudnoff, E. (2015). *Cognitive phenomenology*. Routledge.

Coliva, A. (2006). Error through misidentification: Some varieties. *Journal of Philosophy*, *103*(8), 403–425.

Coliva, A. (2012). Which "key to all mythologies" about the self? A note on where the illusions of transcendence come from. In S. Prosser & F. Recanati (Eds.), *Immunity to error through misidentification*. Cambridge University Press.

Descartes, R. (1984). Meditations on first philosophy. In J. Cuttingham, R. Stoothoff, & D. Murdoch (Eds. and Trans.), *The philosophical writings of descartes* (Vol. II). Cambridge University Press.

Gallagher, S. (2005). *How the body shapes the mind*. Oxford University Press.

Gallagher, S. (2017). *Enactivist interventions*. Oxford University Press.

Husserl, E. (1970). *The crisis of european sciences and transcendental phenomenology* (Translated by D. Carr). Northwestern University Press.

Jackson, F. (1982). Epiphenomenal qualia. *Philosophical Quarterly*, *32*(27), 127–136.

Jackson, F. (1986). What Mary didn't know. *Journal of Philosophy*, *83*(5), 291–295.

Kant, I. (1998). *Critique of pure reason*. P. Guyer and A. Wood (Trans.). Cambridge University Press.

Kaplan, D. (1989). Demonstratives: An essay on the semantics, logic, metaphysics and epistemology of demonstratives and other indexicals. In J. Almog, J. Perry, & H. Wettstein (Eds.), *Themes from Kaplan* (pp. 481–564). Oxford University Press.

Kitcher, P. (2011). *Kant's thinker*. Oxford University Press.

Kriegel, U. (2015). *The varieties of consciousness*. Oxford University Press.

Kripke, S. (1980). *Naming and necessity*. Oxford University Press.

Levine, J. (1983). Materialism and qualia: The explanatory gap. *Pacific Philosophical Quarterly*, *83*, 354–361.

Levine, J. (2001). *Purple haze: The puzzle of consciousness*. Oxford University Press.

Lewis, D. (1979). Attitudes de dicto and de se. *Philosophical Review*, *88*(4), 513–43.

Loar, B. (1997). Phenomenal states. In N. Block, O. Flanagan, & G. Güzeldere (Eds.), *The nature of consciousness* (pp. 597–616). MIT Press.

Longuenesse, B. (2012). Two uses of I as a subject. In S. Prosser & F. Recanati (Eds.), *Immunity to error through misidentification*. Cambridge University Press.

Longuenesse, B. (2017). *I, me, mine*. Oxford University Press.

Merleau-Ponty, M. (2012). *Phenomenology of perception*. D. A. Landes (Trans.).Routledge.

Nagel, T. (1974). What is it like to be a bat? *Philosophical Review*, *83*(4), 435–450.

Perry, J. (1979). The problem of the essential indexical. *Noûs*, *13*(1), 3–21.

Peacocke, C. (2012). Explaining de se phenomena. In S. Prosser & F. Recanati (Eds.), *Immunity to error through misidentification*. Cambridge University Press.

Rosenthal, D. (1997). A theory of consciousness. In N. Block, O. Flanagan, & G. Güzeldere (Eds.), *The nature of consciousness* (pp. 729–754). MIT Press.

Sartre, J. P. (2020). *Being and nothingness*. Routledge.

Searle, J. (1992). *The rediscovery of the mind*. The MIT Press.

Senderowicz, Y. (2010). *Controversies and the metaphysics of mind*. John Benjamins Publishing Company.

Senderowicz, Y. (2021). The "I" that thinks: Kant on self-reference as a thinker. *Iyyun*: *The Jerusalem Philosophical Quarterly*, *69*, 221–247.

Shoemaker, S. (1968). Self-reference and self-awareness. *Journal of Philosophy*, *65*(19), 555–567.

Shoemaker, S. (1996a). Introspection and the self. In *The first person perspective and other essays* (pp. 3–24). Cambridge University Press.

Shoemaker, S. (1996b). On knowing one's own mind. In *The first person perspective and other essays* (pp. 25–49). Cambridge University Press.

Shoemaker, S. (1996c). Self-knowledge and "inner-sense." In *The first person perspective and other essays* (pp. 201–268). Cambridge University Press.

Shoemaker, S. (1996d). Unity of consciousness and consciousness of unity. In Shoemaker, S. (1996). *The first person perspective and other essays*. Cambridge University Press.

Thompson, E. (2007). *Mind in life: Biology, phenomenology and the sciences of mind*. Harvard University Press.

Wittgenstein, L. (1958). *The blue and brown books*. Basil Blackwell.

Zahavi, D. (2006). *Subjectivity and selfhood: Investigating the first person perspective*. MIT Press.

21 Performing Space

Anat Shamgar and Tom Soloveitzik

Personal Reflections

Tom Soloveitzik: *In a moment my time to get on stage will come. Even after years as a performing artist, I still have butterflies in my stomach before I get to meet the critical eyes of the audience. Deep inside, I know that I will cope and that I will do what is necessary for the performance's best. However, I know as well that oftentimes it still leaves me with unanswered questions: Will I be able to express myself freely without the boundaries, restrictions, expectations, and self-censorship that I impose upon myself? Will I be able to truly exist in time and space from an inner place that does not force itself on its surroundings? I do not have satisfactory answers to all these questions from my personal experience over the years. As the years go by I learn that my ability to surprise myself gets slimmer and slimmer. Once in a while a small gap is being revealed behind the wall of habit to which I lean back onto, in a changing degree of variations.*

As I write these lines a memory of a failure from a recent performance lingers. This uncomfortable sensation brings back vivid moments wherein I try to analyze and understand what went wrong. I remember the thoughts inside my head a moment before entering the stage: I was trying to plan in advance what my first action to start the performance with would be. In retrospect, I understand now that the need to choose something to do in this context (and not to wait and see how to best contribute to the situation; in other words, what is needed from me) derives from fear or from the need not to confront the moment in its fullness. By imposing my own will in this instance I caused inaccuracy throughout the performance, a little incitement that created a growing gap during its duration.

I chose to open this short text with a personal statement on the fragile object of the art of improvisation. Often it is said that in improvisation "everything goes" and that one cannot make any mistakes, but it is far from the truth. Like in any other art form, inaccuracies and unintentionality affect the results (of a performance) even if one cannot predict how things will evolve in advance.

I would like to digress for a moment from this personal note and focus on some thoughts from my collaboration with dancer Anat Shamgar through a sound and movement prism.

We meet in different ways, as separate beings who share the same space. Our actions and their by-products, our personal input and output systems, are all based on the unique ecological equilibrium created in the space and time of the performance, each part of which affects the weaving of our (in)visible dialogue. We attempt to cohabitate, side by side, and to attune ourselves to the split second in which time fractures, and its echoes, unsynchronized, are heard and seen.

We hear and perceive sound in space in a different manner than we do movement. Below the surface, sound is ever-changing and in constant flux: multiple sound waves vibrate in space, affected by fluctuating air pressure, their source or cause transparent or hidden from our eyes. Over time, we hear many aural histories mixed together in one space, their interplay undecipherable and unpredictable for us.

Building a common space for performances is at times a ritualistic endeavor, or it can consist of ordinary actions in a space that opens itself up to an ever-unfolding creation. The objects in space are thrown into the multilayered present moment, which holds both the memory of their collaborative past and a speculative present, forever writing itself into being. This act of writing the present etches itself upon the space, resulting in layers of elements that can later be uncovered and rediscovered.

The rift between sound and image, inherent in the performance arts, is usually ignored in an unwritten agreement between spectators and performers. In our mutual work this rift serves as a starting point for our engagement in and with a given space. We strive to create a multidimensional space where there is no one dominant focus, unlike our usual experience in our vision-centered culture. Every space carries its own physical and acoustical characteristics. As Seth Cooke phrased it, "Every contact leaves a trace"; simply dwelling in a place is an intervention. In our performances we find ourselves occupied with (re) designing space: we make it perceivable and audible, using the means available to us. Inhabiting the space is an attempt to create parallel realities, where space, gaze, sounds, and movements intermingle and coexist. The new space created by the performance reaches toward a new experience where, if only for a moment, our gaze dissolves, and our perception expands toward a new presence.

The performance comes to an end. The time has come again to examine the road my consciousness has gone in its course. Have I succeeded in entering a deeper, more synced existence during it? Were there splits in my body-consciousness experience during the performance and what could have caused them? Was I aware of any physical contractions inside myself? Did I keep breathing without interruption?

These questions remain open, and their importance lies within the internal dialogue that they raise and in the way I will continue to reflect and rethink what they may bring.

Evanescence and Paradox

Anat Shamgar

Overview

Being invited to take part in the "Perspectives on Consciousness" conference as artists rather than scientists, performers rather than speakers, was a challenge and an honor. As artists we are interested in the zone of "in-between"—between conscious and subconscious, reality and dream. We aim at opening a broad zone of an "in-between" as a space for the piece to happen. We keep questioning what we can do at every moment of the performance. The performance is an arena for watching and listening to the becoming of our piece, while we are making it, and for inviting a body-mind multilayered conversation.

In our duet, Tom Soloveitzik and myself share a common interest in the art of collaboration, and in improvisation as a performance. Unlike most improvisations that follow schemes, structure, patterns and forms of action-reaction, self-expression, and aesthetics, we aim to maintain and locate ourselves within a state of clearing and availability of our body-mind—to have a "state of nothing" between the narrative self and the performing artists' experiential self. Of course, one cannot escape from expressing one's own personality, but this is not the aim of the performance. We follow principles that help us open ourselves to a deep sensing that is beyond oneself, and find ways to draw our performance from within. We are preparing the body to think, to listen, and to watch how the body and sound become shapes in the space. We lead and watch the tuning of our state, and observe the content of movements and sounds. Our piece is about what is interpellated by itself, revealing its own identity, quality, and texture. We direct and maintain the same principles anew, as they become, while they happen.

Characteristics of silence, stillness, emptiness, and abstraction come from the conditions of the form, and are reflected in the piece. We are moving in simultaneous realities, both movement and sound are sharing the same space and experiencing the same conditions. We build a space for a dialogue as a concept, and as an act.

> It is by lending his body to the world that the artist changes the world into paintings. To understand these transubstantiations we must go back to the working, actual body—not the body as a chunk of space or a bundle of functions but that body which is an intertwining of vision and movement . . . My movement is not a decision made by the mind . . . my body moves itself; my movement is self-moved. It is not ignorance of self, blind to itself; it radiates from a self.
> —Maurice Merleau-Ponty, *Eye and Mind*

Discussion

The act of art is always a movement of opening our subconscious, letting it appear and reveal itself in the body as it observes and writes itself, and aiming to break the limited scope of conscious awareness. In our text we will attempt to articulate the form and the practice of our work, and the means through which staying in a state that is attuned to our body and environment enables embodiment and presence of movement and sound as materials and contents to be seen, heard, and shared.

Tuning is a process in which intention, attention, awareness, and action shift between a state of intuition and a state of cognition. Our awareness witnesses the unknown, reveals the body's desires, deepens the physical experience. We direct ourselves to a process of letting our experience become present and be seen. Clearing our mind and allowing the body-mind to act. We stay alert and open, attend to the information that arrives through our senses. We are in an ongoing act of composition, following the stream of images, sounds, and movements. Our composition is a process of revealing the unknown, letting it appear and transform or disappear. Our process of composition is focused on what emerges, as it happens. We are interested in designing a state of things that can let our artistic desire be fulfilled.

Composition is a practice through which we make our art. It is the way we organize the conditions and define the tools to make art. We find ways to reveal the composition as a process of becoming. It defines the area of creation as a zone of in-between: between intuition and cognition; known and unknown; body and space; sound and body; past and present; subject and object. It is the meeting of the concrete and the abstract. The in-between is a broad zone where many possibilities and combinations emerge. We direct ourselves through the composition process as followers rather than leaders, as receivers rather than instructors. We "forget" the nature of the things we watch and hear in order to be seeing and listening openly. It is as if the act of composition is a "net" through which we fish our patterns and variations. The practice of composition alternates ways to use our senses, and the ways they direct us while we are moving, perceiving our movements and sounds as materials in the act of art. We learn what we do while doing it. We recognize and make choices about how to construct our experience, and the composition itself is the theme we explore.

Our bodies are a source of exploration, and composing of the piece raises our dedication to finding ways to enable the moving body to find its course. The quest is to free the body-mind from inhibitions or expectations and maintain awareness for making choices. Composition is questioning the ways we think and identify materials that are at the core of our artistic practice. We are constantly aware of their reappearance and the way they are restaging themselves. Therefore, the process of composition needs to define the procedures according to which it can take place.

Procedures serve an additional purpose, less concrete and less decisive: they serve as a "net," a fishing net that is thrown into the "sea" of the unknown. Throwing the "net" is not an exploration of something concrete, or something correct—it is rather an expedition to the unknown that will surface and validate itself because it has emerged through such a process. The "catch" is the unknown, and the practice is the use of this tool. It is a conclusive act that synchronizes materials and continuity, shape and sequence, theme and story, research and art. The process of determining such procedures affect the process of creation, and the marks and artistic signature it leaves on the products. Choices of procedures yield artistic results and point to the connection and effect of the process on the artistic act itself.

The integration of the roles of performer and composer, dancer and dance maker, musician and music maker, are the essence of this form.

Movements and sounds are materials in the composition that is always there and is always changing. A practice of composition is opening conscious to the subconscious. It is an interest and focus in the transformation of sensing and "making sense," seeing and forgetting, drawing and erasing. The practice of composition is the thing we discover and explore.

The practice is executed through various mechanisms of communication: deep listening, flow of shift in our attention, availability, and responsivity of our body. We can see these mechanisms through different elements of performance, such as time, space, memory, practice, and communication. We will now expand upon each of these elements as they are reflected in the artistic act.

Time

When we deal with the time dimension within the act, the focus on the in-between is by staying with the "now." The starting and ending points are mapping the middle. The "middle," the expansion of the movement, is the interest. Between the beginnings and endings is the vanishing trace of flow, along which things happen. We need to listen to the continuity of time, and we need to let the trace of things vanish to make space for the "now." We experience "time" as a space in order to measure and direct our attention. We initiate a perception of time as a zone of an "in-between": in between present and future, doing and contemplating. The challenge is to keep the real time (i.e., present time) dominant, while the past is still present in our mind, and the future is on the horizon of our perception. Time becomes space, a container for the piece.

We identify the present while it becomes dominant, as if "before and after" are being diminished. But we also note how every movement and sound comes from the past, and moves to a future. We focus on the moment that we call "present"—but we are also bound by the continuity of the past, present, and future. Conducting time, or just being aware and acknowledging it, gives one an opportunity to utilize it; to have a dialogue within it; to stop the moving the body and observe its stillness; to be a witness and an interferer in the role of "timing the time." Every movement has its beginning, continuation, and end. It can stop and continue at any moment. Tracing the duration and fragmentation, we can watch and follow their shape, direction, sensation, meaning, or else. The movements and sounds that "come to us" are curved within time.

Fragmented perception maintains a possibility of "reading while writing" and "writing while reading." Our attention and intention are like routes to our actions. Listening to the route of time is a tool to maintain the multicity of layers and the zone of the in-between. It makes us slow down, do less, wait, and let go. As it goes, time is curved, curves the composition. Time becomes open and flexible to change. We need to maintain a balance between our body's desire, our ambition, and our judgmental reports, and to follow the rhythm of the piece. When it happens, time can dissolve, disappear, become a timeless space. Space of nothing.

Space

The body encounters space through perception. The way in which we look and see the space enables us to perceive it. Our perception includes the space itself and our presence in it. We use our sight to draw the space, place the body in space, and "spread" the body in space. The eyes are absorbing the space. The body transforms space to movement, and the movements of the body, like a drawing in space, become the body itself.

The skin (like the eyes) opens the possibility for exploring our changing relations with space. Through the sensation of space, we can have an image of space and of us being in it. We see through experiencing. Body and sight become one. The dialogue of body and sight becomes the dialogue of body and space. Separation and unity are simultaneously acknowledged. Reading space makes our actions mark space, be seen in space, become and transform to their reappearances. Building this space is an ongoing act of accepting the limitations of how much we know about what we see and do. The space is a place that is observed, that contains and is held by what is happening or not happening within it, and by what is happening with it.

We think of space as an area that is filled up and is emptied. We are interested in the potentiality of space and in what it contains. Making space present, "spacing" space, experiencing space's multidimensional quality—we move the space, and are moved by space. Space is the container of the two of us, affected by the two of us, and that affects the two of us; or actually, the three of us—Tom, myself, and the space.

We are moved by the need to perceive the dialogue with space, following and joining its continuous movement. Presence of movement depends on the awareness and experience of space. Making it actual by the physical body and by the juxtaposition of the two of us in the same space. We share the space and use it to be aware of each other. Moving in order to sense the space makes the movements happen. Movements are shapes of relocation of the body in space. Being in the "middle" of the movement—in the move—is like joining the movement of space. Space, time, sounds, and movements interplay simultaneous realities. We need to keep spacing them.

Memory

Memory takes time and holds space. It's the most interesting vertebra in the chain of the process—as if unseen but crucial. We use memory to bridge the physical and the visional, the happened and the happens, the shape and the abstraction. It absorbs things that occurred, and marks actions that take place in the present.

Memory combines sensing, identifying, exploring, and storing. The process is letting an action transform into an image—to be identified or recognized by our body-mind, condense it as an image, and archive it. In the chain of identifying, exploring, and memorizing, condensation is our way to absorb and hold large quantities of information into a condensed image, a packet that we can contain. The image is the visual abstraction of the concrete body movement in our mind.

And so, memory is our path to imagination. It is the condition to be in contact with our imagination, and to let imagination be the origin of our choices.

Memory is also the "space within." An inner laboratory through which we can have the images that are stored and be impacted by what is happening in the space around us.

Collecting the information is like a sponge. We let images arrive, be printed in our memory and be "given" to the body's imagination and reaction. The different stages of the process are equally important. They are all relevant to the process.

Memory interplays within awareness and the flexibility of change. We let the image that arrives from the body be "printed" so we can be aware of it—and let go. We keep our interest in the gap we hold between image and the action in order to open a space for awareness and choice. Awareness is our tool to keep us away from automaticity. We need to open this gap for a decision to arrive.

Memory contains the perception of the actual space, its architecture and environmental details, the same way it contains our perception of each other. It structures and stabilizes the communication of the performance.

Memory enables the interplay between the imagined and the actual, between what comes to mind and what happens actually in body and space. The transformation between the body and the image in both ways is a shift, a gap, and a riddle. The difference between the imagined and the experienced feeds the dialogue between our body, our awareness, and the making sense in the act of composition.

Practice

In order to experiment with constant unknowing, we need a lot of care regarding how to experiment. We need to administer or organize our fears, boost our power to act and move.

Let the body become something else, more than itself, to make the action not be bound by the limitation of the conscious.

We need to ask how we can achieve this in between situations at all the conceivable levels of the performance. We see the performance itself as a "zone"—an unknown place, a place of anonymity, a place of drift, a displaced location where things can be thought anew. We direct ourselves to see and hear, re-see and re-hear—watch and listen. Directing ourselves is the practice.

Knowing any decision could have been different, we need to keep the acceptance to what happens and go on with the process of composition. The inner sounds of judgment must dissolve from the body-mind. There is an ongoing act of clearing the mind. Sometimes, at some point, it stays clear; this is a moment of grace. Otherwise, we have the practice to hold on to.

Making art, like sleep, is a state that is beyond our control. We identify it, memorize it, and remember signs from it—but cannot fully control it. We develop a practice that takes us along.

Tuning through the senses is a practice. We direct our senses to read our body and our environment. Letting ourselves open to our body-mind attention, curiosity, passion, and imagination. We start by emptying and listening as a way of opening up our senses, as a practice. Emptying our body and mind needs time and intention. It needs practice not just as a preparation—but also as a procedure. We practice the observation and the shifts of intention and action.

Tuning through our senses is a practice for exploring ways of listening as a mode of being. Tuning is a practice that lays foundations to a practice of directing. It includes the simultaneous action of directing and reacting, organizing and making.

We cannot reach the levels of conception and sensing the act of art without following a practice of letting our subconscious surface, trusting it, and being led by it.

At the heart of the practice is the performance. And the communication that evolves. Our aim is to be one with our environment, a moving part of it. Receive the movement and be moved. "Being free" is an image. In order to make it possible we use the routine of a practice.

Communication

Contemplative performance reaches outward; it includes the participants and the audience. The audience is drawn to experience the transformation in the state of the performers' state of consciousness.

In the performance, we trust the process of "becoming." We keep loyal to being "at the moment," not knowing what will evolve, in order to be able to accept and release in real time. The audience is witnessing the composition in process. Our previous expectations are moved aside for the accuracy and precision of the process. We share and expose our dialogue.

We need to build a language to identify our materials so we can follow them, let them continue, transform, or disappear. We are building ways to separate them—in order to see them, recognize them from a distance, in order to open the space for the in-between—the dialogue—as the zone for the composition. We learn ways to hear our language—like listening to a foreign language—questioning the meaning with an open mind to different potentials of each word. We watch the elements of the piece, the connections and relations between them. We thrive to follow the elements, the relations between them and the juxtapositions between us. We all take part in a practice. We are letting the mind watch and rest. We need alertness as much as restfulness—to witness, be led, and let go.

Beauty comes in from effortless moving. From following the flow of the piece. In movement and sound an opportunity is given to what opens between one and the other. A route is created and discovered together. A joint route. Through witnessing our process, the audience is a key participant in the process of the performance, becomes part of the environment, its memory and experience.

Conclusion

While the audience remains a passive participant in the performance, communication with the audience is essential to the process in which an accumulation of elements and matters come together. The paradoxes are all happening at the same time, in real time, in communication with an audience.

The experience is that of an opportunity to touch the limits of the paradoxes, a quest for moving our act in between the paradoxes, between deep conscious and free flow of matter from the subconscious.

The act of reducing the judgmental voice and internal chatter, entering a quiet space—enabling subconscious content—becomes conscious.

Unlike the process of our performance, in which clarity is abstract and nonverbal, the attempt to illustrate it in words and analyze its ingredients and rules is forever short of capturing the state in which our act of art really takes place. The work itself is forever elusive, dynamic, and changing—and the attempt to mold it into words and phrases is forever frustrating and partial.

We hope we have been able to share with you the sensations, and the elements of our composition process, and to provide you with a glimpse at the process that we go through.

Figure 21.1
An image of the writer, Anat Shamgar from a performance "Second Duet" with Tom Soloveitzik. Photographer: Natasha Shakhnes.

22 The Indian Hard Problem of Consciousness

Eviatar Shulman

Personal Reflections

Consciousness has been a key idea in my thinking for a long time; in academia, it became the locus for the investigation and conceptualization of deeper intuitions that I may have cultivated even as a child. I began my studies at the university in the midst of a fiery search for answers to the big questions of life. These did not seem to be addressed by my studies in psychology, with its experimental, statistical, and theoretically shallow focuses; no one was talking about deep meaning, about life, about soul, or even psyche, not to mention God. Scans of rat brains seemed beside the point. So, when I started studying the Tibetan language and was introduced to Buddhism and Indian philosophy, I suddenly felt that someone was speaking my language. Within my studies in psychology, consciousness emerged as my main interest (special thanks go to Benny Shanon).

With time, my research delved deeper into the ideas and practices of Buddhist religion. My focus became the Pāli texts of the Theravāda tradition, and I grew interested in the way these texts formulated the relation between philosophy and practice. To undertake a serious inquiry into the philosophy I needed to find answers to fundamental questions: What is a Buddha? What are the texts and how and why were they composed?

It seems to me a dangerous sign of our times that materialism is so prevalent. Indian philosophy would consider this an ominous form of nihilism. I feel that a discussion of consciousness that would free itself from the grasp of a dogmatic materialism may help cure society of some of its ailments. True, we must keep dogmatic religious assumptions at bay, but still—must we do away with meaning? Can everything reduce to function? Is God really irrelevant, even when stripped of his or her religious modes and fundamentalist philosophical assumptions?

A confident discussion of consciousness, based on true respect for first-person experience, may or may not allow us, or me, to answer some of these questions. But I do feel it is important, and that the stakes are higher than a few more words printed on the screen or that gather dust on a library shelf. This is how we understand ourselves, and who we expect ourselves to be, what we think is real and possible. Ultimately, I hope that through the study of consciousness we can generate more care in a world that seems to be becoming harsher, and more objectifying, day by day.

Overview

This paper engages with contemporary discussions of the problems inherent within the hard problem of consciousness in order to suggest that the move from consciousness to matter may be an easier one than the one from matter to mind. Furthermore, drawing on insights from Indian philosophy, the idea is raised that mind is not equal to consciousness—the powerful and basic fact of knowing, which remains inexplicable in materialist accounts. In order to take consciousness as seriously as it deserves, which for some reasons has been a struggle in science, the position taken here is that we are better to cultivate the instinct to take consciousness as primary, as happens in central strands of Indian philosophy. Some of these approaches, expressed mainly in philosophical creation myths, are here introduced and discussed.

A heap of bones, smeared with flesh, full of impurity and exceedingly filthy–how can it be the knower, itself always distinct?

śalyarasiḥ māṃsalipto malapūrṇo ’tikaṣmalaḥ / kathaṃ bhaved ayaṃ vettā svayam etad vilakṣaṇaḥ //
—Vivekacūḍāmaṇi 158

Philosophical Introduction

It seems like the eliminativist (see **glossary**) views of consciousness are dying out. At the turn of the millennium positions that denied the very existence of consciousness, or that at least seemed to deny it, were still popular.[1] Today, however, scientists and philosophers combat a mysterianist (see **glossary**) approach that affirms consciousness as an entity that exists beyond its physical, materialist conditions.[2] Materialism—or call it naturalism in a robust, materialist interpretation—reigns, so that consciousness is commonly understood as a mechanism that works in the service of evolutionary survival.[3] Consciousness is thereby interpreted as a mode of presentation of information, as in the global workspace theory developed by Bernard Baars (1997), or in the corresponding global neural network proposed by Stanislas Dehaene and Lionel Naccacche (2001). Commonly, consciousness is said to shape its own picture of the world, an idea that leads to *illusionism.* This view comes in many grades and shades, with prominent voices being those of Daniel Dennett (1991, 2016, 2020), Thomas Metzinger (2009), Keith Frankish (2016), Donald Hoffman (2019), and Anil Seth (2021), as well as Graziano and colleagues (2020).[4] For these thinkers, matter is *real*; consciousness does not provide true access to external reality but creates a powerful illusion that is as close as the brain can get to understanding what is outside the skull (and, in fact, what is in the body). This position does not, in my mind, face up to the real challenge of consciousness: Why do we have it and why do we care so much about it?[5] Why is it that we actually *see* all this beauty and horror?[6]

There is something paradoxical about the illusionist view. On the one hand, evolution produced a mechanism to connect us better to our surroundings and to improve our use of their potentials. On the other hand, this mechanism is thought of as unreal. If consciousness does indeed highlight what is most conducive to survival and to genetic reproduction, that must be reality; maybe not all of reality, but at least something important. Without denying a certain appeal in the illusionist position, and while acknowledging that a coherent formulation for the role of consciousness in its biological context is important, it seems doubtful that such an explanation will encompass all others.

There is an intellectual tyranny embedded within the exclusivity of the naturalist position that results in everyone searching under the same spotlight. Even if consciousness evolved sometime during evolution, perhaps on that day evolution no longer involved only adaptation and survival. If evolution produced consciousness, perhaps it was headed toward something bigger than materialistic ends to begin with, and there were different potentials embedded within matter from the start. Meaning that if consciousness came to be so dominant, perhaps this tells us something with respect not only to what the world is constituted by, but also to what it can possibly be.[7]

Moreover, it seems that the materialist, evolutionary approach is still recalcitrantly trying to bypass the hard problem—How can we get not only the mental, but also the *conscious experience*, from the material or physical? This distinction between the mental and the conscious is important, and may turn out to be Indian philosophy's main contribution to the study of consciousness. The *mind* is explainable within a biological, cognitive approach when it is viewed as an expert organ in digesting information encoded in matter, which interprets sensory data and integrates these with internal information from the body and with self-monitoring capacities. The mere fact of making this information *conscious*, however, and broadcasting the vision of reality we encounter at every moment, is a different issue. That is to say that so long as matter is matter, it's dead, and it's never going to be conscious; so long as information is information, it doesn't have much life in it either (and perhaps matter and information should not be distinguished to begin with). That fact that information, matter, objectivity, and anything we identify are *known* (in the rich sense that consciousness experiences) is different from any functional definition.

Some would like to go beyond the hard problem, speaking of the "hard question,"[8] the "real problem,"[9] "the harder problem,"[10] or by arguing that the very idea that consciousness is anything beyond the material is a mistake.[11] It may be that such attempts to dodge the question are arising because scientists are becoming more confident that the problem is unsolvable.[12] In many ways, such approaches return to the eliminativist notion that just as water is "really" H_2O and heat is molecular movement, conscious processes are this or that neural activity. Yet matter, defined as dead matter, is just not conscious, so that for these thinkers consciousness remains an asset of "folk psychology."[13] The Buddhist sage Nāgārjuna provided the argument almost two millennia ago: any full-fledged, robust, real thing in the world (and for our concerns matter or consciousness) would have to possess *svabhāva,* "own being," which thereby defines it as a truly existent entity with intrinsic characteristics inhering within itself, of and for itself. As a self-defined entity constituted through its own nature, a *svabhāvic* entity is not dependent on other facts and cannot change. In truth, there are no such things, which makes any conceptual definition provisional,[14] although conventionally real. Nevertheless, this is what matter is supposed to supply in a materialist system, an endurable substance that is self-contained. This means that if illusionists want to continue to be materialists in the strong, monist sense that believes matter is real, they end up eliminating consciousness once again.

At the same time, it may be true that there is no such *thing* as consciousness. This, however, can also be said about matter, as idealist philosophers have argued long ago.[15] Ruling one side of the coin out while retaining the other will keep us riding along philosophical tides. Faced with this situation—in which we justifiably take matter very seriously, but are ready to make real room for consciousness—it may seem like dualism is the most realistic position, even if it creates new problems.[16] This is the option adopted by David Chalmers; while formulating his influential definition of the hard problem he opted for a version of property dualism that is less metaphysically committed than a robust substance dualism.[17]

A more intriguing metaphysical option may be panpsychism, which seems to be gaining adherents by the day. Authors like Galen Strawson (2008), Christoff Koch (2012, 2019), and Phillip Goff (2019) have been arguing that the only way to retain a real materialist outlook, while taking consciousness as a fundamental feature of reality in the way that it requires, is to define consciousness as an intrinsic quality of matter.[18] This option may be more attractive than a full-blown idealism,[19] although it is not without problems: if matter is conscious, any distinction between matter and consciousness must go. Panpsychism also tacitly assumes the explanatory priority of materialism, saying that matter is conscious. However, if consciousness is just as fundamental as matter, why not speak of pan-materialism, suggesting that consciousness is the primary fact and matter its manifestation, rather than consciousness being an attribute of matter? Why should it make better sense to have consciousness emerge from matter, rather than the other way around? Is consciousness less fundamental?

Given this state of affairs, it may be good to develop intellectual habits that take consciousness as primary. When the efforts to explain consciousness solely through matter are failing, perhaps we can try the opposite approach, which takes consciousness as fundamental and asks how matter is shaped by it. While taking consciousness as fundamental may lead to the same problems as the ones that materialism involves, this may help us learn to think better about consciousness as a primary feature of reality. To this end, I will here retell a number of Upaniṣadic creation stories that see the world as an emergent property of consciousness. This discussion will be followed by a presentation of the dualist school of *Sāṃkhya,* which offers a conceptualization of related ideas.

Our phenomenology suggests that there is reason to prioritize consciousness, because consciousness appears as the most basic fact of life.[20] Yet there may also be a logical advantage in deriving the objective from the subjective, rather than the more habitual vice versa. When we work from consciousness "outward," we need not explain how "matter" comes into being, but only how the different levels of objective content manifest in experience. This is easier than leaping over the conceptual abyss between matter and consciousness. In an emergent system that begins with matter, it may be possible to explain how higher levels of mind rely on lower biological and chemical levels, yet this investigation has trouble making it all the way up to include the very fact of conscious experience. If knowing is all about information or content, a central locus for integration of information would do; there is no need for consciousness, the rich sense of experience we know so well. If all is chemistry, then leave it at that. Coming at this issue top-down, without needing to ever reach the metaphysical bedrock of *matter,* may therefore be more achievable. At the very least, we should be approaching this problem from both ends.

The Indian hard problem of consciousness is therefore this: Given that consciousness is the basis for all that we know, how did physical (as well as psychological and social) reality emerge from it?[21] More commonly, scholars turn to Buddhist traditions, given their appeal and larger pool of representative voices. Here we are treating the philosophical background from which Buddhist ideas emerged, including that of karma. Naturally, the Indian voices I present here are not framed as a direct engagement with contemporary discussions. Nevertheless, they are interesting as philosophical options that take consciousness as fundamental, and are therefore good to think with.

Upaniṣadic Theories of Creation

We begin with the *Upaniṣads,* texts that conceptualize basic metaphysical assumptions in Indian thought. These texts, and especially the earlier *Upaniṣads* I will discuss here, emerge from the ancient world of sacrifice, while searching for its inner logic and discovering the reality of cosmic connectivity. The *Upaniṣads* perceive that the world is based on links or connections (*bandhu*) between the mental and psychological, the physical and external, the divine, the linguistic, the sociological, and most importantly the sacrificial realms of reality.[22] They intuit that these faces of reality are woven upon each other—the external on the internal, as in the sun (and the sun god) on the eye, the moon on the mind, the wind on breath. Inner levels grow more and more subtle, so that mind is finer than speech, intention finer than the mind, knowing finer than intention, and so on.[23] This leads to the understanding that at the base of the web of myriad cosmic connections is found a primeval conscious reality, the most subtle level of existence, which is called the self (*ātman*), and which is seen by some theorists as the bare fact of subjectivity. Once discovered, the self comes to carry the burden of cosmic coherence, being the condition of all types of existence and what proves to hold them together. Anything real must be known.

Thus, creation tales in the *Upaniṣads* tell us in different ways how the world originated through the deliberations of this fundamental cosmic principle of human subjectivity. Naturally, there are conflicting presentations of the emergence of multiplicity from oneness, each defining a particular mode of reality, so that the theories intersect, at times agreeing or disagreeing on the details, but sharing the same grammar of creation. Here, externality is a function of subjectivity, and material manifestation the result of knowing.

We read in the *Bṛhādāraṇyaka-upaniṣad* ("The hidden teaching [Upaniṣad] of the Great Forest"), chapter 1.4:

> At the beginning, this (world)[24] was only the self, shaped like a man. He looked around and saw nothing but himself. "I am" was the first thing he said, so the term "I" came to be. Therefore, even today, being summoned, a person says first "It is I" and then states his name, whatever it may be.[25]

Notice the beginning with the "self," the conscious principle as the very origin. His original movements condition our own patterns of experience:

> He was afraid. Therefore, whoever is alone is afraid. Then he examined this, (saying) "There is no one other than me, why should I be afraid?" From this, verily, his fear pacified.
>
> He had no pleasure at all. Therefore, one who is alone has no pleasure. He wanted another . . . So he split this very self into two, so husband and wife came into being . . . He united with her, and from this people were born.
>
> Having examined this, she (said)—"How—having created me from himself—does he unite with me?! Well, I'll hide." She became a cow, and he a bull; he united with her and from this, bovine cattle were born. She became a she-horse, he a male; she a she-donkey, he a male one; he united with her, and from this, one-hoofed animals were born. She became a she-goat, he a male, she a

she-sheep, he a male one; he united with her, and from this (small) cattle were born. In this way he created all couples that exist, all the way down to the ants.

Our daily consciousness still behaves according to the more basic movements of the conscious self; alone, we are afraid and have no (sexual) pleasure. Furthermore, external reality and multiplicity result from an inner perception, which generates action on its way to objectivity and the concretization of things.

We learn that creation stories need not be gravely serious. Here, "divinity" is not an abstract, omnipotently imagined creator who knowingly designs cosmos, but an immanent part of reality, its inner conscious force, who is exploring his own being as he creates a reality that is imbued with his inner impulse. More specifically, the world of real, external things is characterized as a splitting up of the original self into parts that continue to shape more and more parts through the ones that first emerged. At the root of this process is an act of self-reflection.

Further along in the story, the *Bṛhādāraṇyaka* explains that external entities are mere "names and forms," *nāma-rūpa*, the objectified ossifications of selfhood. The text also includes the famous statement *ahaṃ brahmāsmi*, "I am Brahman," expressing a moment when a person realizes that they themselves are the original conscious unity, so that even the gods are subservient to them, being effects of creation that the person who knows this can control.

The model of an original, conscious oneness of the self that generates externality, physicality, and multiplicity out of its own inner curiosity and will, is a recurrent feature in ancient Indian creation myths.[26] These do not agree on the details, but are inclined to accept consciousness as the primeval source of existence.

In the last account, the original self split into sets of two. In the classic chapter 6 of the *Chāndogya-Upaniṣad*, the dynamic is threefold. The text begins in a similar tone—"At first, (all) this was the existent; one alone without another." Now, the conscious principle desires to multiply and omits three qualities out of himself: heat, water, and food, which proceed to enter each other and generate further multiplicity. Heat is first, as it reflects basic movement.[27] After heat comes water, which explains why we sweat when we are hot. After water, food emerges, which is why food grows after the rains. Eventually, the activation of externality shapes a physical world.

My point is obviously not to suggest that these theories should be taken verbatim; their authors would probably agree. But the idea that consciousness is fundamental, and externality responds to it, is a serious one. Is it not conceivable that physical reality is thickening of the way it is perceived, that knowing generates its objects rather than the other way around (this is indeed a feature of an intriguing strand in the illusionist view discussed above)? Perhaps matter is the objective condition for the working of the mind, not its sources or underlying reality. We move to examine another related conceptualization.

Consciousness in the *Sāṃkhya* Method of Reasoned Observation

The *Sāṃkhya* school offers a more systematic presentation of similar ideas. This is a tradition of philosophy and practice with deep roots in Indian thought, some of which return well into the first millennium BCE. Here we observe the classic articulation in the *Sāṃkhya-kārikā* ("verses on reasoned analysis"), composed by *Iśvarakṛṣṇa* toward the middle of the first millennium CE.[28]

Sāṃkhya provides a trajectory of analysis that is meant to help one reach a vivid perception of reality, as it is defined within the system. This means first and foremost to understand consciousness as an absolutely pure subjectivity, inherently and categorically distinct from anything objective. In standard presentations of Indian philosophy, *Sāṃkhya* is considered a dualist system that makes a complete, categorical differentiation between consciousness and subjectivity, on the one hand, and matter, objectivity, and the evolved physical world, on the other. The first of these, the ultimately pure and stable subject, is called *puruṣa* ("man"), while the objective sphere is referred to as *prakṛti*, "nature," which refers etymologically to the evolution (prefix *pra*) of that which is made or effected (*kṛ*). A remarkable feature of this system is that our common ideas of subjectivity, including mental activity, a sense of self, and personal identity, as well the first-person perspective, are relegated to the field of objectivity. According to this analysis, anything that can be perceived as an object *is* an object—that is, it relates to the physical field of emergent (or "de-mergent") contingency.

Take the mind as an example—the organ that produces thoughts, cognitive perceptions, emotional preferences, as well as attitudes that judge things as pleasurable and beneficial or unpleasurable, harmful, or dangerous. Any assessment of good or bad, any like or dislike, is always something that we are aware *of*; it does not characterize awareness in itself, which here is continuously, essentially, and fundamentally distinct. Our ideas of personal identity and selfhood, even agency, are all things we can observe, even if this takes some mental effort. The perceiver, however, is different. The emergent, *Sāṃkhya* theorists would say, is material and nonconscious. What knows it is categorically distinct.

In India, dualism suffers from the same problems it has in the West. Thus, *Sāṃkhya* was eventually forsaken and subsumed within different strands of *Vedānta* (classical Indian philosophy), mainly the popular *Advaita* (non-dual) school. Nonetheless, its main concepts and arguments were adopted by *Advaita* theoreticians and employed within a monistic framework that explicitly adopts a metaphysic that takes consciousness as the only substantial reality.[29] Even within the latter, we can identify strands that are dualistic, given that here too there is a dedicated effort to distinguish between consciousness and any form of objectivity, such that the two are said to be differentiated like light and day.[30]

We are thus treating a basic principle of Indian philosophical thinking and praxis, expressed in both dualist and monist/non-dual modes, which speaks of the manner in which consciousness is a fundamental feature of reality, such that objectivity is generated in response to it. Here, we will avoid entering the details of the *Sāṃkhya* picture of reality, and especially so with regard to some of its intriguing and counterintuitive logical arguments. We focus on its explanation of how objectivity—including anything mental or cognitive that is known as object—is generated in response to subjectivity.

Generally, objectivity is said to arise "for the sake of Puruṣa" (*pūruṣārtha eva hetur*), so that the objective elements "present everything for the sake of Puruṣa, lighting it like a lamp" (verse 36). Objectivity is defined as "causal, impermanent, non-pervasive, producing effects, multiple, relying on a material base, having an objective sign, composite, dependent."[31] It is further "characterized by the three qualities (*tri-guṇam*)": the inner inclinations of matter toward lightness and lucidity (*sattva*), heat and activity (*rajas*), and heaviness and inertia (*tamas*). This means that it is conditioned by an interactive multiplicity so as to remain complex and consist of inner conflict. Matter is "objective, general, nonconscious, generative" (*viṣayaḥ sāmānyam acetanaṃ prasavdharmi*). In contrast, subjectivity is the opposite of all this (*tadviparītas tathā ca pumān*; verse 11). There must be a difference between these two spheres, and there must be a true, subjective foundation of *Puruṣa*, "since coming together [of diverse causal factors] exists for the sake of another" (*saṃghātaparārthatvāt*), the ultimate other being pure subjectivity. Such a subjective element must also exist as a controlling faculty (*adhiṣṭhānāt*) and be the enjoyer or experiencer of the objective sphere, which has to exist (*bhoktṛbhāvāt*).[32] Given the absolute distinction between the objective and subjective sphere, the *Puruṣa* is further defined as characterized by being a mere witness (*sākṣitvaṃ*), fully liberated in its own isolation (*kaivalyaṃ*), indifferent (*madhyasthyaṃ*), a seer (*draṣṭṛtvam*), having no activity (*akartṛbhāva*).[33]

Within this picture, the *Sāṃkhya-kārikā* presents a theory for the evolution of life from the basic category of matter, *Prakṛti*. *Prakṛti* has an unmanifest component, referred to as the basis of objectivity, *pradhāna*, from which arises "the great" (*mahat*) element of the intellect, *buddhi*, the first manifestation of objectivity as a basic sense of knowing that is involved with otherness and objectivity. From *buddhi* is then generated a sense of self, of agency and ostensible subjectivity, *ahaṃkāra* ("the I doer"), with which we normally identify ourselves. *Ahaṃkāra* is further creative in producing the third element in the inner-organ (*antaḥkaraṇa*), called mind (*manas*), which is generated together with the five sensory organs (eyes, ears, etc.), and the five organs of action (speech, hands, feet, organs of excretion and sexual pleasure).

Interestingly, together with the last category of mind, the sensory organs and organs of action, arise the five subtle elements (*tanmātras*), which are the basic sensory potentials (sight, hearing, smell, taste, and touch). From these arise the five physical elements that constitute material reality. Matter is thus the end process of an evolution of objective elements from perceptual faculties, which in themselves are objective aspects that respond to the presence of subjectivity and which manifest for the sake of *Puruṣa*.[34]

Few will take this as a viable theory of creation. Nevertheless, the *Sāṃkhya* view has some compelling features, relying on the notion that the harder facts of experience are ingrained in the reality of acts of knowing. "Reality" is here not an independent given, from which we could derive consciousness, but is the outcome of a process that responds to consciousness, the encrusting of subjective vision into objectivity. It is difficult to explain within the *Sāṃkhya* system why the world comes to be when consciousness is "a mere witness, indifferent, having no activity, etc." The orientation of the system toward liberation is also perplexing from a modern standpoint. Still, the turning round of the picture, so that reality begins with consciousness then to be objectified through processes of projective perception and identification, reflects something of the way we know the world. Even if we know illusions, the fact of illusion is real.

Conclusion

The focus of this chapter has been on conceptualizations within Indian philosophy of a view that takes consciousness as fundamental and matter as derivative. The argument is that such a view is no less serious than our habitual modes of thinking that prioritize matter, which have led to a situation in which we cannot explain the most basic fact of our life—experience. After surveying *Upaniṣadic* creation stories that relate how reality emerged from conscious reflection, the *Sāṃkhya* school of thought articulated how objectivity is shaped in response to knowing. Such a view is perhaps more faithful to the way human knowledge works, suggesting that the prioritization of the objective is not as natural and inevitable as we have grown accustomed to believe. Perhaps some would prefer arguments. Here I have supplied these stories not so much in order to convince anyone, but to help us cultivate an intellectual habit that respects the inherent validity of conscious experience, and does not try to derive its explanation from its objective content. Perhaps we need a new creation story.

Glossary

Eliminativism: The theory of consciousness that denies it as a real fact that must be accounted for in a scientific theory, given that it will eventually be explained in material terms.

Mysterianism: The view of consciousness that it is necessarily beyond the understanding of any scientific theory.

Notes

1. Patricia Churchland (1988, p. 129) has defined her position *eliminative materialism*, intending that any fact of consciousness will eventually be *replaced* by a definition of its corresponding neurological processes. The question has been raised whether consciousness is epiphenomenal, meaning most specifically that it has no causal power, as in Churchland (1984).

2. For such critiques, see Frankish (2016), eagerly applauded by Dennett (2016) and Graziano et al. (2020), then again supported by Dennett (2020). Such a position is different than McGinn's (2004) mysterianism, for whom the mystery relates to the potential to be penetrated by science.

3. This would probably even be true of phenomenal realists such as Searle (1992, 1997) and Block (2002).

4. Garfield (2016) believes that Buddhist are also illusionists, which may be true for certain strands of Buddhist philosophy, and especially in the Mahāyāna.

5. See Chalmers's (2018) discussion of the "meta-problem" of consciousness, and the responses to it in the *Journal of Consciousness Studies* (2019, volume 9–10).

6. The fact that different mystical traditions, among them prominent Indian ones, see the discovery of the nature of consciousness as a transformative moment of enlightenment or liberation, is an interesting fact in this respect.

7. For interesting suggestions regarding the teleology embedded within the natural order, see Nagel (2012).

8. Dennett (1991).

9. Seth (2021).

10. Tye (1996); see also Tye (1999) for why he thinks there is no explanatory gap.

11. As in Frankish (2016) and Graziano et al. (2020).

12. The most important advocate of keeping consciousness on the table has been Thomas Nagel (1974, 1986), joined by John Searle (1992, 1997) and David Chalmers (1996).

13. See note 1 above, yet this notion retains its appeal, as in Graziano et al. (2020).

14. For Nāgārjuna, see Garfield (1995, 2008) or Shulman (2007, 2015).

15. Bishop Berkeley's compelling immaterialist views are the classic example. This position also reflects the views of the fourth century Indian Buddhist philosopher Vasubandhu, as discussed in Shulman (2015).

16. For these, see Dennett (1991). A practical application that faces up to the stubborn phenomenological reality of dualism is the neurophenomenological approach of Varela (1996).

17. Chalmers (1996).

18. See Strawson (2008), Koch (2012, ch. 8; 2019), and Goff (2019). The influential integrated information theory (IIT) also has pan-psychist elements.

19. Although inviting, the idealist ideas presented by Kastrup (2019) are more convincing in their critical mode than in their constructive one.

20. Notice that this argument is arguably different than folk philosophy that can, for example, take the world to be flat, given that this is how it appears to us. Here we are speaking not of an interpretation of content, but of the very fact that content is experienced.

21. This is not to suggest that Indian philosophy does not have other things to offer for present theories of consciousness, or that all Indian theories would subscribe to the idealistically inclined ones I present here. For a recent, inspiring realistic formulation, see Chakrabarti (2020).

22. A classic expression of this theme is in the *Puruṣa-sūkta* ("Hymn on man", *Ṛg-Veda* 10.90). For further discussion, see Smith (1989) and Olivelle (1998, introduction).

23. See *Chāndogya Upaniṣad*, chapter 7, for a cogent articulation of this idea. For the notion of weaving worlds or levels of reality upon each other, see *Bṛhādāraṇyaka* 3.8.

24. In the empirical context of the *Upaniṣads*, proximate "this" (*idaṃ*), as it appears to the speaker, relates to what we call "the world."

25. I take the Sanskrit text from Olivelle (1998), whose translations I consulted in offering my own.

26. See, for example, the opening story of the *Aitareya Upaniṣad*. For an interesting parallel see *Bṛhādāraṇyaka* 1.4.10 and 1.4.17, which are contained within the same chapter we discussed. See also *Ṛg-Veda* 10.90.

27. Compare *Ṛg Veda* 10.129, verse 3.

28. For for a Sanskrit edition and translation of the text, see Larson (1979 [1969]); for a concise presentation of *Sāṃkhya*, see Hiriyanna (1995, ch. 6). See further Larson (1969) and Osto (2018).

29. On the potential compatibility between *Sāṃkhya* and *Advaita*, see Nicholson (2007).

30. As in Śaṅkara's introduction to the *Brahma-sūtra-bhāṣya*; see translation in Thibaut (1962).

31. Verse 10: *hetumat anityam avyāpi sakriyam anekam āśritaṃ liṅgam sāvayavaṃ paratantraṃ*. This verse specifically speaks of the manifest level of objectivity; I do not discuss here the distinction between the unmanifest (*avyakta*) and manifest (*vyakta*) levels of objectivity, which is important to the text.

32. These last arguments are taken from verse 17.

33. Verse 19.

34. This understanding is expressed beautifully in verse 59: "Like a dancer who ceases from the dance after having been seen by the audience, so *Prakṛti* turns back, after lighting herself for the sake of *Puruṣa*."

References

Baars, B. J. (1997). *In the theater of consciousness: The workspace of the mind*. Oxford University Press.

Block, N. (2002). The harder problem of consciousness. *Journal of Philosophy*, *99*(8), 391–425.

Chakrabarti, A. (2020). *Realisms interlinked: Objects, subjects and other subjects*. Bloomsbury Academic.

Chalmers, D. J. (1996). *The conscious mind*. Oxford University Press.

Chalmers, D. J. (2018). The meta-problem of consciousness. *Journal of Consciousness Studies*, *25*(9–10), 6–61.

Churchland, P. (1984). *Matter and consciousness*. MIT Press.

Churchland, P. (1988). Can neurobiology teach us anything about consciousness. In N. Block, O. Flanagan, & G. Güzeldre (Eds.), *The nature of consciousness: Philosophical debates* (pp. 127–140). MIT Press.

Dehaene, S., & Naccacche, L. (2001). Towards a cognitive neuroscience of consciousness: Basic evidence and a workspace framework. *Cognition*, *79*, 1–37.

Dennett, D. C. (1991). *Consciousness explained*. Little Brown.

Dennett, D. C. (2016). Illusionism as the obvious default theory of consciousness. *Journal of Consciousness Studies*, *23*(11–12), 65–72.

Dennett, D. C. (2020). On track to a standard model. *Cognitive Neuropsychology*, *37*(3–4), 173–175.

Garfield, J. (1995). *The fundamental wisdom of the middle way*. Oxford University Press.

Garfield, J. (2008). Turning a madhyamaka trick: Reply to Huntington. *Journal of Indian Philosophy*, *36*, 507–527.

Garfield, J. (2016). Illusionism and givenness. *Journal of Consciousness Studies*, *23*(11–12), 73–82.

Goff, P. (2019). *Galileo's error: Foundations for a new science of consciousness*. Pantheon.

Graziano, M. S. A., Guterstam, A. B. B. J., & Wilterson, A. I. (2020). Toward a standard model of consciousness: Reconciling the attention schema, global workspace, higher-order thought, and illusionist theories. *Cognitive Neuropsychology*, *37*(3–4), 155–172.

Hiriyanna, M. (1995). *The essentials of Indian philosophy*. Motilal Banarsidass.

Hoffman, D. D. (2019). *The case against reality: Evolution hid the truth from our eyes*. Penguin Books.

Kastrup, B. (2019). *The idea of the world: A multi-disciplinary argument for the mental nature of reality*. Iff Books.

Koch, C. (2012). *Consciousness: Confessions of a romantic reductionist*. MIT Press.

Koch, C. (2019). *The feeling of life itself: Why consciousness is widespread but can't be computed*. MIT Press.

Larson, G. (1969). Classical sāmkhya and the phenomenological ontology of Jean-Paul Sartre. *Philosophy East and West*, *19*(1), 45–58.

Larson, G. (1979 [1969]). *Classical sāṃkhya: An interpretation of its history and meaning*. Motilal Banarsidass.

McGinn, C. (2004). *Consciousness and its objects*. Clarendon Press.

Metzinger, T. (2009). *The ego tunnel: The science of the mind and the myth of the self*. Basic Books.

Nagel, T. (1974). What is it like to be a bat? *Philosophical Review*, *83*, 435–450.

Nagel, T. (1986). *The view from nowhere*. Oxford University Press.

Nagel, T. (2012). *Mind and cosmos: Why the materialist neo-darwinian conception of nature is almost certainly false*. Oxford University Press.

Nicholson, A. J. (2007). Reconciling dualism and non-dualism: Three arguments in Vijñānabhikṣu's Bhedābheda Vedānta. *Journal of Indian Philosophy*, *35*, 371–403.

Olivelle, P. (1998). *The early upaniṣads: Annotated text and translation*. Oxford University Press.

Osto, D. (2018). No self in Sāṃkhya: A comparative look at classical Sāṃkhya and Theravāda buddhism. *Philosophy East and West*, *68*(1), 201–222.

Searle, J. R. (1992). *The rediscovery of the mind*. MIT Press.

Searle, J. R. (1997). *The mystery of consciousness*. New York Review.

Seth, A. (2021). *Being you: A new science of consciousness*. Dutton.

Shulman, E. (2007–2009). Creative ignorance: Nāgārjuna on the ontological significance of consciousness. *Journal of the International Association of Buddhist Studies*, *30*(1–2), 123–157.

Shulman, E. (2015). Vasubandhu the Mādhyamika? Nāgārjuna the Yogacārin? On the middle way between realism and anti-realism. In G. Jay & W. Jan (Eds.), *Madhyamaka and Yogācāra: Allies or rivals?* (pp. 184–212). Oxford University Press.

Smith, B. K. (1989). *Reflection on resemblance, ritual, and religion*. Motilal Banarsidass.

Strawson, G. (2008). Realistic monism: Why physicalism entails panpsychism. In *Real materialism and other essays*. Clarendon Press.

Thibaut, G. (1962). *The Vedānta sūtras of Bādarāyana with a commentary by Śankara*. Dover.

Tye, M. (1996). The function of consciousness. *Nous*, *30*(3), 287–305.

Tye, M. (1999). Phenomenal consciousness: The explanatory gap as a cognitive illusion. *Mind*, *108*(432), 705–725.

Varela, F. (1996). Neurophenomenology: A methodological remedy to the hard problem. *Journal of Consciousness Studies*, *3*, 330–349.

23 Consciousness and Cognition: A Personal Journey from Function to Phenomenology

Narayanan Srinivasan

My experience is what I agree to attend to.
—William James, 1890

Personal Reflections

As I started my career, my initial studies were predominantly based on the computational view of the mind with the intention of studying the effects of attention on perception, measured objectively using reaction time and accuracy. I found them limited in terms of the properties of conscious experience that could be studied. Over time, I have broadened my research from visual experience to temporal experience based on first-person reports.

Recently, I have been reading phenomenology, especially works focused on time consciousness (Husserl 1991). The work by phenomenologists, especially Husserl, on time consciousness has been inspiring. However, there is a need to integrate phenomenological research with computational and experimental work in the cognitive sciences. I have been impressed by the work on neurophenomenology by Francisco Varela and his colleagues. I think neurophenomenology and the focus on embodiment, combined with the existing methodologies in cognitive science, would be a way forward in understanding subjective experience.

As someone who grew up in India, I am familiar with contemplative practices and Indian philosophy. I am not a meditation practitioner, but I have been thinking about meditation for a long time. It is with no small regret that my own subjective experience is not used in a significant manner in my own scientific research. I do think it is important that our subjective experience is utilized in the scientific study of consciousness, and I look forward to utilizing it more in my future research. I have also been trying to see whether concepts that are used in Indian philosophy can be studied rigorously in a scientific manner. As one of the first steps, I published a paper last year on consciousness without content that drew its inspiration from the Upanishads as well as Buddhist philosophy.

The two well-conducted workshops on "Perspectives on Consciousness" held in Israel in 2019 and 2020 have only strengthened my resolve to find a way to incorporate subjective experience informed by one's contemplative practices into my own future research, and I am thankful to my colleagues who have contributed to this volume and have been helping me in this endeavor.

Overview

Consciousness has been an enduring mystery. My foray into the study of consciousness started with my training as a cognitive psychologist. I essentially perform experimental studies with human participants to understand consciousness by studying how different cognitive processes are causally linked to consciousness. While some have argued that consciousness is an epiphenomenon, my efforts to understand consciousness assumes that consciousness is causally efficacious and tries to show that consciousness is not only influenced by cognitive processes like attention but that "being" aware of something also influences subsequent cognitive processes. This squarely belongs to functionalist approaches to consciousness. I will discuss some studies that we have done focusing on intentions and time perception that hope to demonstrate the role of volition. In addition, I believe it is important to incorporate phenomenology as much as possible. I will discuss studies in which we try to move away from measures like accuracy and reaction time and move toward reports about experience that are simple but not necessarily objectively verifiable as accurate. I will discuss studies demonstrating the effect of attention on visual awareness and how meditation practice could influence our perceptual experience. Finally, I will discuss the importance of control and prediction and how a hierarchical control-system view could enable us to understand consciousness, self, and agency.

Introduction

I became interested in consciousness research as a cognitive psychologist. One of the earliest papers I read about consciousness was published by the late great cognitive psychologist George Mandler (1975). Another significant influence was Daniel Dennett's book, *Consciousness Explained* (Dennett, 1992). During my graduate training, I encountered a new journal titled *Consciousness and Cognition* edited by Bernard Baars and William Banks, which coincidentally was started in the same year I started my PhD in psychology. At that point in time, I was interested in visual perception and attention. Consciousness was not a word commonly used in psychology. Eventually, I read Baars's book on consciousness (Baars, 1988) in addition to the expanding research on consciousness in the 1990s. During my doctoral training, I was interested in figure-ground perception and the effects of attention on visual perception. After a fairly long unplanned detour into engineering, I returned to consciousness and cognitive science research in 2004. I have primarily focused on human visual and temporal experience for the last seventeen years.

The main approach that I have used so far is to consider that the properties of conscious experience are influenced by different mental processes, most notably attention. The empirical approach has been based on designing lab-based experiments to show that processes like attention influence conscious experience or conscious states like intentions have downstream causal effects on perception. I have also studied these relationships with meditators, with the understanding that meditative practice modifies mental processes and hence influences their conscious experience. It has been long known that attention influences performance. In the next section, I will discuss some salient findings on the effects of attention on conscious experience (visual and temporal). One important way in which attention changes due to training is through meditation. The section "Meditation and Conscious Experience" will discuss the research with meditators and non-meditators that shows the effects of meditation on visual and temporal experience. In addition to perceptual experience, I have been studying volition and the sense of agency. The section "Volition" will focus on the work on intentions and their effect on time perception and agency. The final section will discuss the work on hierarchy and how the notion of hierarchy can be used to understand agency and time consciousness.

Attention and Awareness

While there has been agreement that attention and consciousness are different, the relationship between attention and consciousness has been contested (Srinivasan, 2008). Some have argued that attention is necessary for consciousness; others have argued for the opposite (Cohen et al., 2012; Koch and Tsuchiya, 2007). In addition, attention is not a unitary process and can be manipulated in many ways (Srinivasan et al., 2009). For example, we can manipulate spatial attention in terms of a spatial cue or in terms of its spatial scope or size (Srinivasan et al., 2009).

Cuing and visual appearance

A significant advance, going beyond showing attention influences not just performance but also appearance, was made by Carrasco and colleagues (Carrasco, 2011). While they demonstrated the effect of cuing on the appearance of simple visual features, we decided to see whether cuing to a spatial location would influence the appearance of a more complex but socially meaningful aspect like emotional intensity (Mishra & Srinivasan, 2017). Participants indicated that they perceived the emotional face that appeared at the cued or attended location as more intense—that is, the happy face looked happier, and the sad face looked sadder. The results indicate that attention not only influences visual appearance based on early visual features but also more socially relevant complex features like emotional expressions.

Scope of attention and visual appearance

Changes in spatial scope of attention also result in differences in awareness measured via the duration of color afterimages (see **glossary**) (Baijal & Srinivasan, 2009). When people adapt to a blue color square for some time and are shown a gray screen, they see a yellow square. This yellow color square is called an afterimage. In this study, participants performed a letter counting task on a stream of letters/numbers presented at the center of a screen surrounded by a blue square for twenty seconds. A particular letter/number was small, large, or hierarchical. A hierarchical stimulus (see figure 23.1 for an example) is a large letter/number made up of a set of the same small letters/numbers. With a hierarchical letter, the large letter is called the global target and the small letter(s) is called the local target. The results indicated that the duration of the color afterimage and hence awareness was influenced by the nature of the task and more specifically, the nature of attentional processes. Afterimage duration was longer for participants who performed a task with small stimuli (narrow scope of attention) compared to large stimuli or hierarchical stimuli (see **glossary**). In addition, afterimage

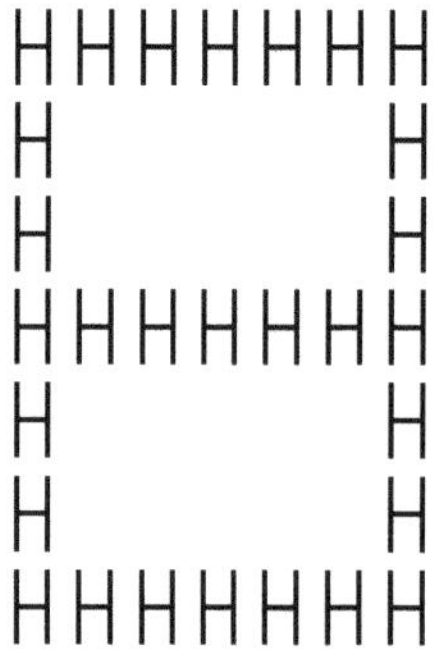

Figure 23.1
An example of a hierarchical stimulus. It shows a large number 8 made up of many small letters ("Hs" in this case).

duration was longer when participants performed a task with the large letters/numbers (global task) compared to a task with a similar sized single large letter/number.

Gradedness of consciousness

Different theories of consciousness have argued that consciousness is discrete (all or none) or graded (Windey & Cleeremans, 2015). Gradedness of conscious experience has been investigated using a perceptual awareness scale (Sandberg & Overgaard, 2015). We have recently investigated the issue in the context of differences in spatial scope of attention with hierarchical stimuli using the perceptual awareness scale (Thiruvasagam & Srinivasan, 2021). In addition to objective performance, a four-point perceptual awareness scale was used to measure gradations in visual awareness. Psychometric functions were constructed for objective performance and awareness ratings. Slope of the psychometric function and percentage of middle awareness scale ratings was taken as a measure of gradedness. The slope was less steep, and the percentage of middle awareness scale ratings was larger for global processing (broad attentional scope; see **glossary**) compared to local processing (narrow attentional scope), indicating that awareness of a global percept is more graded than a local percept.

Attentional blink

Our conscious perception also depends on attention deployed over time. This has been studied by asking people to identify two targets in sequence. For example, participants are asked to identify a target in the left or right location, which is followed by the identification of a second target that appears quickly after the first target in the top or bottom location in the attentional dwell time paradigm (see **glossary**) (Srivastava et al., 2010; Srivastava & Srinivasan, 2010). The duration between the two targets called the stimulus onset asynchrony (SOA) is varied. Generally, people have difficulty identifying the second target when it comes too soon (around 200–300 milliseconds) after the first target. In a different paradigm, called the attentional blink paradigm (see **glossary**), a sequence of stimuli is presented rapidly at the same location (Raymond et al., 1992). Once again, two targets need to be detected, both of which are marked for identification in some way. People once again have difficulty identifying the second target after they identify the first target if the second target appears 200–300 milliseconds after the first target. This phenomenon is called attentional blink.

We used two hierarchical stimuli presented at two different sets of locations (left-right and top-bottom) and asked different groups of participants to identify targets at the global or local level first and identify targets at the global or local level second, resulting in four different situations (Srivastava et al., 2010). There were very few blinks (failures to identify) when participants had to identify a global target followed by another global target. There was a typical blink effect when participants had to identify a local target followed by another local target. Identification of the second target was more difficult when participants had to switch from identifying a target at one level to a target at another level.

In our everyday life, we need to identify emotions expressed by other people (through their facial expressions), and we used a similar attentional dwell time paradigm with one target being an emotional expression and another target being a letter (Srivastava & Srinivasan, 2010). There is also a tight link between broad scope of attention and positive emotions as well as narrow scope of attention and negative emotions (Fredrickson, 2004). Given this link, we found the happy expression was detected much more easily with emotional faces as the second target compared to the sad expression—that is, there was less attentional blink with happy faces compared to sad faces. In addition, a non-emotional target showed less attentional blink when it was preceded by recognition of a happy expression compared to a sad expression. Recently, we investigated emotion recognition in an attentional blink paradigm when both target faces were emotional faces (Ray et al., 2020). We found that happy faces were easier to identify when they were the first or second targets in the stream, consistent with our earlier results. Angry faces, on the other hand, were much harder to see, especially when they appeared after the presentation of a happy face.

Interim summary: Spatial and temporal attention influence awareness

As mentioned in the beginning of this section, the link between attention and awareness is of significant interest to consciousness researchers. Our studies contribute to this debate by showing that attention does influence conscious experience. This has been shown with emotional intensity on the one hand and color afterimages on the other hand (Baijal & Srinivasan, 2009; Mishra & Srinivasan, 2017). The influence of attention on conscious experience can be seen not only in the visual modality but also time as well. More importantly, the spatial and temporal aspects of attention influence what we see and how we see it (Srivastava et al., 2010; Thiruvasagam & Srinivasan, 2021). Finally, conscious experience seems to be graded with gradedness being dependent on attention (Thiruvasagam & Srinivasan, 2021).

Meditation and Conscious Experience

Meditation refers to a set of practices that regulate different mental processes including attention, affect, and awareness. With Antonino Raffone, we tried to develop an understanding of meditation and conscious experience (Raffone & Srinivasan, 2009) based on the global workspace theory (Baars, 2002). According to the global workspace theory, mental contents that could come from different modules of the brain are conscious when they are globally broadcast throughout the brain. We argued that processes that adaptively code in areas like prefrontal cortex are important for consciousness. In addition, we proposed that such adaptive workspaces change due to meditation practices influencing attention and consciousness.

Multiple mental processes are involved in meditation training (Raffone & Srinivasan, 2017), but a critical process of interest is attention. In general, meditation practice influences perception and attention (see Maclean et al. (2010) for an example). In my own research, I have focused on investigating the effects of meditation on perception (specifically visual perception and time) by manipulating attention. These studies have been generally performed with experienced meditators with many years of practice.

In one set of studies, we manipulated spatial scope of attention and showed differences in conscious experience due to meditation practice (Singh & Srinivasan, 2017). We found concentrative meditators experienced the color afterimages for a much longer time than non-meditators. In addition, they also reported that the color afterimages they experienced were much sharper or clearer than those experienced by non-meditators. Thus, there was a difference not only in the duration of their afterimages, but also in the quality of the afterimages they experienced.

Multistable stimuli like the Necker cube provide an important method to study visual awareness. When we see a Necker cube, we perceive the cube in a particular orientation on display and after some time, the perception changes to a cube in a different orientation without any change in the stimulus on the screen. The time taken between two consecutive switches (called dwell time) in experience (two different orientations of the cube) is usually measured. If meditators can control what they experience, they should be able to control the duration for which a particular orientation of the Necker cube is seen by them. We asked both non-meditators and meditators to change what they see as quickly as possible, and in another condition we asked them to hold on to what they see for as long as possible in addition to a baseline neutral condition (Srinivasan et al., 2020). The switches in experience occurred slowly in meditators. Meditators were also able to hold on to what they experienced for much longer durations than non-meditators. The results indicated that meditators could control what they experienced in ambiguous situations by voluntary control.

We also investigated how the perception of a Necker cube changes when the ambiguous stimulus is surrounded by cubes that are oriented in one direction or another in a clear, unambiguous manner (Srinivasan et al., 2020). Generally, when ambiguous stimuli such as the Necker cube are surrounded by unambiguous views of the cube, the dwell times are shorter when the orientation of the cube they were seeing did not match that of the surrounding cubes. With meditators, the surrounding cubes did not affect their experience. These meditators were able to focus better on the target and ensure that conflicting information not relevant to the task did not influence their perception. The results from both experiments show the ability of meditators to control what they consciously experience both volitionally as well as when they were not explicitly told to control their perception in the presence of conflicting information.

The duration between an action and a consequent perceptual outcome has been taken as a measure of temporal binding and hence agency. The shorter the duration, the stronger the temporal binding. We have considered this binding as a measurement of attachment or bias toward stimuli that is associated with our selves. If meditation practice enhances non-attachment or equanimity, then it should reduce binding for information about the self, given that binding has been linked to our experience of agency. We normally process information related to ourselves faster (Chiarella et al., 2020; Makwana & Srinivasan, 2019). Both non-meditators and meditators show this effect. We investigated temporal binding with shapes associated with "self" and "stranger." Participants judged the duration between their keypress and the appearance of a shape on the screen. For the non-meditators, the durations were shorter for the shape associated with self compared to the stranger, indicating stronger binding and more attachment to self-related information. More importantly, the meditators did not show this self-bias binding effect—that is, differential binding or preference for self-related information—indicating that the meditators are less attached to self when it comes to measures associated with agency.

Meditation has been argued to change the way we perceive time as well (Berkovich-Ohana et al., 2011). Given that differences in meditation practices seemed to have different effects on meditation (Berkovich-Ohana et al., 2011), we focused on a concentrative type of meditation. Using the temporal bisection task described earlier, we investigated whether concentrative meditation practice would influence perceived time and whether that influence would depend on the difficulty of the task (Singh & Srinivasan, 2019). We found subjective expansion of time with meditators but only when the task was more difficult. This means that the effects of meditation practice on subjective expansion are visible when the attentional demands are more due to the difficult task.

The studies in totality show that meditation practice influences visual experience, temporal experience, and self-related processing. One aspect to note is that these cross-sectional studies have been performed with expert meditators and the results could be due to other possible differences in those experts on meditation. Longitudinal randomized control studies would be needed to conclusively show that meditation practices influence cognition and consciousness.

Volition

Some have questioned whether consciousness can cause behavior and have termed it as epiphenomenal (Wegner, 2017). To partially address this issue, I had decided to study volition in terms of asking people to consciously intend a particular perceptual outcome and show that our intentions do have a subsequent effect on the way we perceive the world (Makwana & Srinivasan, 2017; Sarma & Srinivasan, 2021). It should be noted that intentions are conscious volitional states and their effect on subsequent outcomes would indicate that consciously intending has causal efficacy. These are contrasted with making the same action through instruction, seeing the same outcome and performing the same task based on the outcome. Once again, the focus has been mostly on visual aspects and temporal properties of our percepts. For example, people were asked to choose whether they want to see a red or green stimulus in a trial and indicate that by pressing an appropriate key. After a brief and variable interval, a stimulus that they intended or not would appear on the screen and participants were asked to perform a task.

In terms of temporal properties of the outcome, we decided to investigate the effects of intentions on time perception (Makwana & Srinivasan, 2017). We varied the duration of presentation of the perceptual outcome (red or green colored stimuli) and asked participants to judge durations. When participants got the intended outcome, they perceived that stimulus as being shown for a longer period (subjective expansion of time) when the interval between the action and the perceptual outcome was short. Such an effect was not present when participants were instructed to press a key specific to a colored stimulus—that is, without making a volitional choice—indicating that having an intention makes a difference to the way people perceive time when they get what they intended or not. Our current explanation is that the subjective expansion of time is due to possible pre-activation of the representations of what was intended before a choice was made.

Our intentions are not just about visual shape or color but can be about many other aspects. One important aspect is emotion. Our social communication comprises of our intentions to elicit certain emotions from the listener and the actions we perform to elicit those intended emotions. We studied intentional binding (see **glossary**) with emotional faces in which participants intended to see a particular emotional face (happy or disgust) (Sarma & Srinivasan, 2021). Judged durations are shorter (stronger binding) for when participants intended negative emotional expressions compared to positive emotional expressions. The emotional content of the actual emotional face that appeared on the screen did not matter, only their intentions. Again, this effect was not present when participants were given instruction rather than making a choice. The studies discussed in this section not only indicated the importance of asking participants to explicitly form intentions, but also showed that prospective mechanisms are involved in this intended emotional effect on intentional binding. Further studies would be needed to understand the way volition influences our actions and perception.

Hierarchy and Consciousness

Sense of agency and control hierarchy

One aspect of my work on the self has focused on our sense of agency through the event-control framework (Jordan, 2003). The event-control approach conceptualizes the mind as a hierarchical control system, with each level interacting with the level above or below it. According to this approach, the self emerges through such interactions between different levels. Specific aspects about the self would depend on the levels that are involved. Unlike theories that categorize self in terms of specific categories like phenomenal self or narrative self, the event-control approach emphasizes the dynamic aspects of self, and the dynamics will depend on the specific levels involved at a particular point in time. The levels generally involve perception-action events that operate at different spatiotemporal scales, with the higher levels being associated with more distal spatiotemporal scales (one could consider them as being narrative self) and the lower levels being associated with proximal spatiotemporal scales (perhaps linked to phenomenal self). The higher, more distal level constrains the lower, more proximal level of event control.

We have shown that sense of agency is not associated with a particular level but changes dynamically as an organism continuously interacts with the environment (Kumar & Srinivasan 2014, 2017). The sense of agency would be associated with the level at which control is successfully achieved. At the level of goals, success of control is defined in terms of whether the goal was achieved or not. At the level of perceptual-motor control, the success of control is defined in terms of how much success people have in performing actions in real time in their interactions with the environment. In one study, we used a computer game with multiple agents (shown as yellow circular disks with numbers on them) that chase another agent (blue disk) to try to catch it (Kumar & Srinivasan, 2014). One of the agents was controlled by the participant. However, they did not know which of the disks they controlled, and they had to figure out the identity of the disk by interacting with the disks on the screen. We also asked about the strength of their identification with the disk they chose and their opinion about the influence of their actions on the yellow disk they thought they controlled using rating scales. When the agent controlled by the participant caught the blue disk, their sense of agency was not dependent on the perceptual-motor control they had over their joystick. However, when the agent controlled by the participant did not catch the blue disk, their sense of agency was dependent on perceptual-motor control with sense of agency increasing with more control over the joystick. The results support our prediction that sense of agency is dependent on control hierarchy. This paradigm allowed us to investigate how sense of agency could develop hierarchically through interactions with the environment.

Temporal hierarchy

In another recent formalization, we have extended the notion of hierarchy to integrate temporal consciousness and cognition in time (Singhal & Srinivasan, 2021). The framework we propose here draws extensively from phenomenological and philosophical work in time consciousness (Dainton, 2018). Philosophical models that have been proposed to explain time consciousness include cinematic, extensional, and retentional models. Our hierarchical framework (figure 23.2) proposes that mental content evolves in tandem over three distinct levels of temporal regularities (approximately at 30, 300, and 3,000 milliseconds). These levels collectively form the basis of our time consciousness. We also propose that different models (cinematic, extensional, and retentional) can be combined and that different scales in our model work according to a different model. The fastest 30-millisecond level acts in a cinematic manner. Our experience of the present moment is linked to the intermediate 300-millisecond timescale. This intermediate timescale integrates and updates information from the fast timescale of 30 milliseconds. It is extensional in nature; that is, our phenomenal experience is extended in time. The contents at the intermediate level are what underlie our conscious experience. This intermediate level is connected to the slower timescale conceptual level. Ultimately, the slow conceptual level "retains" traces of our experiences over time as they slowly fade away over a few seconds (retentional). This level also proactively predicts and anticipates our upcoming experiences. These three levels in unison form the forward-looking and just-passed nature of what is called the "now" or the "psychological

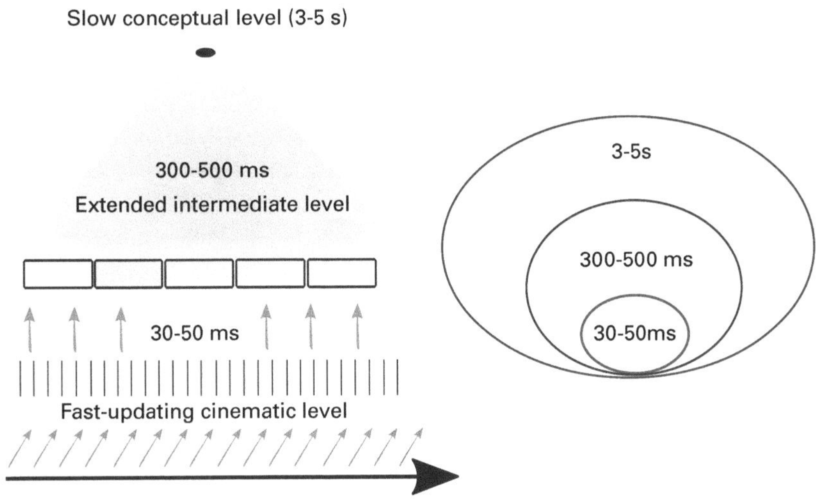

Figure 23.2
A nested multiscale temporal hierarchy consisting of a fast scale (30–50 milliseconds), intermediate scale (300–500 milliseconds), and slow scale (3–5 seconds). Each level is conceptualized to operate differently based on different philosophical models of time consciousness with the fast scale being cinematic, the intermediate level being extensional, and the slow scale being retentional.

present." Complementarily, the framework also allows a common scaffolding for research in timing of cognition. Different cognitive processes operating at specific timescales can potentially be brought together to coexist under a common theory of the timing of cognition. The proposed theory takes the structure of experience seriously and argues that time is represented based on the temporal structure of our conscious experience.

Concluding Reflections

The strength of the current scientific approaches used to study conscious perception is the ability to design controlled lab experiments to study specific aspects of consciousness. The weakness is that the rich structure of conscious experience is still not adequately captured by the dependent variables that are currently in use. Over my career, I have tried to focus more on measurements that are based on experience that are not solely dependent on their "objectiveness," like accuracy or response time. Stimuli like the Necker cube and color afterimages, while still simple, allow one to study aspects of experience that are not just tied to the stimulus presented at that point in time by the experimenter. In addition, we can get measurements that are from discrete trials that are tied to the visual or temporal experience. As a process, attention significantly influences visual and temporal experience. Conscious states like intentions influence what we subsequently perceive. The empirical studies that I have performed over my career contribute to the literature on how different cognitive processes influence what we experience, how practices like meditation influence cognition and consciousness, and how conscious states like intentions influence subsequent perception. We have shown that our minds operate at multiple hierarchical spatiotemporal scales influencing our sense of agency, and any theory of consciousness must meaningfully address the hierarchical nature of our minds and conscious experience that operate at multiple spatiotemporal scales.

In the future, I hope we will take phenomenology, or more specifically, the structure of conscious experience more seriously rather than relying only on simple objective measurements from the lab. The easier quantification of these objective measurements allows us to develop precise models, but they are significantly limited in understanding the wide variety of conscious experiences. More interaction between people whose experiences span a wide range due to multiple factors—including meditation practitioners, philosophers, psychologists, neuroscientists and modelers—will be needed for us to get a better understanding of the fundamental mystery of consciousness.

Acknowledgments

I would like to thank all the students and colleagues who have contributed to the studies discussed in this chapter.

Glossary

Attentional scope: Refers to the region attended in space, which can be broad or narrow.

Attentional dwell time: Poorer identification of a second target that immediately comes after a first target is presented at a different location when participants are asked to identify both targets.

Attentional blink: Poorer identification of a second target that immediately comes after a first target presented at the same location in a rapidly presented stream of targets, when participants are asked to identify both targets.

Color afterimages: The color image that one sees on a blank monitor after staring at a colored stimulus for a period of time; if one stares at a blue square, then one sees a yellow square as the afterimage.

Hierarchical stimuli: Stimuli used to study attentional scope and consists of a large letter/number (called global level) made up of smaller letters/numbers (called local level).

Intentional binding: Perceived shortening of time between a voluntary action and its consequence.

References

Baars, B. J. (1988). *A cognitive theory of consciousness.* Cambridge University Press.

Baars, B. J. (2002). The conscious access hypothesis: Origins and recent evidence. *Trends in Cognitive Sciences, 6,* 47–52.

Baijal, S., & Srinivasan, N. (2009). Types of attention matter for awareness: A study with color afterimages. *Consciousness & Cognition, 18,* 1039–1048.

Berkovich-Ohana, A., Glicksohn, J., & Goldstein, A. (2011). Temporal cognition changes following mindfulness, but not transcendental meditation practice. In D. Algom, D. Zakay, E. Chajut, E. Shaki, Y. Mama, & V. Shakuf (Eds.), *Fechner Day 2011: Proceedings of the 27th annual meeting of the International Society for Psychophysics* (pp. 245–250). International Society for Psychophysics.

Carrasco, M. (2011). Visual attention: The past 25 years. *Vision Research, 51*(13), 1484–1525.

Chiarella, S. G., Makwana, M., Simione, L., Hartkamp, M., Calabrese, L., Raffone, A., & Srinivasan, N. (2020). Mindfulness meditation weakens attachment to self: Evidence from a self vs other binding task. *Mindfulness, 11,* 2411–2422.

Cohen, M. A., Cavanagh, P., Chun, M. M., & Nakayama, K. (2012). The attentional requirements of consciousness, *Trends in Cognitive Sciences, 16*(8), 411–417.

Dainton, B. (2018). Temporal consciousness. In E. N. Zalta (Ed.), *The stanford encyclopedia of philosophy* (Winter 2018 Edition). https://plato.stanford.edu/archives/win2018/entries/consciousness-temporal/

Dennett, D. (1992). *Consciousness explained.* Penguin.

Fredrickson, B. L. (2004). The broaden-and-build theory of positive emotions. *Philosophical Transactions of the Royal Society B, 359,* 1367–1377.

Husserl, E. (1991). *On the phenomenology of the consciousness of internal time.* J. Brough (Ed. and Transl.). Kluwer.

Jordan, J. S. (2003). Emergence of self and other in perception and action: An event-control approach. *Consciousness and Cognition, 12*(4), 633–646.

Koch, C., & Tsuchiya, N. (2007). Attention and consciousness: Two distinct processes. *Trends in Cognitive Sciences, 11*(1), 16–22.

Kumar, D., & Srinivasan, N. (2017). How multi-scale control structures sense of agency: Investigating intentional binding using event-control approach. *Consciousness and Cognition, 49,* 1–14.

Kumar, D., & Srinivasan, N. (2014). Naturalizing sense of agency with a hierarchical event-control approach. *PLoS ONE, 9*(3), e92431.

Maclean, C. A., Ferrer, E., Aichele, S. R., Bridwell, D. A., Zanesco, A. P., Jacobs, T. L., King, B. G., Rosenberg, E. L., Sahdra, B. K., Shaver, P. R., Wallace, B. A., Mangun, G. R., & Saron, C. D. (2010). Intensive meditation training improves perceptual discrimination and sustained attention. *Psychological Science, 21,* 829–839.

Makwana, N., & Srinivasan, N. (2017). Intended outcome expands in time. *Scientific Reports, 7,* 6305.

Makwana, M., & Srinivasan, N. (2019). Self-associated stimuli produce stronger intentional binding. *Journal of Experimental Psychology: Human Perception and Performance, 45*(11), 1436–1442.

Mandler, G. A. (1975). Consciousness: Respectable, useful, and probably necessary. In R. L. Solso (Ed.), *Information processing and cognition.* Lawrence Erlbaum.

Mishra, M. V., & Srinivasan, N. (2017). Exogenous attention intensifies perceived emotion expressions. *Neuroscience of Consciousness, 3*(1), nix022.

Raffone, A., & Srinivasan, N. (2009). An adaptive workspace hypothesis about the neural correlates of consciousness: Insights from neuroscience and meditation studies. *Progress in Brain Research, 176,* 161–180.

Raffone, A., & Srinivasan, N. (2017). Mindfulness and cognitive functions: Toward a unifying neurocognitive framework. *Mindfulness, 8,* 1–9.

Ray, S. B., Mishra, M. V., & Srinivasan, N. (2020). Attentional blink with emotional faces depends on emotional expressions: A relative positive valence advantage. *Cognition & Emotion*, *34*(6), 1226–1245.

Raymond, J. E., Shapiro, K. L., & Arnell, K. M. (1992). Temporary suppression of visual processing in an RSVP task: An attentional blink? *Journal of Experimental Psychology: Human Perception and Performance*, *18*(3), 849–860.

Sandberg, K., & Overgaard, M. (2015). Using the perceptual awareness scale (PAS). In M. Overgaard (Ed.), *Behavioural Methods in Consciousness Research* (pp. 181–195). Oxford University Press.

Sarma, D., & Srinivasan, N. (2021). Intended emotions influence intentional binding with emotional faces: Larger binding for intended negative emotions. *Consciousness and Cognition*, *92*, 103136.

Singh, A., & Srinivasan, N. (2019). Concentrative (Sahaj Samadhi) meditation training expands subjective time. *PsyCh Journal*, *8*, 28–35.

Singhal, I., and Srinivasan, N. (2021). Time and time again: A multi-scale hierarchical framework for time-consciousness and timing of cognition. *Neuroscience of Consciousness*, *2021*(2), niab020.

Srinivasan, N. (2008). Interdependence of attention and consciousness. *Progress in Brain Research*, *168*, 65–75.

Srinivasan, N., & Singh, A. (2017). Concentrative meditation influences visual awareness: A study with color afterimages. *Mindfulness*, *8*, 17–26.

Srinivasan, N., Srivastava, P., Lohani, M., & Baijal, S. (2009). Focused and distributed attention. *Progress in Brain Research*, *176*, 87–100.

Srinivasan, N., Tripathi, S., & Singhal, I. (2020). Meditators exercise better endogenous and exogenous control of visual awareness. *Mindfulness*, *11*, 2705–2714.

Srivastava, P., Kumar, D., & Srinivasan, N. (2010). Time course of visual attention across perceptual levels and objects. *Acta Psychologica*, *135*, 335–342.

Srivastava, P., and Srinivasan, N. (2010). Time course of visual attention with emotional faces. *Attention, Perception, & Psychophysics*, *72*, 369–377.

Thiruvasagam, S., & Srinivasan, N. (2021). Gradedness of visual awareness depends on attentional scope: Global perception is more graded than local perception. *Consciousness & Cognition*, *94*, 103174.

Wegner, D. M. (2017). *The illusion of conscious will*. MIT Press.

Windey, B., and Cleeremans, A. (2015). Consciousness as a graded and an all-or-none phenomenon: A conceptual analysis. *Consciousness and Cognition*, *35*, 185–191.

24 Open-Ended Inquiry of Psychedelic Experiences via First-Person Methodologies and Neurophenomenology

Christopher Timmermann

Personal Reflections

The incorporation of lived experience has always had a place in psychopharmacology. There is a long tradition of investigators trying compounds as an important complement to their research with humans and animal subjects. Famous examples are the case of Sir Humphry Davy who experimented with nitrous oxide in the eighteenth century, and later Freud, who developed an extensive practice of ingesting cocaine during hist studies with the substance.

Psychedelics are no exception in this regard, as historically these have been tested by the investigators on themselves before performing experiments on others. In some cases this self-experimentation has led to meaningful insights regarding the scientific examination and practical application of the substances. For example, after intentionally trying LSD for the first time Albert Hoffman recommended the use of psychedelics to psychiatrists to understand the inner worlds of their patients, which then led to significant research in both the fields of neuropharmacology and psychiatry.

On the other hand, some forms of self-experimentation can contribute to biases, which may arise because the experiences induced by psychedelics can be both idiosyncratic and personally meaningful. This may lead researchers and users to hasty convictions related to what psychedelics reveal about consciousness and reality, as well as biased appraisals regarding the safety and efficacy of psychedelics. While these convictions are not necessarily wrong, they should inform unbiased forms of research where these ideas are properly tested.

The fields of cognitive neuroscience and pharmacology have tended to resist incorporating lived experience into research as anything else than an optional complement to objective measures of brain activity or behavior. However, this incorporation is essential to advance our understanding of the brain and mind, and what the modification of consciousness can teach us about the experience of being alive. A careful and humble approach to lived experience is necessary to make the steps required to incorporate measures of subjectivity as an essential part in neuroscience and pharmacological research. Such an approach should consider practices where researchers and participants examine and reflect on their own experiences in order to reveal biases and opportunities in the study of consciousness.

Overview

Psychedelic compounds act on specific brain processes, which can lead to rich experiences of wide phenomenological breadth. Because of this, psychedelic phenomena present a unique opportunity to explore the tension between the objective and the subjective. However, in-depth first-person approaches fostering inquiry into the psychedelic experience are still underdeveloped in psychedelic science and clinical practice. This limitation can lead to a reduced awareness of how contextual and cultural factors may shape the experience and its recollection, which in turn may lead to users and researchers to forms of subjective realism in which preexisting culturally-loaded categories are reified, thereby hindering a broader, safer, and more efficacious exploration of psychedelic experiences. This chapter proposes that first-person and neurophenomenological approaches may significantly aid in these issues by fostering a careful and mediated examination of psychedelic experiences, and their relationship to brain dynamics. Empirical examples are provided to support this proposal based on our research involving psychedelic compounds, and suggestions are made to foster an open-ended inquiry in the emerging field of psychedelics in consciousness science and clinical practice.

Introduction

Psychedelics are compounds of increasing interest in consciousness and mental health research. The enthusiasm for the potential of psychedelics has been propelled by two intertwining areas of research: consciousness science seeking to understand the unusual effects of psychedelics in the human brain and mind (Carhart-Harris et al., 2016; Griffiths et al., 2006), and clinical research into the potential application of psychedelics for the treatment of mood, anxiety, and substance abuse disorders (Andersen et al., 2021). However, the subjective effects induced by psychedelic experiences appear to be variable and challenging to predict. This variability present challenges when attempting to link features of psychedelic experiences to brain processes, and outcomes in mental health benefits. As with other areas in neuroscience, a plausible reason behind these issues is that the methods and frameworks used to explore practice, and address psychedelic experiences are heterogeneous, and at times lack a method for reflecting on them.

The shortcomings of studying psychedelic phenomena is contrasted by how their appeal revolves around the extraordinary quality of the experiences they induce. For example, recreational users (so-called "psychonauts") are interested in psychedelic compounds because of their psychological effects. Furthermore, psychedelic-induced experiences of catharsis, psychological insights, and mystical-type effects appear to mediate reductions in depression and addictive behaviors (Yaden & Griffiths, 2021). Individuals interested in the potential for psychedelics to increase creativity and well-being are looking for novel insights that are made available principally during the psychedelic experience. Finally, consciousness researchers are interested in brain and mental phenomena related to the experience evoked by these compounds. Thus, it is not unreasonable to assume that the fundamental appeal behind most of today's interest in these compounds revolves around their experiential properties.

Despite this appeal, it seems that interest in psychedelic experiences has only been partially addressed by current scientific research about psychedelics. Central concepts that the scientific communities use to describe and measure psychedelic experiences share a rather specific set of influences, which may arguably bias the characterization of the psychedelic experience in specific frameworks of mysticism (see **glossary**). This chapter argues that lacking an open-ended inquiry into psychedelic phenomena (which fosters an awareness of the embeddedness of psychedelic experiences within wider cultural and historical contexts) may limit both the understanding and content of these experiences.

First-person and neurophenomenological approaches (see **glossary**) seeking to relate mind and brain via reciprocal constraints may be able to tackle some of these issues at the root (Varela, 1996; Varela & Shear, 1999). The rich (and sometimes ineffable) psychedelic phenomenology results from the ingestion of a molecule acting in the brain. This fundamentally highlights the core issues related to the explanatory gap (i.e., the intuition that experiences cannot be fully accounted for by the physical properties of the brain) that propelled the neurophenomenological program (Varela, 1996; see **glossary**).

Within this context, psychedelic science can be understood as an open space for practice to develop novel methods that employ first-person approaches that take the view from within the experiencer as the starting point of inquiry. This starting point could then be employed to explore the relationship between the psychedelic experience and brain dynamics. Furthermore, perceiving psychedelic phenomena under the light of the theoretical complement of neurophenomenology, the enactive approach (Varela et al. 1991), provides an intuitive framework to understand and practice psychedelic phenomena. The situated character of psychedelic experiences to social, cultural, and historical contexts is increasingly supported by research (Aday et al. 2021; Hartogsohn 2016) and is thus resonant with the enactive approach.

The main proposal of this chapter is that first-person approaches may provide fruitful avenues for disciplined inquiry into consciousness and clinical practice involving the use of psychedelics. The first part of this chapter argues that a naive immersion into the psychedelic experience (i.e., subjective realism) may be the result of limited awareness of the historical, contextual, and cultural determinants of such experiences and could hinder the exploration of the wide phenomenological breadth associated with these experiences. The second part addresses what could constitute first-person and neurophenomenological approaches for psychedelic phenomena and attempts to address the issues outlined in the first part. The third part provides preliminary examples of psychedelic neurophenomenology in science and practice, inspired by recent work performed using approaches employing first-person methods. The final part of the chapter highlights the importance of incorporating open-ended inquiry via skillful know-how for researchers and clinicians who work with these experiences in order to foster an open-ended approach to psychedelic phenomena.

Subjective Realism in Psychedelic Practice

Psychedelics are of interest in consciousness research because of the unusual phenomena they induce in users. The effects of psychedelics are commonly described in the scientific literature as follows: sensory phenomena (e.g., eyes open visual distortions, eyes-closed visual imagery, modulations of bodily awareness, and sometimes auditory hallucinations); hyperassociation of ideas; changes in meaning; modulations of time perception; experiences of so-called ego-dissolution; and mystical-type experiences (see **glossary**) (Barrett et al., 2015; Dittrich, 1998).

While these experiential categories have become commonplace in the literature that describes the effects of psychedelics, many of them are deeply influenced by work that relates to specific frameworks. It is possible that this obscures other experiences that may arise during these states. For example, the popular belief that psychedelics induce spiritual-type experiences is influenced by the work on mysticism (see **glossary**) by William Stace, and on perennial philosophy (see **glossary**) by Aldous Huxley. These frameworks claim the existence of a common experiential core that may be reported across all religions and therefore places these experiences apart from other (contextually bounded) experiences, granting them a form of exceptionalism or truth (Taves, 2020). Furthermore, ethnographic work suggests that these views have pervasively influenced psychedelic scientific practice and non-academic users alike (Langlitz, 2012), and this is illustrated in psychedelic studies by the use of self-reported questionnaires such as the mystical experience questionnaire (MEQ).

The MEQ results in subscales or factors such as *transcendence of time and space, ineffability*, and *mystical*. However, interpreting these categories as features of high-dose psychedelic experiences independent of the cultural milieu in which they happen may arguably be misleading. Even if, from the standpoint of the user, these experiences appear to be unmediated by cultural contingencies, they may be unconsciously assembled by processes of mediation and interpretation in collaboration with the surrounding context before, during, and after the experience (Taves, 2020). Furthermore, the use of retrospective questionnaires may influence participants to re-interpret their experience in light of the items in the questionnaire, and reifying these terms as a "black box," inaccessible to naturalistic inquiry (Sanders & Zijlmans, 2021). While discussing the metaphysical implications of these experiences is beyond the scope of this chapter, here practical issues are outlined when these frameworks are used to understand psychedelic phenomena.

Evidence indicates that psychedelics increase suggestibility (Carhart-Harris et al., 2015) and psychedelic experiences are highly susceptible to contextual and cultural influences (Hartogsohn, 2016). Therefore, it is reasonable to suspect that popular narratives about the psychedelic experience have a significant impact on the actual experience (and not just its interpretation) of individuals who have been exposed to those narratives (a similar process has also been described as a form of *socialization of hallucinations* induced by psychedelics) (Dupuis, 2020). Considering the significant influence that psychedelic science has had in the mainstreaming of psychedelics in the past ten years, these issues may result in a scenario in which the practices and methods of science have an impact on psychedelic phenomenology and the interpretation of the psychedelic experience after it has taken place (Gearin & Devenot, 2021).

It could be argued that limited awareness of the contextual, cultural, social, and historical processes that mediate the contents of psychedelic experiences may lead users, facilitators, and researchers to promote an approach that favors a naive interpretation of psychedelic phenomenology, both during and after the experience. The process by which the experience may be conflated with culturally primed categories and with reduced awareness of the contextual devices that may shape the interpretation and phenomenology of these experiences

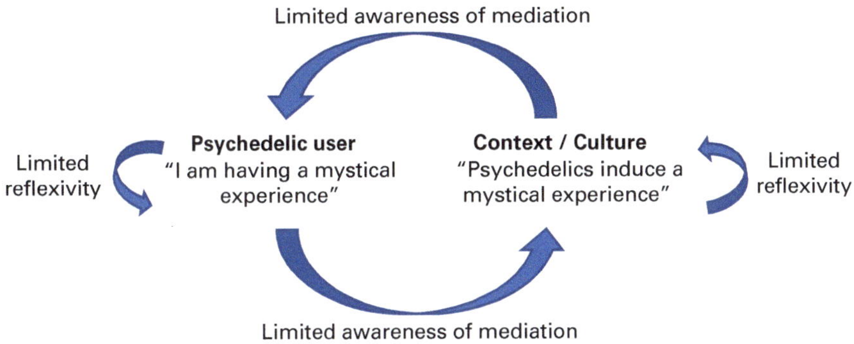

Figure 24.1
Subjective realism in psychedelic practice. Limited mediation and reflexivity concerning psychedelic experiences may enable a naive immersion in that experience and its interpretation. An approach that disregards the role of contextual, social, cultural, and historical mediation, as well as reflexivity of the experience tends to reify preexisting experiential categories (e.g., mystical experience) and limits the exploration of psychedelic phenomenology beyond these.

is resonant with the notion of "subjective realism" that describes a naive immersion in an experience (Lebois et al., 2015).

Forms of subjective realism during psychedelic experiences may also lead to long-term consequences. A survey study in naturalistic retreat contexts found that participants often adopted beliefs such as the transcendence of reality, universal consciousness (panpsychism), and the dualistic nature of the universe, while rejecting physicalism. These belief shifts were influenced by factors like age, peer-conformity, and emotional connectedness during the ceremony (Timmermann et al., 2021). Although preliminary, these findings suggest that psychedelics may facilitate the transmission of beliefs, a process that may be exacerbated in the absence of social instances that foster a reflective examination of these experiences and their impact on worldviews (with ethical implications related to consent, persuasion, and even mind-control) (Timmermann et al., 2022).

Neurophenomenological and Enactive Approaches in Psychedelic Science

While this chapter is not intended to debate the ontological validity of concepts related to mysticism or perennialism (see **glossary**), it aims to establish how their use in the context of psychedelic use may have consequences for our understanding of human consciousness, as well as for their safety and utility in clinical practice. Conversely, adopting a phenomenological approach for psychedelics has as a starting point in the careful examination of the experience of participants, users, and patients. More specifically, by fostering a phenomenological disposition, users may become aware of the processes and structures of the psychedelic experience rather than naively being immersed in that experience (i.e., in a natural attitude), which may lead to issues related to subjective realism as stated above.

First-person and neurophenomenological approaches to psychedelic experiences could also bring awareness to the contextual and historically embedded nature of these experiences. The contents of psychedelic experiences and the clinical outcomes that may be gained from them are known to be highly dependent on the *set and setting* (see **glossary**): the physical and social context in which they occur, the surrounding culture, as well as the psychological characteristics of the individual having the experience (e.g., personality traits) (Hartogsohn, 2016). The enactive approach emphasizes that knowing the world is inseparable from a history of coupling (or active engagement) between the perceiver and the world (Varela et al. 1991), and thus provides an intuitive framework to understand the psychedelic experience as not only brainbound, but also bodily, socially, culturally, and historically bounded.

Implementing First-Person and Neurophenomenological Approaches in Psychedelic Science and Clinical Practice

Psychedelic phenomena may present a unique opportunity to explore the possibilities of the enactive approach and neurophenomenological methods as psychedelics inherently exploit the tension between subjective and objective. For example, it has been shown that psychedelics rely on activation of the serotonin 2A receptor. When this receptor is blocked, both subjective effects and alpha brainwave (see **glossary**) suppression (the most reproduced finding in psychedelic neuroscience) are negligible (Vollenweider et al. 1998). This highly specific natural mechanism is contrasted by intimate subjective experiences of rich phenomenological depth, at times labeled as "ineffable" (i.e., the difficulty of translating the experience into words), "intuitive," and "noetic" (i.e., the feeling of gaining information suddenly and without mediation). Psychedelics thus present researchers an opportunity to methodologically explore the hard problem of consciousness from both ends of the subjective–objective continuum.

In our initial experiments we have pursued this opportunity by employing neurophenomenological methods in the context of psychedelic research. In an preliminary study we studied the effects of the psychedelic N,N-Dimethyltryptamine (DMT) (see **glossary**) in brain activity and subjective experience by using electroencephalograms (EEG) and microphenomenological interviews (see **glossary**). DMT is a short-acting (i.e., lasting approximately ten minutes) psychedelic that induces a range of subjective effects, consisting of modulations in bodily awareness, intense eyes-closed visual imagery (described at times as an immersion into alternate dimensions), and communicating with being or "entities." While the contents of these experiences have been described in-depth in naturalistic contexts, the structural features of these experiences and their dynamics have not been explored exhaustively in scientific contexts.

In our study we employed interviews inspired by the microphenomenology method (see **glossary**), which seeks to gather first-person data via mediation by an experienced interviewer (in the second person) who guides the interview by allowing a process of embodied re-evocation of the experience while helping participants to redirect their attention away from categorization and reflective dimensions (i.e., interpretations) of experience, thus favoring prereflective features instead. Microphenomenology fosters a disciplined report of the temporal dynamics (diachronic) and multidimensional (synchronic) structural features of an experience, therefore emphasizing *processes,* as opposed to *contents* of experience (Petitmengin, 2006).

Once the experiential data was gathered, a member of the research team extracted structural categories of experience, and assigned ratings of intensity for each of these features at every minute following the administration of DMT. This kind of experiential temporal resolution was not possible to obtain in a standard neuroscientific approach, which employs commonly used psychometric questionnaires (as these are completed only once after the experience has subsided) (see top left of figure 24.2). Once we had these ratings for each participant and for each minute (first-person data) we used them to constrain the analysis of brain activity (EEG data; third-person data) (Timmermann et al., 2019; see figure 24.2).

These findings allowed us to present evidence of how brain function was related to immersive experiences induced by psychedelics in which individuals experience a world of rich visuals, while being partially detached from the environment (via reduced bodily awareness). This finding led us to infer that similar neuronal and experiential mechanisms may be at play during other states of consciousness that share this immersive quality, such as dreaming.

To test the advantage of using phenomenological methods over conventional tools, we also used self-reported psychometric questionnaires taken at the end of the session (the intensity of the DMT experience precludes exploration of detailed phenomenology within the session). No findings were found relating the immersive quality of sensory experiences of DMT and brain activity, possibly due to the lack of mediation in obtaining the self-reported questionnaires and the complete absence of temporal resolution in psychometric questionnaires of psychedelic experiences (Timmermann et al., 2019; see figure 24.2)

This example highlights the advantages of using phenomenological techniques to gain a higher resolution of reports related to psychedelic experiences, as well as striving to redirect these reports from the contents of the experience (e.g., seeing entities) to their structural features (e.g., sensory dimensions), with the latter arguably being more general across psychedelic experiences and less affected by contextually contingent categorizations (in contrast to features of mystical-type experiences). While the interview process and analysis of the data are always influenced by the context and culture of the researchers, a phenomenological approach could arguably render these biases more visible to both research subjects and researchers themselves and might reduce their impact.

First-person approaches could be taken into more pragmatic applications such as clinical practice involving psychedelic compounds. Recent evidence suggests that experiential features are key drivers in the improvements of clinical outcomes and have been seen in recent studies employing psychedelics for the treatment of depression and substance abuse (Yaden & Griffiths, 2021). At the same time, the use of "integration" practices (see **glossary**), in which patients can revise these experiences, make meaning of them, and develop bridges between the experience and their lives, may play an important role in sustaining the safety and efficacy of these therapies over time (Earleywine et al., 2022). Microphenomenology or other phenomenological techniques—mediated by a skilled facilitator—could be inserted during these integration practices to explore the richness of phenomenological features encountered in the psychedelic experience and connect them to their therapeutic process. Furthermore, adopting phenomenological methods could allow clinicians to determine what kind of interventions are better suited for different acute psychedelic experiences, as well as consider details of prior experiences of each patient to determine a specific course of action within the process of psychedelic therapy.

The mediation of first-person methods could also be used during integration practices for the purposes of enhancing safety because psychedelics may elicit experiences of distressing

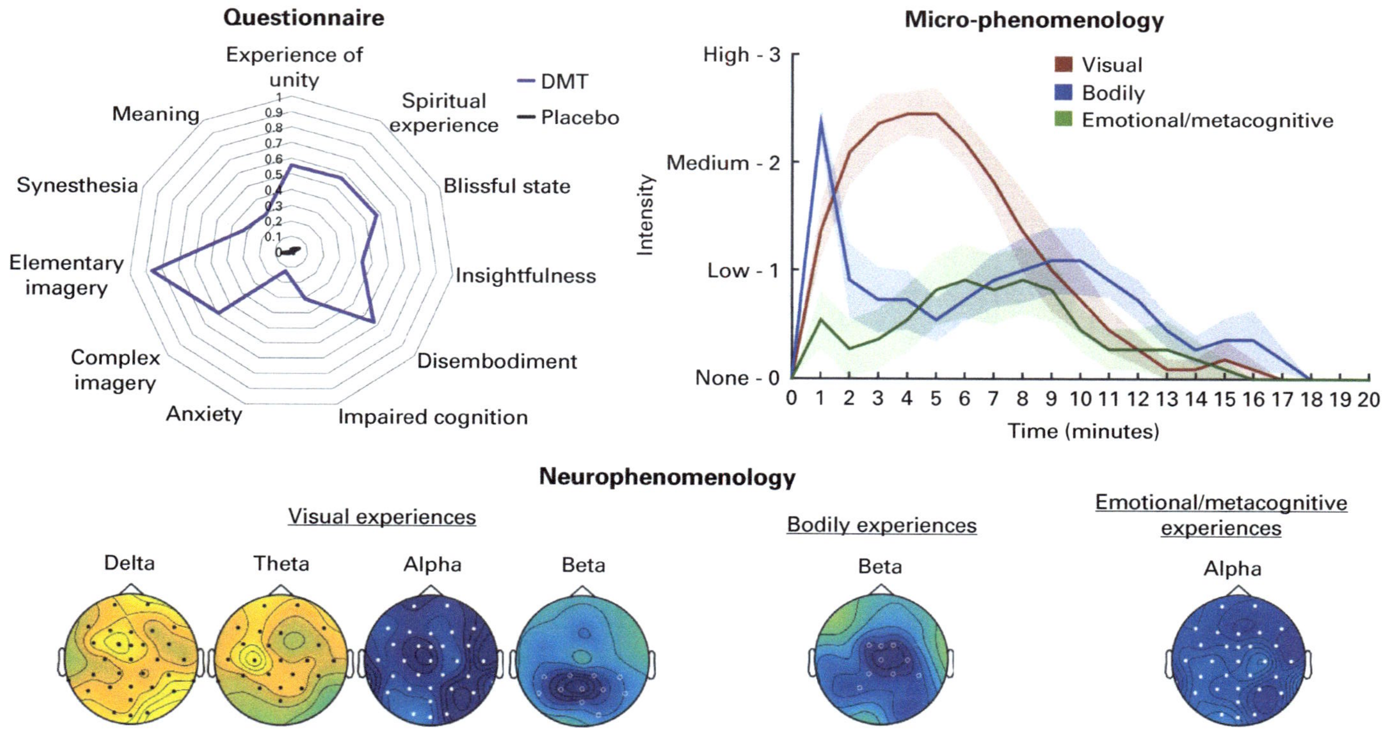

Figure 24.2
Example of a comparison between standard and neurophenomenological methods to account psychedelic experiences. The use of standard questionnaires (e.g., top left) only provided a single score for each predefined category of experience, accounting for the whole duration of the DMT experience. Conversely, microphenomenology (top-right) consisted of the extraction of phenomenological categories, allowing for a richer understanding of the temporal dynamics associated with the DMT experience. This analysis then allowed us to establish the neuronal counterparts of the DMT experience using a neurophenomenological approach (bottom). This showed how the visual component of this experience was linked with increases in power in lower frequencies (yellow) and decreases in alpha and beta brainwaves (blue), while bodily experiences were associated with reductions in beta brainwaves, and emotional/metacognitive experiences were found to be linked to reductions in the alpha band (see **glossary** for brainwaves).

revelations and psychological insights—not uncommonly reported during processes of psychedelic therapy (e.g., challenging autobiographical material, false memories, apparent metaphysical revelations). The revelations and insights occurring during psychedelic experiences are usually characterized by noetic feelings of enhanced significance and attribution of truth, which may exacerbate these complications. Via progressive orientation by an experienced facilitator, the use of intersubjective mediation could aid psychedelic users and patients in order to become aware of their experiential processes and foster a light touch regarding psychedelic insights and revelations (Timmermann et al., 2022).

Finally, the established importance of context (i.e., set and setting; see **glossary**) for the safety and efficacy of psychedelics harbors promise for a neurophenomenological framework to be implemented within the context of psychedelic therapy. Psychedelic-assisted therapy consists of a combination between the use of chemical compound and a context surrounding the experience induced by that compound—this is a model for treatment that differs significantly from standard approaches in psychiatry, which seek to act fundamentally on chemical processes to treat mental illness (Schenberg, 2018). Psychedelic therapy could thus be oriented in such a way that baseline assessments of brain activity could be combined with interventions in which drug, dose, psychological support, therapy, music, and the surrounding setting is specifically tailored to the patient and their therapeutic needs when undergoing the psychedelic session. In this way, the subjective-objective tension of psychedelic action finds a meaningful expression in its application in psychedelic therapy by highlighting the potential gains to be made when adopting biopsychosocial and enactive approaches to health and disease.

Fostering an Open-Ended Inquiry for Psychedelic Phenomena

The enactive approach highlights that the world (and what can be learned from it) is not discovered by the agent, but rather enacted and situated in an intersubjective medium within a larger history of coupling, and thus emphasizes the importance of how valuable knowledge resides in spontaneous and skillful action and perception (i.e., know-how) (Depraz et al., 2003). In a similar fashion the proposal of this chapter stresses how intersubjective mediation and validation of psychedelic phenomena in the context of science and clinical practice can be viewed under this lens of skillful know-how. This approach highlights how valuable methods involved in the use of psychedelics are the result of the embodied practice of facilitators, therapists, and users who have developed structured instances of psychedelic practice, and not solely the result of strict manualized interventions.

An important example of know-how can be found in the case of specific traditions of Indigenous and mestizo shamans in South-America who have used psychedelics in rituals for hundreds of years. In these cases, careful attention is placed on the experience of participants as well as on the elements of context that are used to direct the experience toward specific states of consciousness. The accumulation of these practices over time can be thought of as the historical development of know-how, embodied in lineages of psychedelic practice. Branches of this lineage can be found in contemporary psychedelic therapy, which (similar to its role in traditional contexts) uses music as an essential contextual device to guide and elicit experiences that may be central to the efficacy of psychedelics in the treatment of mental health (Timmermann et al., 2022).

From the perspective of scientific communities interested in consciousness research, psychedelics may be employed to investigate fundamental structures of experience, especially when considering that the use of these substances has been linked to ego-dissolution experiences (see **glossary**). The disruptions in the cognitive processes of the self could be understood as a perturbational tool for researchers aiming to understand these processes and their fundamental contribution to human experience. However, as mentioned above, experiences of psychedelic-induced ego-dissolution may be easily assimilated by unconscious frameworks of mysticism or perennialism, which may limit the exploration of these experiences. It therefore seems that a skillful process of intersubjective mediation of these reports is required in order to extract meaningful data concerning perturbations of the sense of self (Timmermann et al., 2023).

In this context, mediated first-person approaches could be used to explore possible perturbations to the multidimensional facets of the sense of self (e.g., bodily and narrative aspects) and how these relate to brain dynamics. Thus, an open-ended approach may be required to explore the possibilities of phenomenological characteristics that make up this sense of self. While the biasing of reports by researchers' implicit frameworks still remains a possibility in this endeavor, the use of explicit methods (such as the microphenomenological interview) fosters an awareness into these biases and may significantly ameliorate them.

Fostering an approach to the psychedelic experience with explicit awareness of its embeddedness in wider cultural and historical contexts may therefore be fundamental for the study of consciousness and for clinical uses of psychedelics. Importantly, in order to leverage the potential of psychedelics in these domains a central focus resides in forms of intersubjective mediation and validation of psychedelic experiences by an experienced facilitator, guide, shaman, therapist, or researcher who has assimilated the opportunities and pitfalls of that experience in their own practice (Timmermann et al., 2022). The hope is that processes of mediation, validation, and awareness of the embeddedness of these experiences in wider contexts enables an open-ended character of investigation, and the utility of psychedelic experiences in consciousness science and clinical practice.

One of the most intriguing applications for which psychedelics are currently being investigated is their use for the treatment of existential distress in end-of-life terminal patients (Griffiths et al., 2016). Evidence indicates that the degree to which participants experience mystical experiences leads to reduced levels of existential anxiety, leading some to argue that psychedelics might confer a *comforting delusion* to the dying by fostering beliefs of transcendence. This hypothesis has been contested by stating that rather than inducing a comforting delusion, psychedelics would enable therapeutic transformations by unbinding of usually fixed models that constitute the sense of self, which then enables the development of newer, healthier self-models (Letheby, 2021).

Varela (1999) conceptualized the (virtual) sense of *self* as dynamically emerging from cognitive and neural properties of distributed local *selves*. In this context, ethical know-how would consist in the realization of this process of *virtuality*. This may be resonant with the proposal that psychedelics may foster a better understanding of the processes that give rise to the sense of self (Letheby, 2021), thereby allowing users and researchers to inquire into this virtuality. However, as long as we do not carefully consider the experiences of

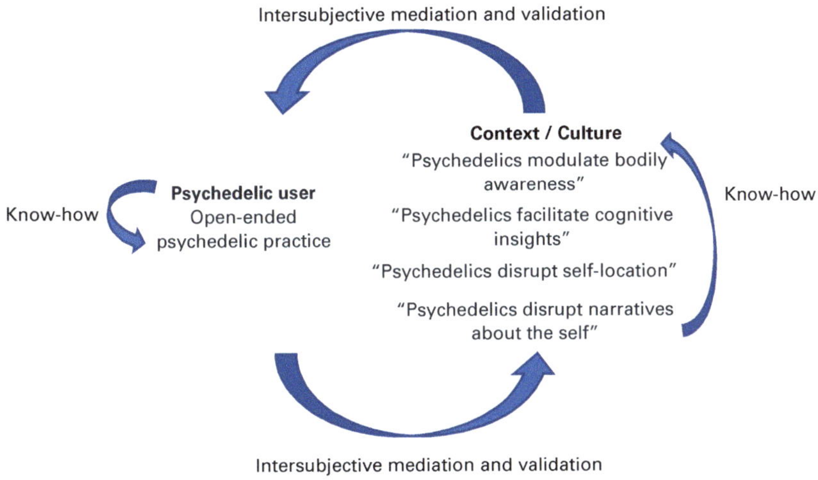

Figure 24.3
Open-ended inquiry in psychedelic practice. Disciplined approaches to psychedelic experiences may facilitate an open-ended approach to the psychedelic experience by fostering an awareness on the embeddedness of the psychedelic experience in larger cultural and historical backgrounds. This awareness may enable psychedelic users develop and use mechanisms of intersubjective mediation and validation of their experiences.

psychedelic users in psychedelic science and clinical practice, the existential opportunities of psychedelics may remain nebulous.

Acknowledgments

The author would like to thank James Sanders and Leor Roseman for useful comments in earlier drafts of this chapter.

Glossary

Ayahuasca, DMT, LSD, mescaline, psilocybin: Serotonergic psychedelic compounds that share a primary pharmacological action in the serotonin-2A receptor in the brain. With the exception of LSD, all these compounds can be found in plants or fungi and have been used in ceremonial contexts for hundreds of years by Indigenous or mixed-race populations in the American continent. LSD was first synthesized by the Swiss chemist Albert Hoffman, who also accidentally discovered its experiential properties in 1943. This led to a wave of psychedelic research in the following three decades. It also had a profound impact on the counterculture in United States and Europe in the 1960s and 1970s.

Brainwaves: Rhythmic activity of neuron firing that is captured by techniques such as the electroencephalogram (EEG). These rhythms (or oscillations) are usually divided into delta, theta, alpha, beta, and gamma rhythms. Each of these rhythms has been linked to cognitive functions and states of consciousness, and thus are common outcomes of investigation in consciousness research.

Ego-dissolution: Umbrella term referring to various forms of disruption to the sense of self, occurring usually during non-ordinary states of consciousness. Ego-dissolution is a heterogeneous category and may refer to the blurring of bodily boundaries or affective processes of the sense of self (i.e., disrupting the minimal sense of self) or the disruption of narrative aspects (e.g., traits, goals, and beliefs) with which a subject identifies.

Microphenomenological interviews: Interview process that seeks to mediate first-person reports and limit their bias by redirecting attention from reflective to prereflective (or embodied) characteristics of experience. In addition to its use for the intrinsic purposes of experiential research, it has been tailored to facilitate the study of reciprocal constraints between subjective and objective data in the context of neurophenomenology (see below).

Mysticism: In this context, mysticism refers to a framework that was strongly influenced by the work of the philosopher Walter Stace. Stace described the phenomenology of mystical experiences as consisting of facets, such as feelings of revelation, unity, paradoxicality, and ineffability, among others.

Neurophenomenology: Research program proposed by Francisco Varela aiming to pragmatically bridge the explanatory gap separating the irreducible domains of mental and biological phenomena. It places a central emphasis on the role of first-person methods inspired by phenomenology to incorporate subjective phenomena in the study of consciousness.

Integration practices: Instances occurring after psychedelic sessions during which users' experiences are reflected upon and thus may serve as opportunities to develop guidance from therapists or peers in the meaning-making process associated with psychedelic phenomenology.

Perennialism: A perspective in philosophy popularized by Aldous Huxley, and influenced in part by his use of psychedelics, that states that all religions share a common core or truth.

Set and setting: A concept emerging in recreational psychedelic use in the 1960s that highlights how extra-pharmacological factors shape psychedelic experiences and how people interpret them. It is commonly employed in the framework of harm reduction practices and emphasizes the importance of being aware how psychological states and traits (the mindset), as well as the physical context (setting), influence the safety of a psychedelic experience.

References

Aday, J. S., Davis, A. K., Mitzkovitz, C. M., Bloesch, E. K., & Davoli, C. C. (2021). Predicting reactions to psychedelic drugs: A systematic review of states and traits related to acute drug effects. *ACS Pharmacology and Translational Science*, *4*(2). https://doi.org/10.1021/acsptsci.1c00014

Andersen, K. A. A., Carhart-Harris, R., Nutt, D. J., & Erritzoe, D. (2021). Therapeutic effects of classic serotonergic psychedelics: A systematic review of modern-era clinical studies. *Acta Psychiatrica Scandinavica*, *143*, 101–118. https://doi.org/10.1111/acps.13249

Barrett, F. S., Johnson, M. W., & Griffiths, R. R. (2015). Validation of the revised mystical experience questionnaire in experimental sessions with psilocybin. *Journal of Psychopharmacology*, *29*(11). https://doi.org/10.1177/0269881115609019

Carhart-Harris, R. L., Kaelen, M., Whalley, M. G., Bolstridge, M., Feilding, A., and Nutt, D. J. (2015). LSD enhances suggestibility in healthy volunteers. *Psychopharmacology (Berl)*, *232*, 785–794. https://doi.org/10.1007/s00213-014-3714-z

Carhart-Harris, R. L., Muthukumaraswamy, S., Roseman, L., Kaelen, M., Droog, W., & Nutt, D. J. (2016). Neural correlates of the LSD experience revealed by multimodal neuroimaging. *Proceedings of the National Academy of Sciences*, *113*, 4853–4858. https://doi.org/10.1073/pnas.1518377113

Depraz, N., Varela, F., & Vermersch, P. (2003). *On becoming aware: A pragmatics of experiencing*. John Benjamins Publishing Company.

Dittrich, A. (1998). The standardized psychometric assessment of altered states of consciousness (ASCs) in humans. *Pharmacopsychiatry*, *31*, 80–84.

Dupuis, D. (2020). The socialization of hallucinations. Cultural priors, social interactions and contextual factors in the use of ayahuasca. *Transcultural Psychiatry*, *59*(5).

Earleywine, M., Low, F., Lau, C., & de Leo, J. (2022). Integration in psychedelic-assisted treatments: Recurring themes in current providers' definitions, challenges, and concerns. *Journal of Humanistic Psychology*, 002216782210858. https://doi.org/10.1177/00221678221085800

Gearin, A. K., & Devenot, N. (2021). Psychedelic medicalization, public discourse, and the morality of ego dissolution. *International Journal of Cultural Studies*, *24*, 917–935. https://doi.org/10.1177/13678779211019424

Griffiths, R. R., Johnson, M. W., Carducci, M. A., Umbricht, A., Richards, W. A., Richards, B. D., Cosimano, M. P., & Klinedinst, M. A. (2016). Psilocybin produces substantial and sustained decreases in depression and anxiety in patients with life-threatening cancer: A randomized double-blind trial. *Journal of Psychopharmacology*, *30*, 1181–1197. https://doi.org/10.1177/0269881116675513

Griffiths, R. R., Richards, W. A., McCann, U., & Jesse, R. (2006). Psilocybin can occasion mystical-type experiences having substantial and sustained personal meaning and spiritual significance. *Psychopharmacology (Berl)*, *187*, 268–283. https://doi.org/10.1007/s00213-006-0457-5

Hartogsohn, I. (2016). Set and setting, psychedelics and the placebo response: An extra-pharmacological perspective on psychopharmacology. *Journal of Psychopharmacology*, *30*, 1259–1267. https://doi.org10.1177/0269881116677852

Langlitz, N. (2012). *Neuropsychedelia*. University of California Press.

Lebois, L. A. M., Papies, E. K., Gopinath, K., Cabanban, R., Quigley, K. S., Krishnamurthy, V., Barrett, L. F., & Barsalou, L. (2015). A shift in perspective: Decentering through mindful attention to imagined stressful events. *Neuropsychologia, 75*, 505–524. https://doi.org/10.1016/j.neuropsychologia.2015.05.030

Letheby, C. (2021). *Philosophy of psychedelics*. Oxford University Press.

Petitmengin, C. (2006). Describing one's subjective experience in the second person: An interview method for the science of consciousness. *Phenomenology and the Cognitive Sciences, 5*, 229–269. https://doi.org/10.1007/s11097-006-9022-2

Sanders, J. W., & Zijlmans, J. (2021). Moving past mysticism in psychedelic science. *ACS Pharmacology and Translational Science, 4*, 1253–1255. https://doi.org/10.1021/acsptsci.1c00097

Schenberg, E. E. (2018). Psychedelic-assisted psychotherapy: A paradigm shift in psychiatric research and development. *Frontiers in Pharmacology, 9*, 1–11.https://doi.org/10.3389/fphar.2018.00733

Taves, A. (2020). Mystical and other alterations in sense of self: An expanded framework for studying nonordinary experiences. *Perspectives on Psychological Science, 15*, 669–690. https://doi.org/10.1177/1745691619895047

Timmermann, C., Kettner, H., Letheby, C., Roseman, L., Rosas, F. E., & Carhart-Harris, R. L. (2021). Psychedelics alter metaphysical beliefs. *Scientific Reports, 11*, 1–13. https://doi.org/10.1038/s41598-021-01209-2

Timmermann, C., Roseman, L., Schartner, M., Milliere, R., Williams, L. T. J., Erritzoe, D., Muthukumaraswamy, S., Ashton, M., Bendrioua, A., Kaur, O., Turton, S., Nour, M., Day, C., Leech, R., Nutt, D. J., & Carhart-Harris, R. L. (2019). Neural correlates of the DMT experience assessed with multivariate EEG. *Scientific Reports, 9*, 16324. https://doi.org/10.1038/s41598-019-51974-4

Timmermann, C., Watts, R., & Dupuis, D. (2022). Towards psychedelic apprenticeship: Developing a gentle touch for the mediation and validation of psychedelic-induced insights and revelations. *Transcultural Psychiatry*, 136346152210827. https://doi.org/10.1177/13634615221082796

Timmermann, C., Bauer, P. R., Gosseries, O., Vanhaudenhuyse, A., Vollenweider, F., Laureys, S., Singer, T., Antonova, E., & Lutz, A. (2023). A neurophenomenological approach to non-ordinary states of consciousness: Hypnosis, meditation, and psychedelics. *Trends in Cognitive Sciences, 27*, 139–159. https://doi.org/10.1016/j.tics.2022.11.006

Varela, F. J. (1996). Neurophenomenology: A methodological remedy for the hard problem. *Journal of Consciousness Studies, 3*, 330–349.

Varela, F. J. (1999). *Ethical know-how: Action, wisdom and cognition*. Stanford University Press.

Varela, F. J., & Shear, J. (1999). First-person methodologies: What, why, how? *Journal of Consciousness Studies, 6*, 1–14.

Varela, F. J., Thompson, E., & Rosch, E. (1991). *The embodied mind: Cognitive science and human experience*. MIT Press.

Vollenweider, F. X., Vollenweider-Scherpenhuyzen, M. F., Bäbler, A., Vogel, H., & Hell, D. (1998). Psilocybin induces schizophrenia-like psychosis in humans via a serotonin-2 agonist action. *NeuroReport, 9*, 3897–3902. https://doi.org/10.1097/00001756-199812010-00024

Yaden, D. B., & Griffiths, R. R. (2021). The subjective effects of psychedelics are necessary for their enduring therapeutic effects. *ACS Pharmacology and Translational Science, 4*(2). https://doi.org/10.1021/acsptsci.0c00194

25 Four Components of Subjectivity and the Projective Consciousness Model

Kenneth Williford

Our thought or perception is essentially reflexive upon itself; or as it is expressed more happily in Latin, *est sui conscia*, for I never think without knowing that I think. . . . In addition to this reflection which can be called virtual and which is found in all our perceptions, there is another, more explicit kind by which we examine our perception by another perception, as everyone can easily verify.
—Antoine Arnauld, *Des vraies et des fausses idées* (1683)[1]

Personal Reflections

In my view, one's access to one's subjective experience relates to the scientific study of consciousness in an indispensable way (cf. Ataria, and Demšar and Kordeš, this volume). It is our primary source of data about consciousness and its aspects, structures, and possibilities. Without such access, there could be no such thing as a theory of consciousness—nor many other, more beautiful things: the work of Proust or Virginia Wolf, the poetry of Pessoa or Ginsberg, the music of Bach or Mahler.

The cultivation of knowledge of one's consciousness—its patterns, forms, dynamics—is necessary for articulating the "pre-comprehension" we all rely on in our intersubjective concourse into something more precise, into models. This cultivation, by no means a perversely private enterprise, can also increase one's empathy; it is not purely theoretical. It does have pitfalls, but these can be corrected. A tendency toward anti-naturalism is one pitfall. But there is nothing in phenomenology that is at variance with naturalism or even materialism; it can be an ontologically minimalistic discipline, metaphysically neutral. But thinking that consciousness has complete self-access can lead one down the wrong path here. And it is well-known that practitioners of meditation can sometimes mistake flights of metaphysical fancy (not to mention large-scale ego trips) for experiential data. But there is no doubt that meditative practice provides one more important mode of phenomenological cultivation highly relevant to understanding consciousness (cf. Fulder, Ott, Pagnoni and Guareschi, and Srinivasan, this volume). Much the same can be said for the study of variations of consciousness induced by psychoactive substances and other means of altering experience (cf. Mayseless, Ott, and Timmermann, this volume).

All of these paths, properly pursued, can lead one not to an inflated conception of consciousness's powers of self-observation but to a sense of humility and surprise. It is an endless wonder that consciousness can vary in so many dimensions and that even the most obvious things (time consciousness and intentionality; see **glossary***) still seem so amazingly opaque and mystifying, while perfectly familiar! This sense of awe should not surprise us: after all, consciousness is a natural phenomenon too; and every natural phenomenon outstrips our minds in one way or another. No need to invent supernatural worlds behind the scenes, and no need to delude ourselves about our own powers (cf. Segev, this volume). The natural world and we, as some of its "finite modes," are awe inspiring enough. In metaphysics, Husserl and Spinoza were about as far apart as one could imagine, but both would be able to agree, I think, that phenomenology offers a domain of "infinite tasks." If we are to tame any of this domain, we must first look at it honestly and rigorously with whatever tools, "subjective" or "objective," we have at our disposal.*

Overview

After laying out some of my assumptions, I argue that subjectivity has (at least) four key components: (1) a projective structure or point of view, (2) prereflective self-acquaintance, (3) a sense of individuality or concrete specificity, and (4) self-locational capacity (which admits of degrees of elaboration) (see **glossary**). I argue that (1) provides the foundation for (2), and in part for (3) and argue that the first three components together provide the foundation for (4). I conclude with some reflections on future directions and disciplinary limitations.

Introduction: Some Assumptions

I assume,[2] first, that consciousness is a real phenomenon and that with some important qualifications it *is* as it *seems* to us to be upon introspection. Its seeming to be a certain way (e.g., intentional, temporal, self-aware) is compatible with it also being ways it *fails* to seem to be. Consciousness may have many properties, including metaphysically essential ones, that one has no introspective access to. It is introspectively *incomplete*. But this does not prevent it from being infallible with respect to the properties it does positively seem to have (see Williford, 2007, 2020).

The thesis of introspective incompleteness (see **glossary**) is hardly controversial among most neuroscientists, psychologists, and philosophers of mind. But it is important to note that a vocal minority of philosophers of mind accept some version of a completeness thesis (advertently or inadvertently), and this leads them to the paths of neodualism (see **glossary**) and panpsychism (see Chalmers, 2020; Goff, 2017; Williford, 2020) (see **glossary**). Inversely, some philosophers of mind and neuroscientists have been tempted to reject not only the completeness thesis but even qualified infallibility theses too. They have been tempted to think that consciousness not only fails to have introspective access to everything essential about itself, but that it systematically *misrepresents* itself; that introspection, phenomenology, and meditative practices are basically useless in its study—save in the sense that they may define a set of erroneous, intuitive beliefs about consciousness that the theory of consciousness ought to *explain away*.

This view sometimes goes by the name of *illusionism* (see **glossary**) (see Dennett, 2016; Frankish, 2016; Graziano, 2016 [Graziano rejects the label]). I maintain that sometimes illusionism is spawned by conflating the infallibility thesis with the completeness thesis (as, I would argue, Dennett essentially does in *Consciousness Explained* [Dennett, 1991]). I would thus offer a counterexplanation in psychological terms of the grip of illusionism on some theorists: conflating infallibility and completeness, but rejecting the metaphysically extreme consequences of the latter, they end up with the view that consciousness itself is, in some sense, an illusion (see Williford, 2020).

Respecting the distinction between infallibility and completeness, however, could not be more important. If consciousness does not provide some reliable information about its own properties, the theory of consciousness is epistemically untethered. Even the proposals of illusionists and their allies typically attempt to preserve at least *some* connection to *prima facie* phenomenological plausibility.[3] But it is hard to understand why, given their assumptions. Their own proposals about the real nature of consciousness (or its non-illusory correlate) seem no better off in this respect than realist theories (like Integrated Information Theory or the Projective Consciousness Model). Moreover, it is the standard practice of many of the mind sciences (e.g., cognitive neuroscience, psychophysics, and even those very parts of psychology that have to do with self-deception) to rely, at some level, on introspective reports interpreted in a realistic—and not "heterophenomenological" (cf. Dennett, 1991)—way.

On the other hand, imagining that consciousness is completely self-revealing (as Goff [2017] does), that it provides introspective access to *all* of its essential properties, is not only absurd from a biocognitive point of view: it is absurd from the point of phenomenology itself! Nothing could be clearer than that, at multiple levels, consciousness is limited by a world it has access to but does not create. This is so even under a methodologically individualist working assumption—that is, an assumption on which we bracket any presuppositions about the existence of every sort of intentional object of consciousness (the "external world" included): the determinate properties of consciousness itself, as well as its qualitative character (the "what-it-is-like" aspect of consciousness), are *given* to consciousness, not *made* by it, and are not under its voluntary control. There is thus an introspectively accessible limit to one's subjectivism, as even the supposedly idealist Husserl himself understood (see Williford, 2013).

Moreover, one's consciousness is "thrown" into existence and is constrained by its past in ways that it has no control over. And it is impinged upon by its embodiment in ways that make little sense on the assumption of introspective completeness. The operation of psychoactive substances on consciousness is a case in point. Further, if having access to all its essential properties were itself grounded in one of its essential properties (which it would need to be), it should then be able to isolate and identify that property; this is entailed by the completeness thesis itself. But we find no such thing. If that is right, then we can know *by introspection* that consciousness is not completely self-revealing! Finally, if all of the essential properties of consciousness were accessible to introspection, the metaphysics of consciousness should not be phenomenologically undecidable; but it is. We can't decide on the basis of phenomenology alone whether consciousness is a Cartesian substance, a Humean bundle of property tropes, a Spinozistic finite mode of God, a Leibnizian monad, an emergent property of organized matter, or a mathematico-computational form realized in neuronal hardware. Phenomenological realism (see **glossary**), without the completeness thesis, is compatible with any of these metaphysical schemes.

It is thus fully compatible with materialism. And I do maintain some version of materialism about consciousness. In fact, I believe that there is still good enough reason to think that consciousness is some sort of mathematico-computational form realized in neuronal hardware (and operating under some as yet unspecified constraints; see Williford et al. [2018]).[4] Indeed, the version of materialism just suggested is much stronger than what a scientific theory of consciousness requires. All that such a science requires, and this is my second assumption, is that there be some thoroughgoing and in-principle detectable correlations between the real properties of consciousness (its structures and functions) and some processes of the brain or potentially another substrate (cf. Lamy and Micher, Malach, and Segev, this volume). That weaker assumption is compatible with most nonmaterialist metaphysical positions as well.

My third assumption is that phenomenology (broadly construed, but certainly including the work of Husserl and colleagues and Varela's neurophenomenology, as well as Berkovich-Ohana, Gallagher, Rudrauf, and Schmidt, this volume) can give us a good idea of the set of (possibly essential) manifest features of consciousness. Among these features are temporality (or time consciousness); projective lived spatiality (or point of view); prereflective self-acquaintance (see **glossary**); multimodal sensory, affective, algedonic, cognitive, and conative differentiation and integration; synchronic unity; and intentional /representational/ conceptual capacity.

My fourth assumption is that consciousness resides entirely in its realizing substrate (in humans and animals, it resides in their brains). That is to say, I accept a kind of internalism (see **glossary**) about consciousness and reject direct realism (see **glossary**) or an externalism that would have conscious (directly) reach beyond its realizing materials. One's phenomenal world does indeed *seem* to be outside of one's cranium; but this is an adaptive illusion (see Lehar, 2003; Metzinger, 2004; Revonsuo, 2006). The brain realizes a virtual phenomenal space (Rudrauf et al., 2017). However, on the further assumption that the brain and its consciousness are connected to the surrounding world in multiple ways (causal and constitutive), a form of indirect realism (see **glossary**) becomes defensible. It is indirect realism alone, in my view, that allows one to make sense of both the relevant "internalist" data (e.g., dreams, hallucinations, perceptual variation, perceptual illusions, misrepresentation) and the adaptive character of consciousness (and, by extension, the rise and success of the sciences). In the sequel, it will be important to bear this denial of direct realism in mind.

My fifth assumption is that prereflective self-acquaintance is to be understood as a kind of non- or pre-representational self-awareness that is reflexive (in the logical sense of that word). The word "self-acquaintance" is used here only to indicate that this is not a matter of representation or conceptual thinking about oneself (Williford, 2015). That is, an episode of consciousness is acquainted with itself prior to any psychological reflection upon or any other kind of representation of itself (see the epigram from Arnauld above). This means that it has an irreducibly circular structure.[5]

I have shown elsewhere that this assumption does not lead to an infinite regress; and indeed, it leads to some strikingly phenomenologically and metaphysically plausible conclusions about consciousness: that homuncularism is false (there is no "observer in the Cartesian Theatre," separable from consciousness itself [Dennett 1991]); that consciousness is both like a relation and like a relatum of that relation; and that consciousness "has an outside" that it cannot see directly but nevertheless has some implicit awareness of (e.g., it must be structured by a projective origin that is outside of phenomenal space; see below and see Rudrauf et al. [2017] and Williford [2019, et al. 2022]).

My sixth assumption is that projective lived spatiality or point of view is to be understood in terms of consciousness being structured by a three-dimensional (3D) projective space with its attendant geometry. This is elaborated in our projective consciousness model (PCM) (see Rudrauf et al., 2017, 2020, 2022, 2023; Williford et al. 2018, 2022). This means that consciousness has an essentially spatial aspect that can be oriented around a "lived origin" or point of view. However, this assumption also entails that consciousness has an aspect that it cannot directly access, for a 3D projective space requires a fourth dimension that is outside of the other three. This fourth dimension need not be thought of as a "Lovecraftian" physical dimension; it could be a computationally realized vector dimension. The point-of-view structure (the "lived origin") is thus built into the space of conscious experience, but the point of view itself remains elusive—something that can never be a perceptual object in the space of conscious experience.

On the view assumed here, the fact that certain sounds, smells, and so on might be thought of as not being located in the space of vision (the space that most obviously has a projective structure, with its perceptual origin and horizons) is quite beside the point. Even a smell with a nebulous location is manifest with an "appearance-to" structure (the odor is manifest *to* you). And on the view assumed, all "appearance-to" is understood as having a projective presentational aspect—it is the appearance *of* something *to* someone. That relational structure, even in its most minimal form, is a projective one, implying a point of view (and a projective geometry; see **glossary**).

In what follows, I develop some thoughts about the nature of subjective character (see **glossary**), beginning from these assumptions. The assumptions about phenomenological realism and about the usefulness of phenomenology to a scientific theory of consciousness bear on the discussion below in the following way: if the phenomenology of subjectivity has a mathematical model, one could use it as a guide for and constraint on the development of both computational and neuroscientific approaches (in basic agreement with Varela's [1996] "neurophenomenological" program). Issues pertaining to time consciousness and differentiated quality and attitudinal spaces will, of necessity, be left aside. Any full theory of consciousness would, no doubt, involve a synthesis of mathematical models of all of these (apparently) essential structures.

Prereflective Self-Acquaintance, Subjective Point of View, and Intersubjectivity

We have developed an account of prereflective self-acquaintance or self-consciousness (PRSC) that relates PRSC to the thesis that consciousness is structured by a projective 3D space with its attendant geometry (see Williford et al. 2022). The latter is the key thesis of the PCM. In brief, certain properties of PRSC can be understood in terms of the projective geometrical structure of consciousness. For the present purposes, I will focus on only two closely related ones: "witness consciousness" (see **glossary**) and what I shall call "proto-intersubjectivty."

By "witness consciousness" I mean the sense that one is always, in a way, one's own observer (prior to any reflection), one is a "witness to oneself" (even when one "loses oneself").[6] This is understood in projective geometrical terms in the following way. In virtue of being structured by a projective 3D space subject to a projective transformation group (immanent rules, so to speak, that limn and delimit the possible "moves" available within phenomenal space), consciousness involves an inherent perspectival ambiguity. This ambiguity is *au fond* irremovable, though it is dampened when particular perspectives are taken on objects in the phenomenal field. For example, this means that while looking out the window from my third-floor office at a tree below, there remain alternate perspectives immanent in the very experience itself. In looking out the window, there remains, in particular, its reciprocal—my being looked back at from the position of the tree. And there are all the other, peripheral perspectives that form a kind of "halo" around the one actually taken; its reciprocal is, however, privileged.

Imagine now the pure space of consciousness, without any specific perspective being taken on anything in the phenomenal field—something like the state of pure awareness that is the goal of some meditative practices (see, e.g., Metzinger, 2020). Such a space is, in a sense, both aperspectival and omniperspectival. It is aperspectival in the sense that no specific perspective is being taken on anything, but it is omniperspectival in the sense that it is the common ground for all possible perspectives open to it.

The adoption of a specific perspective, we might say, involves the "breaking" of this perfect symmetry. Prior to the breaking of symmetry, no particular perspective is privileged. But even upon the breaking of symmetry, the phenomenal space retains its essential character as a space of multiple possible perspectives. This gives phenomenal space a kind of ineradicable perspectival ambiguity. Speaking very loosely then, no matter what perspective one takes one is "being observed" from all the other possible perspectives immanent within one's phenomenal space itself. The reciprocal perspective, which is ineradicably bound up with a taken perspective, further highlights this sense of being observed because it encodes the possibility (and often actuality; see below) of being "looked back at." But the sense is there, inherent in phenomenal space, even in the "pure" state; it is a feature of a projective phenomenal space as such. That is the theory, roughly stated. (For more precise formulations involving the projective geometrical notions of duality, reciprocity, perspectivity, frame, projective transformation, closure, and closedness, see Williford et al. [2022]).

It was customary among the classical phenomenologists to say that consciousness was conscious of itself in PRSC without "profiles." This means that consciousness does not experience itself in the way that it experiences physical objects—over time (perspectival aspect by aspect, frame by frame; just think of walking around a table). In that sense, consciousness takes no point of view on itself, though it is immediately conscious of itself. According to the model described here, self-consciousness without profiles is a matter of consciousness being, irreducibly, a space of multiple possible perspectives, no matter which perspective is actually realized at a time. In looking at the tree from my window, I do not *actually* need to be looking back at myself from the tree. In looking at the tree, I am looking back at myself looking in a *virtual* fashion; this sense is grounded in the pre-perspectival structure of consciousness *qua* projective phenomenal space.

This "virtual reflection" (as Arnauld called it) is intrinsic to the space of consciousness as a space of possible perspectives. This virtuality is grounded in the *actual* and irreducible structure of the space. One could say that in looking at the tree I am, *ipso facto*, conscious of myself looking at the tree because such looking necessarily carries its reciprocal with it due to the *actual* structure of the space. There is a symmetry between the looking at and being looked at that is also provisionally broken, but not completely eliminated, in my conscious orientation toward the tree (from "me to the tree and back" so to speak). And the fact that it can't be completely eliminated is grounded in the very structure of the projective space. Again, that this is *not* a matter of consciousness "representing," "perceiving," or "thinking about" itself. It is, rather, a feature built into the very structure of phenomenal spatiality (see **glossary**). Thus, it is better here to speak of "self-acquaintance" rather than "self-representation," "higher-order thought," and so on.

If phenomenal space, the space of consciousness, is a space of many possible perspectives in which any actually taken perspective contains an inherent ambiguity between the perspective taken and its reciprocal (or, more generally, the inner sense of being observed "from anywhere and everywhere"), then the sense of always being quasi-observed (or "witness consciousness") is accounted for. But one can also see here one of the bases of *intersubjectivity* (see **glossary**) (and cf. Gallagher, this volume).

Notice the curious fact that consciousness, at one and the same time, is both an interface for intersubjective action and a space of privacy (see **glossary**). There is a shared world and a hidden domain that one may choose to reveal or to conceal. Often the shared world is called "public," "objective," or "observable." However, recall my earlier denial of direct realism: the domain of the publicly observable is just the typically most coordinated domain of many individual, phenomenal spaces *modulo* perspectival framing. And, for evident reasons, these are typically the subdomains of individual phenomenal spaces most closely connected to basic sensory perception and shared priors about the nature of objects and the ways they can appear from various points of view. It is for this reason that, for example, we feel rather confident that our perception of a ball is similar enough to a dog's perception of a ball to make playing fetch together a possibility. But consider that dogs share phenomenal sub-subdomains with one another that humans do not share with them—namely, parts of the auditory and olfactory domains. Again, indirect realism makes the best sense of the fact that much but not all is thus shared.

When it comes to an individual's thoughts, ones not expressed in action or in language, there are simply too many degrees of separation from the "publicly observable"—that is, ultimately, from the ways in which the sensory and sensory-phenomenal systems of two or more individuals in the same place and time are being commonly affected—to expect coordination. We may be moving a table together, and we can expect of each other a common idea of where the table is in relation to one another and to the surrounding room and its other furnishings. But I may be thinking of mathematical developments in thirteenth-century China, while you may be wondering whether you should tell me the *real* reason you invited me over to help you move furniture.

The point here is to illustrate how the *private sphere* of one's own conscious experience must carry within it the notion of multiple possible points of view and thereby possibly actually realized points of view that one does not currently occupy. I know, roughly, how the table must look to you from where you have grasped it. I know that you know, roughly, how it must look to me from where I have grasped it. I know, roughly, how I and the surrounding furnishings must look to you from where you are; and you know how you must look to me from where I am. But this sense of a shared world with multiple possible and actual points of view immanent in it is, in reality, just a necessary part of two private but well-enough coordinated phenomenal spaces, virtual projective spaces realized in two separate brains. I could be dreaming the whole scene. This means that private consciousness has the resources for the sense of intersubjectivity *within itself.* We can call this its "proto-intersubjectivity" (see **glossary**). And this is intimately connected to its prereflective self-acquaintance, its "witness consciousness." It is not unreasonable to think that this connects also to theological, psychological, and psychoanalytic extrapolations like the "voice of conscience," the "eyes of God," the "last judgment," the "law of karma," the "superego," and the "big other."

What I am calling proto-intersubjectivity is, it should be noted, not dependent on "theory of mind" or the ability to recognize faces or eyes. Theory of mind and such abilities are normally integrated with it. But proto-intersubjectivity is operative even when one is all alone and even if one could not tell the difference between a person and a hat. When one navigates a space, one is constantly making implicit predictions about how things should appear from the various points of view one anticipates occupying.

Suppose you are going to the kitchen to make tea. You expect to see the teapot from certain angles, from various distances; the same goes for the tea, the cups, the water, the stove. If any of these expectations are violated, you will become distinctly aware of your implicit prediction about perspectival character: for example, suppose that, unexpectedly, you begin to undergo a bout of macropsia (or micropsia, telopsia, or pelopsia). Your first instinct might be to attribute the distortion to some unexpected change in the objects or your objective relation to them. These implicit predictions that guide and regulate our action planning and execution are made possible by the transformation group immanent in our projective phenomenal space (viz., PGL(4), the projective linear group).

This group provides the laws that allow us to predict changes of perspective and thereby to detect deviation from prediction. It is in virtue of this group that one can engage in moment-to-moment navigation, deliberatively imagine more or less advantageous perspectives on the world relative to a given project, engage in more or less explicit intersubjective simulation, and conjure more "distanced" or "objective" perspectives on the whole.

To sum up, "witness consciousness" is a key aspect of PRSC. This "witness consciousness" can be (at least partly) understood, according to the PCM, in virtue of the fact that, as structured by a projective space, consciousness is *ipso facto* a space of multiple possible perspectives, and any actually taken perspective is ineradicably ambiguous—it is entwined with its reciprocal perspective, and at a deeper level with a whole space of possible ways to break global perspectival symmetry. In that sense, any perspective *from A to B* is simultaneously a perspective *from B to A*, as well as, more fundamentally, a perspective "seen from everywhere." And this means that in taking a perspective, one is taking a virtual perspective on the perspective (A "sees" B who "sees" A "seeing" B). The "breaking of symmetry" that gives a single directionality a kind of dominance is not sufficient to eliminate the reciprocal or opposite directionality (or the other possible directionalities immanent in the space). These cannot be eliminated because it they are of the nature of the space itself. The claim is that this yields the sense of being observed in observing or "witness consciousness."

Moreover, this reciprocity of perspectives (along with the more general perspectival ambiguity) is but one aspect of the fact that the projective space of conscious experience is a space of possible perspectives equipped with immanent rules governing transformations from perspective to perspective. This proto-intersubjectivity constrains and enables our quotidian navigational abilities, but it also makes the sense of full-blown intersubjectivity possible from within what is in fact a private domain, encased, more or less, in one's cranium.

The Sense of Individuality

There is, however, more to subjectivity than PRSC, proto-intersubjectivity, and privacy. There is also the sense that one is *this very unique individual and no other*. Even if we imagine multiple universes containing endlessly many exact duplicates of ourselves (mine are each writing a paper qualitatively identical to "this one right now"), I know that I am *this one* in *this world* and not any of my modal duplicates. This is to be distinguished from merely knowing the logico-metaphysical truth that everything is self-identical. My knowledge of my own individuality is not, first and foremost, a matter of knowing the truth of some proposition (though it enables me to know certain propositions). It is more like a direct acquaintance with my own being, my own individual existence.

How is it that I can know myself *as myself*? The ability to answer this question is what Manfred Frank calls meeting the "*de se* constraint" on any adequate theory of self-consciousness (see, e.g., Frank, 2019). This too seems to be an essential component of subjectivity (cf. Senderowicz, this volume). My view is that prereflective self-*acquaintance* is the key. One can come to represent or frame beliefs about anything with which one is acquainted. The field of acquaintance (see **glossary**) is a field of similarities and differences and the various relations that obtain between the items in the field. If consciousness is acquainted with itself, then it can, in principle, distinguish itself from its components and the components from one another; and with cognitive machinery operative, it can represent and express beliefs about these similarities and differences (see Williford [2019, 2022] for details).

If self-acquaintance, as I have argued, can be understood within the framework of the PCM and is integrally related to the projective spatial structure of consciousness, then the PCM is sufficient for capturing the basis of this sort of *de se* self-knowledge. If there is a residual sense of puzzlement, it may stem from two sources. First there is the curious fact that by being realized in the world, a consciousness knows pre-representationally the very fact of its realization. Most realizations of properties are not like that. But the hypothesis here is that the realization of a virtual projective space (under, perhaps, certain yet unspecified conditions) in a brain (or potentially another substrate) is, among other things, simultaneously the realization of direct (pre-representational) knowledge *of that very realization*.

Second, there is the nagging general metaphysical issue of individuation. Why is *this* particular realization of a projective phenomenal space *me* and not someone else? But this worry applies to *every* individual thing. Why wasn't I *that potato* instead of *this human being*? Why is proton A the one it is and not proton B? If we trace this problem to its roots, we find that there must be some brute, primordial differentiation in the cosmos. You are you and not me *ultimately* because of that differentiation, even if we must first begin talking about the particular bits of matter that realize us now and then our particular histories of form-preservation through material flux. And I suggest that this differentiation is brute because I do not think that *anything* can explain it further. The world simply contains real differences—differences within a unity and unities within difference, to be sure. But we cannot eliminate the many from the one. And this is ultimately reflected in one's puzzlement about one's own apparently contingent facticity. If prereflective self-acquaintance allows us to know directly (and later represent to ourselves and others) our own individuality or concrete realization, then that is sufficient for this component of subjective character. That it does not also allow us to solve the general metaphysical riddle of primordial individuation is not a cause for further concern, and the riddle should not be confused with a problem specific to the theory of consciousness and self-consciousness.

Expandable Self-Locative Capacity and the Function(s) of Consciousness

Once, however, we have the machinery of perspective-taking and *de se* cognition in place, we have the basic tools we need for locating ourselves in a variety of spaces and planning accordingly. We have already discussed navigating ambient physical space and coordinating actions with others, but the more cognitive capacities we develop the more "spaces" we can locate ourselves in.

We can "know where we are" in a brief dialogue or long correspondence; "know where we are" in a book we are reading or paper we are writing; "know how we compare" to others in our field; "know what we need to do" to grow and improve in this or that; and we can know "where we were" and "how far we have come"—in bodily, psychological, intersubjective, or social terms. If one had no abiding sense of self this would be impossible. And if one had no basic self-acquaintance to ground this "extended" sense of self (see Damasio, 1999), one could not have that sense. Notice that this does not mean that the sense of an "extended self" with a history and socially mediated identity involves some connection to a substantial entity. The self or ego here could be the object of distorting and even fictional representation. What I am saying is compatible with narratival and partially narratival views of personal identity. The self, in this extended and "autobiographical" sense, may indeed be a useful fiction (and sometimes useless or harmful) (see, e.g., Williford 2011; cf. Dreyfus, this volume). But the self-acquaintance and perspective-taking abilities that enable it cannot be. They are real. Realism about consciousness and self-consciousness is compatible, in other words, with fictionalism about the extended self; it is compatible with varieties of realism as well, and certainly does entail realism about a "thin," "minimal," or "core" self (see Dainton [2008], Damasio [1999], Ganeri [2012], and Strawson [2009] for a range of interesting views.)

What consciousness enables us to do, I would argue, is to locate ourselves (itself) in a variety of spaces—the spaces of perception, action, and basic intersubjectivity being the most fundamental. By enabling this, consciousness can serve as a kind of virtual "user interface" (see Metzinger, 2004). This, I maintain, is its primary bio-cognitive and bio-social function, but there is no space to develop this thought much further (see Rudrauf et al., 2023; Williford et al., 2018). Suffice it to say that even if things other than consciousness *could* enable the relevant behavioral-functional aspects here, that in no way means that consciousness is not what has *in fact* come, via evolution, to enable them.

Concluding Reflections

Finally, a few words about future directions and disciplinary limitations. For a long time, the idea that the philosophy of mind and phenomenology could not proceed in good conscience in isolation from developments in psychology, neuroscience, and AI has held considerable sway, and rightly so, in my opinion. There has always been an influential minority of philosophers and phenomenologists who still maintain what we might call non-naturalist positions; these often end up defending positions like neodualism and panpsychism in philosophy of mind (Chalmers, 1996; Goff, 2017) and anti-realist views in phenomenology (see, e.g., Thomasson, 2007; Zahavi, 2016). And it must be conceded that these positions do often highlight remaining conceptual aporia, points recalcitrant to the development of a fully naturalistic, scientific realist, "unified" theory of mind and consciousness.

In my view, mathematics holds the promise of showing us the way forward (cf. Maimon and Klein, this volume). And the PCM is an example of precisely the sort of modeling I have in mind. It allows us to understand hitherto puzzling aspects of consciousness once thought to be theoretically intractable. Many problems remain. We still need a mathematical theory of qualia, for example, and a theory of representation (about which I am less sanguine). The further elaboration of the PCM and the development of these needed theories will require the combined efforts of mathematically minded phenomenologists and philosophers and phenomenologically and philosophically minded mathematicians, neuroscientists, psychologists, roboticists, and software engineers. If experimentation and simulation continue successfully apace, eventually the mathematical and empirical sophistication of the theory of consciousness will make it harder to access, more a domain for certain highly qualified and rigorously trained specialists. That day may be closer than we would like to think, and we would do well to reflect seriously, as many are, on its potentially Faustian moral aspects (cf. Noah Harari and Segev, this volume). If the day does come, then Bertrand Russell's idea that it is the job of philosophers to work themselves out of job might come, once again, to fruition. But we need not fear the darkening of the world or its complete disenchantment. There will always be beautiful things we do not (yet) understand.

Acknowledgments

Above all, I must thank David Rudrauf, Daniel Bennequin, and Manfred Frank. I would also like to thank Bjorn Merker, Georges Dreyfus, Stefan Lang, Gerhard Preyer, Marc Borner, Davide Bordini, audiences at the "Bordini's Experiences" lecture series, and Marc Borner's 2022 seminar at Goethe-Universität Frankfurt. Thanks also to Aviva Berkovich-Ohana for her patience and encouragement, and to Eviatar Shulman and Or Rapahel for very useful feedback on this chapter.

Glossary

Acquaintance: The direct knowledge consciousness has of itself and its internally differentiated manifold. Acquaintance is to be distinguished from representation. In contradistinction to the objects of representation (e.g., a unicorn), the objects of acquaintance (e.g., a flavor) cannot fail to exist (when one is acquainted with them).

Antirealism: Here the view that consciousness (or the mind) cannot be understood as an objective process (object, set of events), whether physical or not, for fundamental epistemological and semantic ("transcendental") reason that render such a scientific or metaphysical realist view ill-founded or nonsensical.

***De se* attitudes:** Beliefs, desires, memories, and other "propositional attitudes" about oneself *qua* oneself (i.e., not merely accidentally about oneself). Attitudes *de se* depend on self-acquaintance and the sense of individuality it enables.

Direct realism: The (erroneous but highly intuitive) view that, via perception, consciousness has direct and typically veridical access to the world that surrounds it.

Illusionism: A family of views according to which consciousness (as it seems to us) is a kind of illusion or otherwise very self-misleading about its own properties. According to most illusionists (and allies), the problem of consciousness reduces to *explaining away* the *problem* of consciousness in standard psychological or neuro-scientific terms.

Indirect realism: The view that, via perception, consciousness has indirect, causally mediated (but typically veridical enough) access to the world that surrounds it.

Intentionality: The capacity for mental states to be *about* or *of* other things or states of affairs. In analytical philosophy the word is basically synonymous with the word "representation." But in phenomenology it often has a broader, more ambiguous meaning such that it is sometimes taken to include both representation and acquaintance.

Internalism: Here the view that consciousness is localized in and limited to its realizing basis (e.g., the brain or CNS) and does not literally extend out into the world beyond that basis.

Intersubjectivity: Our general capability of "sharing" a world (experiences, thoughts) with other conscious beings. Includes both "theory of mind" and "simulation" and well as our access to the domain of the "publicly observable."

Introspective completeness: The thesis that consciousness has introspective access to *all* of its essential properties.

Introspective incompleteness: The thesis that consciousness *does not* have introspective access to all of its essential properties.

Introspective infallibility: The thesis that *with respect to those properties consciousness does have introspective access to* introspective judgment is not prone to error.

Neodualism: A view, associated with David Chalmers's 1996 book, *The Conscious Mind*, that consciousness must be a fundamental feature of reality perhaps contingently supervening, in a nomologically regular way, upon complex, functionally organized, material entities (e.g., brains) but not reducible thereto. It is called "neodualism" to distinguish it from the "paleo-dualism" of Descartes.

Panpsychism: Here the view that consciousness is literally ubiquitous in the physical world and an irreducible feature of reality.

Perspectival character (projective character): The "point of view" structure immanent in consciousness. One experiences the world from a situated point of view; and one is aware of this structure without being able to make the point of view into a perceptual object. According to the projective consciousness model, this is accounted for by the fact that consciousness is structured by a 3D projective space.

Phenomenal spatiality: The projective spatial character of conscious experience with its perspectival character and horizonal structure.

Phenomenological realism: The view that consciousness is a real phenomenon and (with some appropriate qualifications) is the way it seems to itself to be.

Prereflective self-consciousness (prereflective self-acquaintance): The reflexive (and not necessarily reflective or introspective) consciousness every episode of consciousness has of itself. All consciousness is directly acquainted with itself, independently of introspective or reflective attention.

Privacy: Herein this word refers to the fact that one's individual conscious experiences are not (or not normally, in any case) immediately sharable with other conscious beings. This is so in spite of a functional illusion of immediate shareability vis-à-vis the domain of the "publicly observable" and common overt emotional expressions.

Projective Consciousness Model: A theory of consciousness, originated by David Rudrauf, according to which consciousness is structured by a 3D projective space/geometry operating under a Free Energy minimization principle (or a similar "optimization" process).

Projective geometry: The nonmetric geometry of incidence relations and properties invariant with respect to projective transformations. Intuitively, a projective geometry adds to a familiar Euclidean geometry a set of points "at infinity" that delimit the horizon (line or plane). Such a geometry is able to characterize the transformations induced by changes of points of view on a field of objects (think of drawing a 3D object on 2D surface as seen from various angles). See any standard text on projective geometry for a more precise and detailed characterization.

Proto-intersubjectivity: The fundamental structure of consciousness, *qua* phenomenal projective space, in virtue of which it can be or become aware of actual and possible points of view other than the one it currently occupies.

Self-location: Here the capacity conscious beings have of locating and situating themselves in a variety of spaces—physical, social, semantic, historical. The capacity expands and contracts with the expansion and contraction of cognitive, conceptual, and memorial capacities but presupposes prereflective self-acquaintance, point of view, and a sense of individuality.

Subjective character: Those aspects of consciousness in virtue of which consciousness could be said to involve subjectivity. They include perspectival character, prereflective self-consciousness, a sense of individuality, a sense of self-location (in multiple spaces), a sense of agency, privacy, a sense of concrete embodiment, and a sense of an extended self, either an autobiographical self or a personal identity.

Witness consciousness: A term for prereflective self-consciousness that derives from the Advaita Vedanta tradition. Here it is chosen simply because it nicely highlights one feature of prereflective self-consciousness—namely, the sense that one is always one's own observer.

Notes

1. See Arnauld (1990/1683, pp. 25–26). I have slightly modified the translation. For the original, see Arnauld (1986/1683, p. 52). Emphases original.

2. Here I can only briefly articulate some assumptions; the interested reader is invited to follow the reference trail for more depth.

3. Graziano et al. (2019) argues for reconciling his own attention schema theory with global workspace theories and higher-order thought theories. But the latter two theories begin from realistically construed phenomenological data. Theories that (more or less) equate consciousness with attention (like Graziano's; see also Prinz [2012]) themselves get their plausibility from *prima facie* phenomenological data concerning the importance of attention to conscious experience (which does not hold up *secunda facie*; see, e.g., Jennings [2015] and Kentridge [2011]).

4. Such a position is not without its difficulties, but nothing much in the remainder of this essay will hang on this issue.

5. This is a view that has a long history in philosophy, going back to Aristotle, the Stoics, and classical Indian and Buddhist philosophy, with representatives in medieval philosophy, early modern philosophy, German idealism and phenomenology—as well as a few in analytic philosophy. For entry points, follow the reference trails in Frank (1991), Williams (1998), Caston (2002), Williford (2003), Yao (2005), Kriegel and Williford (2006), Kriegel (2009), Thiel (2011), Dreyfus (2011), Faschung (2011), and Zahavi (2020).

6. The phrase "witness consciousness" derives, ultimately, from discussions of consciousness and self-consciousness in Advaita Vedanta (see Albahari 2009; Faschung 2011; Gupta 1998). In using this phrase drawn from an Indian philosophical tradition, I am not committing myself to any particular (especially non-materialist) metaphysical doctrines (see Shulman, this volume, for a contrasting point of view in a related context). I mean only to emphasize a certain aspect of the phenomenology of prereflective self-consciousness, one brought out well in many of Sartre's discussions; Sartre, quite independently, also uses the word "witness" (*témoin*) in many of these discussions (see, e.g., Sartre [2018/1943, p. 123]: "[Consciousness] exists for a witness, even though this witness for whom consciousness exists is itself").

References

Albahari, M. (2009). Witness-consciousness: Its definition, appearance and reality. *Journal of Consciousness Studies*, *16*(1), 62–84.

Arnauld, A. (1990/1683). *On true and false ideas* (Elmar J. Kremer, Trans.). The Edwin Mellen Press.

Arnauld, A. (1986/1683). *Des vrais et des fausses idées*. Fayard.

Caston, V. (2002). Aristotle on consciousness. *Mind*, *111*(444),751–815.

Chalmers, D. J. (1996). *The conscious mind: In search of a fundamental theory*. Oxford University Press.

Chalmers, D. J. (2020). How can we solve the meta-problem of consciousness? *Journal of Consciousness Studies*, *27*(5–6), 201–226.

Dainton, B. (2008). *The phenomenal self.* Oxford University Press.

Damasio, A. R. (1999). *The feeling of what happens: Body and emotion in the making of consciousness*. Houghton Mifflin Harcourt.

Dennett, D. C. (1991). *Consciousness explained.* Little, Brown & Co.

Dennett, D. C. (2016). Illusionism as the obvious default theory of consciousness. *Journal of Consciousness Studies*, *23*(11–12), 65–72.

Dreyfus, G. (2011). Self and subjectivity: A middle way approach. In M. Siderits, E. Thompson & D. Zahavi (Eds.), *Self, no self?: Perspectives from analytical, phenomenological, and Indian traditions* (pp. 114–156). Oxford University Press.

Fasching, W. (2011). "I am of the nature of seeing": Phenomenological reflections on the Indian notion of witness-consciousness. In M. Siderits, E. Thompson & D. Zahavi (Eds.), *Self, no self?: Perspectives from analytical, phenomenological, and Indian traditions* (pp. 193–216). Oxford University Press.

Frank, M. (Ed) (1991). *Selbstbewußtseinstheorien von Fichte bis Sartre* [Theories of self-consciousness from Fichte to Sartre]. Suhrkamp.

Frank, M. (2019). From 'Fichte's original insight' to a moderate defence of self-representationalism. *ProtoSociology*, *36*, 36–78.

Frankish, K. (2016). Illusionism as a theory of consciousness. *Journal of Consciousness Studies*, *23*(11–12), 11–39.

Ganeri, J. (2012). *The self: Naturalism, consciousness, and the first-person stance*. Oxford University Press.

Goff, P. (2017). *Consciousness and fundamental reality*. Oxford University Press.

Gupta, B. (1998). *The disinterested witness: A fragment of Advaita Vedānta phenomenology*. Northwestern University Press.

Graziano, M. S. (2016). Consciousness engineered. *Journal of Consciousness Studies*, *23*(11–12), 98–115.

Graziano, M. S., Guterstam, A., Bio, B. J., & Wilterson, A. I. (2019). Toward a standard model of consciousness: Reconciling the attention schema, global workspace, higher-order thought, and illusionist theories. *Cognitive Neuropsychology*, *37*(3–4),155–172.

Jennings, C. D. (2015). Consciousness without attention. *Journal of the American Philosophical Association*, *1*(2), 276–295.

Kentridge, R. W. (2011). Attention without awareness: A brief review. In C. Mole, D. Smithies & W. Wu (Eds.). *Attention: Philosophical and psychological essays* (pp. 228–246). Oxford University Press.

Kriegel, U. (2009). *Subjective consciousness: A self-representational theory*. Oxford University Press.

Kriegel, U., & Williford, K. (Eds.) (2006). *Self-representational approaches to consciousness*. MIT Press.

Lehar, S. M. (2003). *The world in your head: A gestalt view of the mechanism of conscious experience*. Psychology Press.

Metzinger, T. (2004). *Being no one: The self-model theory of subjectivity*. MIT Press.

Metzinger, T. (2020). Minimal phenomenal experience: Meditation, tonic alertness, and the phenomenology of "pure" consciousness. *Philosophy and the Mind Sciences*, *1*(I), 1–44.

Prinz, J. (2012). *The conscious brain*. Oxford University Press.

Revonsuo, A. (2006). *Inner presence: Consciousness as a biological phenomenon*. MIT Press.

Rudrauf, D., Bennequin, D., Granic, I., Landini, G., Friston, K., & Williford, K. (2017). A mathematical model of embodied consciousness. *Journal of Theoretical Biology*, *428*, 106–131.

Rudrauf, D., Bennequin, D., & Williford, K. (2020). The moon illusion explained by the projective consciousness model. *Journal of Theoretical Biology*, *507*, 110455.

Rudrauf, D., Sergeant-Perthuis, G., Belli, O., Tisserand, Y., & Serugendo, G. D. M. (2022). Modeling the subjective perspective of consciousness and its role in the control of behaviours. *Journal of Theoretical Biology*, *534*, 110957.

Rudrauf, D., Sergeant-Perthuis, G., Tisserand, Y., Poloudenny, G., Williford, K., & Amorim, M. A. (2023). The projective consciousness model: Projective geometry at the core of consciousness and the integration of perception, imagination, motivation, emotion, social cognition and action. *Brain Sciences*, *13*(10), 1435.

Sartre, J.-P. (2018/1943). *Being and nothingness* (S. Richmond, Trans.). Routledge.

Strawson, G. (2009). *Selves: An essay in revisionary metaphysics*. Oxford University Press.

Thiel, U. (2011). *The early modern subject: Self-consciousness and personal identity from Descartes to Hume*. Oxford University Press.

Thomasson, A. L. (2007). In what sense is phenomenology transcendental?. *The Southern Journal of Philosophy*, *45*(S1), 85–92.

Varela, F. J. (1996). Neurophenomenology: A methodological remedy for the hard problem. *Journal of Consciousness Studies*, *3*(4), 330–349.

Williams, P. (1998). *The reflexive nature of awareness: A Tibetan Madhyamaka defence*. Routledge.

Williford, K. (2003). *The structure of self-consciousness: A phenomenological and philosophical investigation* [PhD Dissertation]. The University of Iowa.

Williford, K. (2007). The logic of phenomenal transparency. *Soochow Journal of Philosophical Studies*, *16*, 181–195.

Williford, K. (2011). Pre-reflective self-consciousness and the autobiographical ego. In J. Webber (Ed.), *Reading Sartre: On phenomenology and existentialism* (pp. 195–210). Routledge.

Williford, K. (2013). Husserl's hyletic data and phenomenal consciousness. *Phenomenology and the Cognitive Sciences*, *12*(3), 501–519.

Williford, K. (2015). Representationalisms, subjective character, and self-acquaintance. In T. Metzinger & J. M. Windt (Eds), *Open MIND*. MIND Group/MIT Press.

Williford, K. (2019). Self-acquaintance and three regress arguments. *ProtoSociology*, *36*, 368–412.

Williford, K. (2020). Headlessness without illusions: Phenomenological undecidability and materialism. *Journal of Consciousness Studies*, *27*(5–6), 190–200.

Williford, K. (2022). Irreducible circularity and attitudes *de se*: Fichte, Henrich, and self-acquaintance. In M. Frank & J. Kuneš (Eds.), *Selbstbewusstsein: Dieter Henrich und die Heidelberger Schule* (pp. 131–152). J.B. Metzler Verlag/Springer Nature.

Williford, K., Bennequin, D., Friston, K., & Rudrauf, D. (2018). The projective consciousness model and phenomenal selfhood. *Frontiers in Psychology*, *9*, 2571.

Williford, K., Bennequin, D., & Rudrauf, D. (2022). Pre-reflective self-consciousness and projective geometry. *Review of Philosophy and Psychology*, *13*(2), 365–396.

Yao, Z. (2005). *The Buddhist theory of self-cognition*. Routledge.

Zahavi, D. (2016). The end of what? Phenomenology vs. speculative realism. *International Journal of Philosophical Studies*, *24*(3), 289–309.

Zahavi, D. (2020). *Self-awareness and alterity: A phenomenological investigation* (2nd ed.). Northwestern University Press.

Contributors

Yochai Ataria is a full professor at Tel-Hai College, Israel. He conducted his postdoctoral research in the neurobiology department at the Weizmann Institute of Science. He is the author of the following books: *The Structural Trauma of Western Culture* (2017); *Body Disownership in Complex Post-Traumatic Stress Disorder* (2018); *Primo Levi and Ka-Tzetnik* (2021); *Consciousness in Flesh* (2022); and *Genes, Technology, and Apocalypse* (2024).

Shantena Augusto Sabbadini is a physicist, philosopher, and a scholar of Chinese classics. As a physicist he worked at the University of Milan on the foundations of quantum mechanics and at the University of California on the first identification of a black hole. He directs the Pari Center for New Learning, an international institute located in the small medieval village of Pari, Tuscany. He is the author of *Pilgrimages to Emptiness: Rethinking Reality through Quantum Mechanics*; *Tao Te Ching: A Guide to the Interpretation of the Foundational book of Taoism*; and *The Original I Ching: The Eranos I Ching Project.*

Aviva Berkovich-Ohana is a professor of neuroscience at the Edmond J. Safra Brain Research Center, in the Faculty of Education at the University of Haifa, Israel. She is currently head of the Integrative Psychotherapy Program in the Department of Counseling and Human Development. She received her PhD in neurobiology from the Gonda Multidisciplinary Brain Research Center, Bar-llan University, Israel, and did her postdoctoral studies at the Weizmann Institute, Israel. Her major research interests include the effects of various contemplative mental training, including meditation; psychedelics; lucid dreaming; and channeling on consciousness, cognition, and sense of self. Her lab utilizes a broad spectrum of neuroimaging tools, behavioral and cognitive tasks, and phenomenology, with an emphasis on neurophenomenology in collaboration with proficient contemplative practitioners.

Ema Demšar is a cognitive scientist interested in the study of lived experience within the enactive and neurophenomenological framework. She obtained an MS in cognitive science from the University of Ljubljana, Slovenia, where she then went on to work as a researcher and teaching assistant at the Center for Cognitive Science. Demšar's empirical and theoretical work at the Center has focused on advancing methods and designs for first-person research and examining its epistemological and methodological challenges in the context of cognitive science. She is currently a PhD candidate in the Centre for Consciousness and Contemplative Studies at Monash University, with her doctoral research centering on the neurophenomenology of meta-awareness in lucid dreaming.

Georges Dreyfus spent fifteen years in Buddhist monasteries before receiving in 1985 the title of Geshe, the highest degree conferred by Tibetan monastic universities. He is currently professor of religion of the Department of Religion at Williams College.

Stephen Fulder was born in London in 1946 and has a PhD in molecular biology. He has been practicing Buddhist-related spiritual practice for over forty years. He is the founder and senior teacher of the Israel Insight Society, the largest meditation and mindfulness organization in Israel, and has been teaching large numbers of practitioners for twenty-five years. Besides fourteen academic and popular books on health-related subjects, he has two books on his teachings: *What's Beyond Mindfulness: Waking Up To This Precious Life* (2019), and *The Five Powers* (2020).

Shaun Gallagher is the Lillian and Morrie Moss Professor of Excellence in Philosophy at the University of Memphis, and professorial fellow at the School of Liberal Arts, University of Wollongong. He was a Humboldt Foundation Anneliese Maier Research Fellow (2012–2018), and is currently a Humboldt Fellow (2024–2026). He has held honorary professorships at Tromsø University (Norway); Durham (UK); and Copenhagen (Denmark), as well as invited visiting positions at Cambridge, Lyon, Paris, Berlin, Oxford, and Rome. His areas of research include phenomenology, philosophy of mind, embodied cognition, social interaction, concepts of self, and hermeneutics. His most recent book, *The Self and its Disorders* was published by Oxford University Press in 2024. Other publications include *Action and Interaction* (2020); *Performance/Art: The Venetian Lectures* (2021); *The Phenomenological Mind* (2021, 3rd ed); *Enactivist Interventions: Rethinking the Mind* (2017); *How the Body Shapes the Mind* (2005); is editor of the *Oxford Handbook of the Self* (2011); and is coeditor of *The Oxford Handbook of 4E Cognition.* He is editor-in-chief of the journal *Phenomenology and the Cognitive Sciences.*

He is currently a professor in the Department of History at the Hebrew University of Jerusalem.

Moshe Klein received his PhD from Tel Aviv University and is currently a postdoc in the TAU Laboratory for Consciousness and Soft Logic Research. He is also a lecturer at Tel Chai College.

Urban Kordeš is professor of cognitive science and first-person research at the University of Ljubljana where he currently heads the Center for Cognitive Science, Laboratory for Empirical Phenomenology, and the joint master's program in cognitive science. He holds a bachelor's degree in mathematical physics and doctorate in philosophy of cognitive science. His research interests include in-depth empirical phenomenological research, neurophenomenology, second-order cybernetics, as well as epistemological and methodological issues in the research of nontrivial systems. In 2016, Kordeš founded the research group "Observatory" as a platform for contemplative, non-naturalized, and existentially meaningful research of consciousness.

Dominique Lamy was born in Morocco, grew up in France, and emigrated to Israel in 1987. She completed her PhD in psychology at Tel Aviv University in 2000 and was a postdoctoral fellow at Johns Hopkins prior to her appointment at Tel Aviv University, where she is currently a full professor in the Department of Psychological Sciences. She served as the head of the cognitive psychology program for ten years and currently co-heads the consciousness research track of the philosophy-psychology program. Her current research focuses on visual attention, conscious access and working memory, and how these interact.

Oded Maimon is the Oracle chair professor at Tel Aviv University and was formerly at MIT. His areas of research are algorithmic artificial intelligence and consciousness. He has published over one hundred refereed papers and fifteen award-winning academic books. In 2019 he established the TAU Laboratory for Consciousness and Soft Logic Research.

Rafael Malach did his PhD in physiological optics (1982) at University of California, Berkeley and postdoc at MIT. In 1985 he joined the Department of Neurobiology (now the Department of Brain Sciences) at the Weizmann Institute of Science. Malach's group explores the link between physical activity in the human brain and its psychological and mental manifestations. To that end Malach combines functional brain imaging of the human brain using magnetic resonance with invasive electrophysiological recordings, performed for diagnostic purposes in patients. Malach aims to uncover the principles that link human brain activity patterns to conscious and subconscious processes—mainly in visual perception, memory recall, and in free and creative behaviors. Most recently, Malach has been exploring the neuronal mechanisms organized by structuralist principles and the spontaneous waves of biased randomness as central drivers of subjective experience and creativity.

Zohar Maliniak has a BS in physics and mathematics, and an MS in theoretical physics (1987–1993) from the Hebrew University of Jerusalem. He wrote his master's thesis on Bohm's interpretation of quantum theory under the supervision of the late Professor Itamar Pitowsky and the recently deceased Professor Issachar Unna. He then served as the principal of several schools and taught physics and mathematics in high schools and colleges. In 2011, he began writing a doctoral dissertation at the University of Haifa under the supervision of Professor Meir Hemmo, which he completed in 2020. The subject of his doctorate is Bohm's interpretation of quantum mechanics, its history, metaphysics, and the reciprocal relationship between them.

Ofra Mayseless is a professor of developmental psychology in the Faculty of Education (and former dean), at the University of Haifa, Israel. She has published over one hundred articles and several chapters, edited two books, and authored one. Currently, she is the head of the multiuniversity Center for the Study of Human Spirit. Her research has focused on attachment, caring motivation, transition to adulthood, and spiritual development.

Nitzan Micher was born and raised in Israel. He completed his BS in biology and psychology at the Sagol School of Neuroscience at Tel Aviv University in 2020. He is currently an MS student in the neuroscience research program at Sagol and he is also a member of Professor Dominique Lamy's visual cognition lab. His current research focuses on the role of conscious perception in high-level cognitive processes.

Yuval Noah Harari is the bestselling author of *Sapiens: A Brief History of Humankind*; *Homo Deus: A Brief History of Tomorrow*; *21 Lessons for the 21st Century*; and *Nexus: A Brief History of Information Networks from the Stone Age to AI*. His current research focuses on macro-historical questions and the relations between history, biology, and computer science.

Ulrich Ott is a psychologist heading the working group "Altered States of Consciousness—Meditation Research" at the Bender Institute of Neuroimaging. He studies effects of meditation and yoga on mind and brain using interviews, questionnaires, physiological measurements, biofeedback, and neuroimaging methods.

Giuseppe Pagnoni is associate professor at the Department of Biomedical, Metabolic and Neural Sciences, University of Modena and Reggio Emilia, Italy. After a masters in physics, he completed a PhD in neuroscience and has worked for several years in the Department of Psychiatry and Behavioral Sciences, at Emory University, Atlanta, Georgia. He has led and collaborated in neuroimaging studies on diverse topics including reward processing, the interaction of immune and brain function, social cognition, intrinsic brain activity, pain processing, mental effort, and meditation. He is currently involved in the application of the theoretical framework of active inference to the study of contemplative practices.

Or Raphael is a graduate of the Grand Technion Energy Program in Israel, specializing in renewable energy engineering. He holds a BS in physics from the Hebrew University of Jerusalem. Alongside his academic pursuits, Raphael has been deeply engaged in meditation and self-reflection, in particular following the "direct path" teachings of Sri Ramana Maharshi. Since 2019, he has coauthored dozens of scientific articles for kids and teens with Nobel laureates for the journal *Frontiers for Young Minds*. As of 2023, he serves as a scientific engagement coordinator at the "Frontiers Planet Prize," a global prize mobilizing innovative solutions in earth systems science.

A psychologist and neuroscientist, **David Rudrauf** holds two PhDs in cognitive science and neuroscience. He was an assistant professor of neurology and radiology at the University of Iowa, where he directed the Laboratory of Brain Imaging and Cognitive Neuroscience. Rudrauf was also associate professor of psychology at the University of Geneva, where he directed the Laboratory of Multimodal Modelling of Emotion and Feeling. After years of research in neuroscience, neuropsychology, electrophysiology, and multimodal neuroimaging, he focuses on synthetic psychology, combining mathematical modeling of consciousness, virtual reality, and robotics. He is now professor of cognitive science at the University Paris–Saclay.

Stefan Schmidt studied psychology in Konstanz and Freiburg and then embarked on research into consciousness. He worked as a postdoc at the Medical Center, University of Freiburg, Germany, followed by an endowed chair in Utrecht, Netherlands, and then an assistant professorship in Frankfurt (Oder), Germany. Since 2018, he has been a professor of systemic family counseling in Freiburg, Germany. His main topics of research are contemplative science, systemic counseling, and placebo treatments.

Idan Segev earned his PhD in mathematics and neurobiology from the Hebrew University in 1982 and completed a postdoctoral fellowship at the NIH in Bethesda, Maryland. In 1985, he joined the Department of Neurobiology at the Hebrew University of Jerusalem. His team uses computational and theoretical tools to study how neurons and the circuits they form compute and dynamically adapt to our ever-changing environment. Segev's team has recently begun exploring the uniqueness of human neurons and neuronal circuits using large-scale simulations ("The Blue Brain Project") and machine learning techniques. These neurons, still alive and electrically active, are obtained from fresh tissues following brain surgeries. He serves as chief editor of *Frontiers in Neuroscience* and co-chief editor of *Frontiers for Young Minds*—an open access journal featuring articles by leading scientists, including Nobel laureates, reviewed by kids aged ten- to fifteen-years-old. Segev is also deeply interested in the interplay between art and the brain. He coedited two books inspired by collaborations between artists and brain researchers: one (*Brain and Art*) featuring original brain-inspired etchings, and another (*Neuropoetry*) containing brain-inspired poems with reflections from scientists on their research journeys, including their failures and successes.

Yaron Senderowicz teaches philosophy in the Department of Philosophy at Tel Aviv University. He heads, together with Dominique Lamy, the interdisciplinary program for the study of consciousness and cognition. He was also the associate editor of *Pragmatics & Cognition*. His main research fields include philosophy of mind, phenomenology, and German idealism.

Anat Shamgar is a dancer and a dance maker based in Jerusalem, Israel. She is an active performer of solo works and collaborations. She finds collaboration an opportunity for experiencing the "other," merging languages and contradicting the familiar. Shamgar performed in the Montpelier Dance Festival, the Podevil Festival in Berlin, the New York Improvisation Festival, the Israel Festival, the Room Dance Festival, and many other venues. She has been a faculty member of the Jerusalem Academy of Music and Dance in Jerusalem since 1991 and head of the Choreography Track and Movement Department at the Jerusalem Academy. She teaches composition, choreography, body work, and multidisciplinary courses.

Eviatar Shulman teaches Buddhist studies and Indian religion and philosophy at the Hebrew University of Jerusalem. He is chair of the Department of Comparative Religion, and member of the Department of Asian Studies. He has published extensively, including two scholarly monographs—*Rethinking the Buddha: Early Buddhist Philosophy as Meditative*

Perception (2014), and *Visions of the Buddha: Creative Dimensions of Early Buddhist Scripture* (2021).

Tom Soloveitzik is a musician and sound artist based in Tel Aviv, Israel. He works with the materiality of sound, the practice of listening, and the potentiality of recording (a "field") through performance and writing. He codirects the Holon Scratch Orchestra and was, for the past decade, a co-artistic director of Ensemble Musica Nova for experimental music. More information is present at https://www.tomsoloveitzik.net.

Narayanan Srinivasan is currently with the Department of Cognitive Science at the Indian Institute of Technology Kanpur, India. He spent 17 years building the Centre of Behavioural and Cognitive Sciences, University of Allahabad, India. Before wandering into psychology, he studied physics at the University of Madras and electrical engineering at the Indian Institute of Science. He obtained his PhD in psychology from the University of Georgia. He is interested in understanding mental processes, especially consciousness, attention, emotions, self, and decision-making using multiple methodologies. Dr. Srinivasan is a fellow of the Association for Psychological Science, the Psychonomic Society, and the National Academy of Psychology (India).

Fausto Taiten Guareschi is the abbot emeritus of the Zen Temple and Monastery of Fudenji, (Salsomaggiore Terme, Italy), which he founded in 1984. The temple is a center of spirituality that has always sought to promote an active dialogue with the contemporary religious, philosophical, and scientific culture. In 2009, Fudenji was recognized by the Sōtōshū Shumuchō (Soto Zen Administrative Headquarters, Tokyo, Japan) as an official temple with special statute (*Sōtōshū Tokubetsu jiin*), and in 2016 as an affiliate temple (*Ojikimatsu*) of Daihonzan Eiheiji, Fukui prefecture, Japan. Reverend Taiten Guareschi has been leading the Seminar of Theological and Catechistical Studies since 1989 and has published several books with the publishing houses Il Cerchio (Rimini) and Casadeilibri (Padova): the most recent ones are *Fatti di Fuoco* (2015) and *Fatti di Acqua* (2021).

Christopher Timmermann, PhD, is a postdoctoral research fellow at the Centre for Psychedelic Research in the Department of Brain Sciences at Imperial College London, where he leads the DMT Research Group. He is also the director of the Foundation for the Study of Human Consciousness (ECOH) from his native Chile. Dr. Timmermann performs psychology and neuroimaging research associated with psychedelic drugs, and for his PhD, led the first brain imaging studies of DMT (a potent psychedelic). He is interested in the use of neurophenomenological approaches for the study of consciousness bridging human experience with neural dynamics. Dr. Timmermann also studies the long-term effects that psychedelics have on mental health and belief systems, as well as the ethical implications of their use.

Since 2009, **Kenneth Williford** has been a professor of philosophy and the chair of the Department of Philosophy and Humanities at the University of Texas at Arlington. He completed his PhD in philosophy at the University of Iowa in 2003 and a postdoctoral fellowship at the CREA in Paris in 2005. He works primarily in philosophy of mind, phenomenology, and the history of modern philosophy. Since 2007, he has been part of an interdisciplinary collaboration centered on understanding consciousness and subjectivity. With David Rudrauf, Daniel Bennequin, and others, he is one of the architects of the projective consciousness model, a fruit of this collaboration.

Index

Note: Photos, Figures and Tables are indicated by italicized page numbers.

Publisher contact:
The MIT Press
Massachusetts Institute of Technology
77 Massachusetts Avenue, Cambridge, MA 02139
mitpress.mit.edu

EU Authorised Representative:
Easy Access System Europe, Mustamäe tee 50,
10621 Tallinn, Estonia
gpsr.requests@easproject.com

Printed by Integrated Books International,
United States of America